TEXTS AND READINGS IN MATHEMATICS **26**

Differential Calculus in Normed Linear Spaces

Second Edition

Differential Calculus in Normed Linear Spaces

Second Edition

Kalyan Mukherjea
Indian Statistical Institute
Kolkata

HINDUSTAN
BOOK AGENCY

Published in India by

Hindustan Book Agency (India)
P 19 Green Park Extension
New Delhi 110 016
India

email: hba@vsnl.com
http://www.hindbook.com

Produced from camera ready copy supplied by the Author.

ISBN-10 81-85931- 76-3
ISBN-13 978-81-85931-76-0

Contents

Preface

From the usual theory of partial differentiation, no large amount is needed in the present book. Most of that is given here in a manner, not using coordinate systems, which keeps the full meaning always clear.

from *Geometric Integration Theory* (p.58)
by Hassler Whitney,
Princeton University Press, 1957.

This book arose out of a course on Advanced Calculus I taught several years ago to second year students of the B. Stat. programme at the Indian Statistical Institute, Calcutta. The contents of the course had been determined, to a very large extent, by my experience of teaching Differential Geometry to students at the M. Stat. level. I had found that when dealing with functions between open sets of Euclidean spaces, even the best of our students had difficulty in treating the derivative as a linear map (rather than the Jacobian matrix) and that their grasp of the Inverse/Implicit Function Theorems lacked conviction. As a result the first few weeks of my Differential Geometry lectures were always devoted to a crash course on Advanced Calculus with a geometric viewpoint; that is, without using the usual apparatus of partial derivatives, Jacobians, etc. I had found that students found it easier to grasp the essence of results like the Inverse Function Theorem in this kind of geometric treatment of Advanced Calculus.

When the opportunity to teach "Multivariate Calculus" presented itself, I decided to try and give a course which, I hoped, would cover the higher dimensional versions of most of the standard theorems of Differential Calculus (including Taylor's formula) and do so in a basis-free manner. The advantages of such a treatment is succinctly explained in the quotation from Hassler Whitney's celebrated monograph.

I realized that my target could only be achieved if the students turned out to be well prepared and motivated; something which would become known only after some weeks. As a result I was somewhat loth to announce any official text-book for the course. (The only suitable books were, anyway, rather expensive for the average Indian undergraduate.) Instead, I made LaTeX-ed notes of the material I had covered and distributed photo-copies of these to the students. This book is largely based on these notes.

This book provides a self-contained treatment of the Differential Calculus of functions between open sets of Banach spaces. The student is expected to possess a basic (but thorough) background in Analysis, as would be provided by the first eight chapters of Rudin's *Principles of Mathematical Analysis*, ([Rud61]) and a first course in Linear Algebra. Any good undergraduate textbook of Linear Algebra would give sufficient background for reading this book. My favourite, in spite of its antiquity, remains Halmos' classic, *Finite Dimensional Vector Spaces* [Hal61]: Helson's book [Hel94] would be a nice alternative.

There are three exceptions to the claim of self-sufficiency I have made.

1. I have used without proof (but plenty of motivating remarks) the theorem that the elementary symmetric functions generate the ring of all symmetric polynomials. The proof of this result is too far removed from the subject matter of this book to have permitted an adequate treatment in a *short* appendix.

2. I have used without giving a proof, the Open Mapping Theorem for bounded linear operators between Banach spaces. This is used only in a special formulation of the Implicit Function Theorem for functions defined on an infinite-dimensional Hilbert space and in a subsequent application to Lagrange's Theorem of Undetermined Multipliers. It seemed that a long section on the Baire Category Theorem (which would be required for proving the Open Mapping Theorem) was not justified.

3. I have not tried to discuss Zorn's lemma or other variants of the Axiom of Choice. I have, at certain places, simply stated that from what has been done the desired conclusion would follow by the use of the Axiom of Choice. When I was teaching a class of second year undergraduates this seemed the only way to proceed. I have not discussed this point at any length while writing this book since it would have to be a fairly long digression if it were to be useful for neophytes. I have simply stated the Axiom together with a few comments and a reference which should be easily understood by all readers.

Chapter 1 of the book is a quick review of the terminology (mainly concerning Linear Algebra) that will be used throughout the book.

It can be argued that to do a coordinate-free treatment of the higher derivatives, all that is needed is a knowledge of the definition of continuous multilinear functions. While logically correct, from the pedagogical standpoint this is not very helpful. The higher derivatives, by definition, are elements of

$$L(E, L(E, F)), L(E, L(E, L(E, F))) \dots \text{ etc.}$$

It is better to clarify how they become identified with multilinear functions. This has been done by stressing the "Law of Exponents" right from the start.

In Chapter 2, I have spent a fair amount of time dealing with "spaces of linear transformations" including dual spaces and the spaces "where higher derivatives live" and on multilinear functions. This chapter serves another purpose: it should help acclimatize the students to thinking about Linear Algebra without the constant use of matrices. In addition, the "old-fashioned" definition of the tensor product which is presented, might be useful for students going on to study differential geometry.

Chapter 3 covers the basic material on the topology of normed linear spaces. Since most of the results would hold in the more general setting of metric spaces, I have, while writing the book, formulated the statements and proofs, whenever possible, in this wider context. The Principle of Irrelevance of Algebraic Inequalities, which appears in §3.5, is the only nonstandard material that is presented here.

Chapters 4 and 5 constitute the heart of the book. The material here is quite conventional: the textbooks of Dieudonné [Die57] and Cartan [Car67] are the standard references for the subject.

Two topics not covered in my lectures have been introduced in these two chapters.

1. Some of my colleagues, correctly, pointed out that students at this level ought to know how to express important Partial Differential Operators in terms of "curvilinear coordinates" and that students learning Advanced Calculus from this book may have difficulty understanding the traditional treatments of this topic.

Since this topic, by definition, is antithetical to the approach laid out by Whitney, a "quick fix" is not easy. In section 4 of Chapter 4 I have discussed the slightly more abstract way of treating the derivatives of smooth functions via derivations of the ring of smooth functions. This motivates the definition of vector fields and leads quickly to differential operators. In a short subsection I have indicated, using the Laplacian as an example, how to go from one coordinate system to another.

2. The section on ordinary differential equations (in Chapter 5) had not been part of the lecture course, but I felt that even such a cursory introduction might be helpful for some students. A knowledge (or, at least, a firm belief [O'N78]) in the existence and uniqueness of local solutions of differential equations is a prerequisite for most books on Differential Topology/Geometry.

Chapter 6 has been written up for the first time for the book; I had covered the same material (somewhat hastily) in my lectures. My aim here was primarily to show the wealth and variety of subjects to which the tools of Calculus may be applied.

Most sections of the book end with exercises which are not really very difficult: they have been, almost without exception, set as home-work assignments or as problems in tests given during the course.

Acknowledgments

1. The students who attended the course showed great enthusiasm and appreciation. To all of them my thanks. Some of them corrected or improved upon my presentation: particularly, Siddhartha Gadgil, Lakshmi Iyer and Amritangshu Prasad come to mind. I should also thank Anindya Sen, a former student of I. S. I., Calcutta, for contributing two vignettes to this book; they are acknowledged as they arise.

I was fortunate in having three very able and hard-working teaching assistants for my course: Mahuya Dutta, Swagato Ray and Rudra Sarkar.They had proof-read the original notes I had distributed, helped in constructing exercises and in many other ways contributed to the course and this book. It is a pleasure to thank them for their efforts.

I thank my colleagues in the Stat–Math Unit, who have always provided a stimulating and supportive atmosphere for my work.

2. It goes without saying, that there is nothing original in this book, except perhaps, the errors. My debt to the textbooks of Dieudonné [Die57] and Lang [Lan65] should be obvious to anyone familiar with these books. In the section on infinite series I have found K. R. Parthasarathy's notes, ([Par80]) very useful.

On a more personal note, I would like to add that when KRP and I were colleagues in the Delhi Centre of the Indian Statistical Institute, we had a great deal of mathematical interaction. While this influenced my research interests only temporarily, it permanently, and for the better, changed the way I teach and write Mathematics. I am very conscious of the debt I owe him.

3. I would like to thank the Editorial board of the TRIM series for engaging two extremely conscientious referees who have, by pointing out an embarrassingly large number of errors in the original manuscript, helped bring this book to its present form. Thanks also to my friend Bireswar Chattopadhayay, who helped me in scrutinizing the final version of the manuscript.

4. This book has been written using LaTeX and a host of other Free Software programs. My thanks to Donald E. Knuth (the creator of TeX) and the large community of programmers who make their work freely available. Two individuals, deserve special mention.

(i) The macro package diagrams.sty has proved invaluable in producing the commutative diagrams which appear in this book. I would like to thank its author, Dr. Paul Taylor of the Department of Computer Science, Manchester University.

(ii) Some time ago I wrote to the author of Xpdf viewer of pdf-files Derek Noonberg (noonbergd@foolabs.com) requesting him to modify it so that my cataract-enfeebled eyes could use it with greater comfort. Not only did he send me the necessary modifications in a matter of days but then for a week or so reached across cyberspace and held my hand (with upto 4 emails a day!) as I stumblingly but in the end, successfully, built his program. Such astonishing generosity can hardly be repaid by a mere "Thank you"!

Above all, I would like to thank my wife, Lalita, without whose care, love and support the writing of this book would not even have begun.

Kalyan Mukherjea,
Calcutta, Deepavali 2001.

Preface to the second edition

Many changes and corrections of minor nature have been made throughout the book. Chapters 5 and 6 have undergone major overhauls to make the exposition more transparent. In addition, an egregious technical blunder has been corrected in Section 2 of Chapter 6.

In spite of these amendments, the basic structure and purpose of the book remains unchanged: it is a book on Differential Calculus of functions between open subspaces of vector spaces.

I would like to thank all those who have pointed out errors and obscurities in the text; particularly useful have been comments by V. Sunder and Harish Seshadri.

The preparation of a second edition would have been impossible without Emacspeak (a complete Audio Desktop) which enables visually disabled persons to use the personal computer with comfort and profit. No words of thanks suffice for Dr. T. V. Raman (of IBM and Google) for creating this wondrous tool and making it freely available.

Calcutta, December 2006.

How to use this book

A few words about the layout of the book are in order.

1. Throughout this book, § $m.n$ refers to section n of Chapter m. Theorems, lemmas etc. are numbered consecutively within each section. Thus, if a Lemma immediately follows Proposition $x.y.z$ then it will be labelled Lemma $x.y.(z + 1)$ and both these belong to section y of Chapter x. Certain results are not numbered since they are never referred to except in their immediate vicinity or because they are referred to elsewhere by a descriptive name.

2. Throughout "iff" is used as an abbreviation for the phrase "if and only if". The end of a proof is indicated by the symbol □. If this occurs in the statement of a theorem, it means that the result is easy to establish from what has been done already or the proof is omitted (by policy) since it involves the Axiom of Choice. The words "analytic" or "analytical" simply mean 'pertaining to analysis' and have no special mathematical connotation.

Two kinds of people are likely to read this book:

- Those wishing to learn all or part of the material presented here. They will collectively, be referred to as **students**.
- Those who are considering the use of this book to teach a course on Advanced Calculus. They will be referred to as **teachers**.

Here are some suggestions for people in each category.

SUGGESTIONS FOR STUDENTS:

1. Each chapter begins with a preamble where I briefly describe the contents of the chapter. In these preambles you will often encounter terms which have not yet been defined. I will then use the following kind of `typewriter`-like font to draw your attention. If I were to not allow myself these deviations from strict logical progression, these preambles would lose their motivating effect and hence their pedagogical *raison d'etre.*

2. Although basic definitions of Linear Algebra are given in § 1.3, it is *not* suitable reading for someone who has not already had a course in Linear Algebra.

3. With the exception of the section on Tensor Products (§ 2.3) there is very little in the earlier chapters that a student may skip without creating difficulties later.

§ 4.4 may be omitted on a first reading but it might make understanding the motivation behind the definition of ordinary differential equations in § 5.2 a little difficult. The proof of the uniqueness of Implicit Functions in § 5.4 *should* be omitted on first reading.

SUGGESTIONS FOR TEACHERS:

This book could have had the subtitle:

Elementary Mathematics from an Advanced Viewpoint.

What I mean is that while the material of the book should be part of the advanced undergraduate syllabus, I have indicated (explicitly or implicitly) how the same material might appear from a more advanced point of view.

One example of each kind may help clarify what I mean.

1. Although I have not used the appropriate technical term, Remark 1.3.9 can be regarded as hinting at the fact that the set of bases of $\mathbf{k}^n$ is a *torsor* of the group $\mathrm{GL}_n(\mathbf{k})$.

2. In Chapter 2, I have talked rather informally (but explicitly) about functors. Whether or not an instructor should mention such things in a Calculus course is upto his taste and judgement of the ability of the audience to assimilate such ideas. I found my second year undergraduates quite enthusiastic and capable in this particular direction. I might mention here that the rather unusual and functorial proof of the fact that the dual of an injective linear transformation is surjective (see Remark 2.1.6b on page 42) was a part of the solutions provided by one of the students during a 1 hour quiz given during the course. I have been told that I was fortunate in having a particularly receptive bunch of students. I hope others will also sometimes be as lucky!

In the sections on Linear Algebra in Chapter 1, I have made a conscious effort to show how group-theoretic notions involving the classical groups can illuminate many of the more humdrum aspects of Linear Algebra. Whether or not the instructor wishes to do the same is again a matter of taste alone.

The section on Multilinear Functions contains more material than is strictly necessary for the study of Calculus. It would have sufficed to discuss multilinear functions defined on the n-fold tensor product of llinear spaces rather than those defined on spaces such as $E_1 \times \cdots \times E_n$. But then it is difficult to produce really interesting examples of multilinear functions (such as $E^* \times F \approxeq L(E, F)$) and to make the topic "come alive" for the students.

This book contains more material than can be comfortably taught in a 50 hour lecture course. The main reason for this is the minimal prerequisites demanded of the student. If it is felt that the students are familiar with the relevant bits of Linear Algebra, then most of Chapter 1 and § 2.1 may be omitted. If their background includes the basic topology of metric spaces, then the first two sections of Chapter 3 may be omitted.

Otherwise, to fit into a 50 hour course, perhaps the best option would be to decide *ab initio* to omit the section on Higher Derivatives (§ 4.3): this will also enable the teacher to omit multilinear functions, various *avatars* of the "Law of Exponents" and help speed up some parts of Chapter 3 as well.

1

Preliminaries

In this chapter I am going to catalogue some facts, most of which you have encountered already, perhaps in a different form. Many assertions will not be proved: either because their truth will be evident upon a few moments' reflection or because they follow easily from results you have seen earlier in your mathematical education. My main purpose here is to set up the notations I will use and to familiarize you with my eccentricities.

Section 1 is where some basic definitions are set out. There is a *brief* discussion of the Axiom of Choice. The method of "commutative diagrams" is explained and the Law of Exponents is introduced. This last is an elementary device which is useful in dealing with functions defined on Cartesian products.

Section 2 sets out the standard notations, definitions etc. concerning vectors and matrices.

Section 3 is devoted to the basic concepts of Linear Algebra. Among other things, there is a result (Theorem 1.3.3) concerning the basis of a vector space, which might not be familiar to all readers. The process of reviewing the notion of a basis is carried on in Section 4 in the context of orthonormal bases and the Gram-Schmidt procedure. In Section 5, I review some ideas centred around the notion of eigenvalues of matrices. Here I also use a result on `symmetric polynomials` to prove a somewhat surprising result concerning the **similarity classes** of complex matrices.

1.1 Sets and Functions

If A, B are sets, then a function f from A to B will, as is usual, be represented as $f : A \to B$ or as $A \xrightarrow{f} B$. If $f : A \to B$ and $g : B \to C$ are functions, the function which takes $a \in A$ to $g\big(f(a)\big)$ will be said to be the *composition* of g and f and denoted $g \circ f$ or more simply gf.

I will say that a function $f : A \to B$ is:

- *injective* (or that f is an *injection*) if $f(a) = f(a')$ implies $a = a'$;
- *surjective* (or that f is a *surjection*) if for all $b \in B$ there is an element $a \in A$ such that $f(a) = b$;
- *bijective* (or that f is a *bijection*) if f is surjective and injective.
- If $f : A \to B$ is a surjection, a function $s : B \to A$ is called a *section of f* if $f \circ s = 1_B$, the identity map of B.

Proposition (The Axiom of Choice).
Every surjection admits a section. □

As already mentioned in the Preface, I will not enter into a serious discussion of the Axiom of Choice and its many variants. A lucid (and relatively elementary) account may be found in [Sim63]. The following remarks may indicate the subtleties involved in this area.

REMARKS 1.1.1 (ON THE AXIOM OF CHOICE)

Remark 1.1.1a (The usual form of the Axiom of Choice).
Let me rephrase the above proposition. If $f : X \to Y$ is a surjection, then as y varies over all the points of Y, the sets $S_y = f^{-1}\{y\} \doteq \{x \in X : f(x) = y\}$ are nonempty and form a partition of X; that is $\bigcup_{y \in Y} S_y = X$ and $S_y \cap S_{y'} = \emptyset$ if $y \neq y'$. Thus to produce a section we need to:
define a function s from Y to the union of a collection of nonempty sets, sets indexed by the elements of Y in such a way that the value of s at y lies in the set S_y.

The more usual form of the Axiom of Choice is to assert the possibility of defining such a function from Y to an arbitrary union $\bigcup_{y \in Y} S_y$ of nonempty sets S_y.

Remark 1.1.1b. The difficult thing is to do this when one knows nothing specific about either X or Y.

For example, if Y is finite we could simply select arbitrary elements one after the other from each of the S_y's.

If all the S_y's were groups, we could simply select for each $y \in Y$ the neutral element of S_y. It is the lack of information that makes it necessary to invoke an axiom to get around this difficulty.

The following charming example due to Bertrand Russell might clarify the point further.

Remark 1.1.1c (Russell's Example).
If one is confronted with an infinite collection of pairs of shoes and asked to select precisely one shoe out of each pair, one can proceed by picking up all the left shoes. But if shoes were to be replaced by socks, then one would need the Axiom of Choice to perform the same task!

Now I will discuss "commutative diagrams" — a pictorial device which is very helpful. The fact that English is written from left to right and that the value of a function f on an element $a \in A$ is written $f(a)$ causes a little bit of a problem when dealing with compositions of functions. If $f : A \to B$ and $g : B \to C$ are functions, then the symbol '$g(f(a))$' which is quite often more simply written as '$gf(a)$' is read as "Jee Ef of a" when it really is obtained by first applying f to a and then applying g to $f(a)$.

This can cause errors and things becomes worse when we are trying to assert that the composition of two functions is equal to the composition of two or more other functions.

I will throughout use the following pictorial device to keep out of this kind of trouble. If $f : A \to B$, $g : B \to C$ and $h : A \to C$ are functions such that $g \circ f = h$ the functions and sets will be represented pictorially as shown below

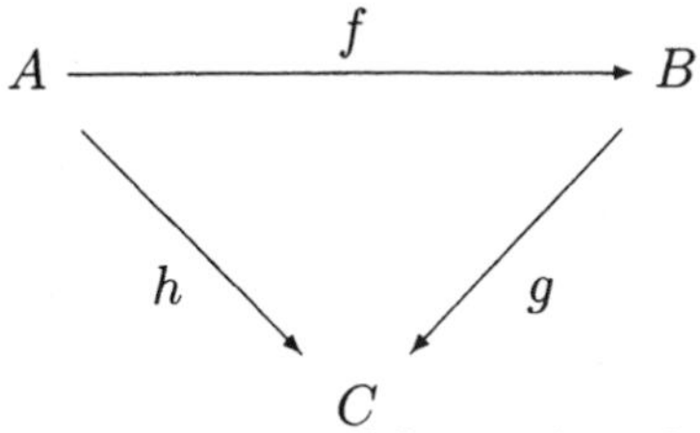

and the condition $gf = h$ expressed by saying that this diagram *commutes*.

Similarly if $f : A \to B$ and $g' : B \to B'$, $g : A \to A'$ and $f' : A' \to B'$ are functions such that $g'f = f'g$ I will represent this pictorially as shown below and say that the diagram.

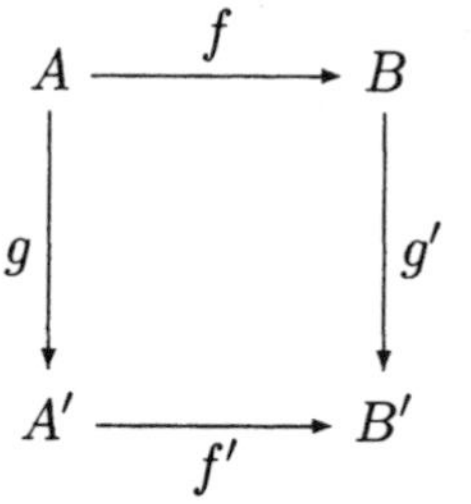

commutes.

The idea is that a function $\phi : X \to Y$ from X to Y, is thought of as a "transfer" of the elements of X to elements of Y via $x \mapsto \phi(x)$. The commutativity of the two diagrams above is supposed to signify that if an element $a \in A$ is moved to C by the arrows along two different routes the end-result is the same element of C. This device of using commutative diagrams is extremely effective as an aid to understanding — indeed, there are fields of mathematics (algebraic topology, for example) which would be almost impossible to learn without the use of commutative diagrams.

A small example will bring the message home. Suppose we have two commutative diagrams

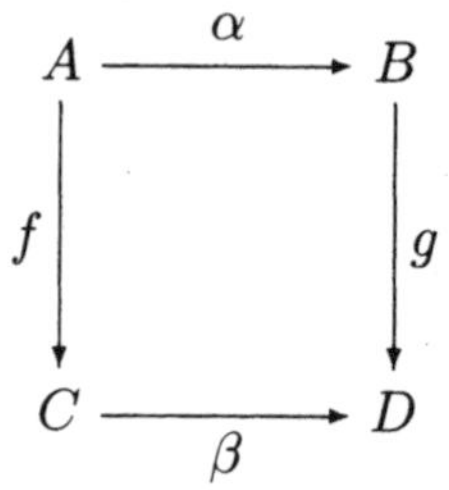

and

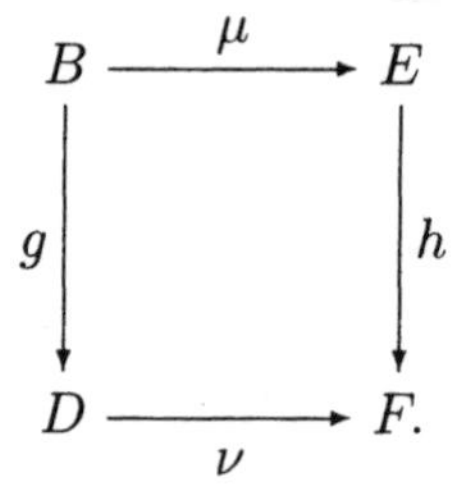

Bringing them together — "concatenating them" — we get a diagram:

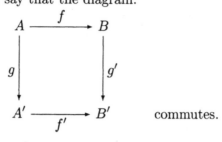

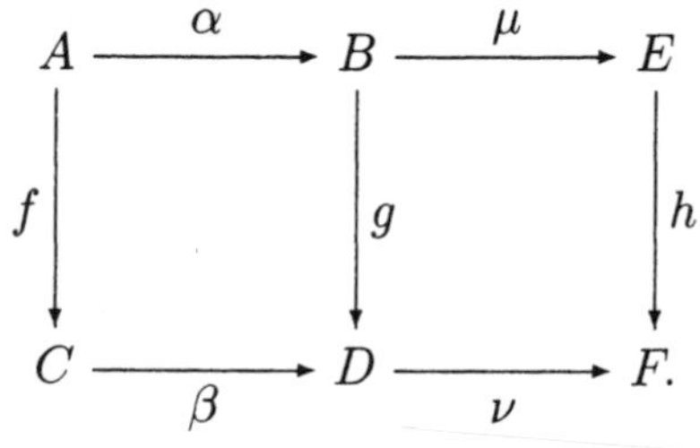

and it is easy to see from the picture above that if each individual square commutes then so does the larger rectangle: that is $h\mu\alpha = \nu\beta f$. This is surely easier than trying to establish this relation by staring hard at the two relations:

$$\begin{aligned} g\alpha &= \beta f \\ h\mu &= \nu g. \end{aligned}$$

I now turn to the so-called Law of Exponents which yields an alternative interpretation of functions of two (or more) variables, that is, functions whose argument is an ordered pair (or ordered p-tuple) of elements. Recall that if A and B are two sets the set of all functions from A to B is denoted by B^A.

Proposition 1.1.2 (The law of Exponents).
Let A, B and C be any sets. Then there are bijections between the sets $A^{(B\times C)}$ and $(A^B)^C$.

Proof. For each $c \in C$ one can identify the set B with the set:

$$B_c = \{(b,c) \,:\, b \in B\} \subset B \times C.$$

Given any function $f : B \times C \to A$ let f_c be its restriction to B_c. This yields, for each $c \in C$, a function from B to A. Thus I have defined a function $\xi : A^{B\times C} \to (A^B)^C$; $\xi(f)(c) = f|B_c$. ξ is a bijection. I define its inverse as follows.

If $G \in (A^B)^C$ define $\Xi(G) : B \times C \to A$ by $\Xi(G)(b,c) = G(c)(b)$. You should check that ξ and Ξ are inverses of each other. □

Remark 1.1.3. Notice that $f(b,c)$ corresponds to $f(c)(b)$. This feature will be even more pronounced in the generalizations to more than two variables.

I now wish to generalize this to functions of p variables and immediately we face a problem concerning notation: even with positive integers, $(a^b)^c \neq (a)^{b^c}$ and this is emphatically more so when a, b, c are sets. To continue with the "exponential notation" B^A for all functions from A to B would be impractical when more than two sets are involved. So I introduce some new notation; though I will continue to speak of "the law of exponents" for the final result. If $\mathbf{A} = (A_1, \ldots, A_p)$ is an ordered p-tuple of sets and $1 \leq m \leq p$, then I will denote by $\mathbf{A}(\widehat{m})$ the ordered $(p-1)$-tuple obtained by omitting A_m.

I will write $\times\mathbf{A}$ for $(A_1 \times \cdots \times A_p)$. Also, if each $A_i = A$, I will write $\mathbf{A}[p]$ for the p-tuple $(A, \overset{p \text{ factors}}{\cdots\cdots}, A)$ and $A^{[p]}$ for the p-fold Cartesian product of A.

For any p-tuple, $\mathbf{A} = (A_1, \ldots, A_p)$, define $\mathfrak{F}_p(\mathbf{A}, B)$ inductively as follows.
For $p = 1$, if $\mathbf{A} = \{A\}$, let $\mathfrak{F}(A; B) = B^A$, the set of all functions with domain A and range B.

If $p > 1$, then define $\mathfrak{F}_p(\mathbf{A}; B) = \mathfrak{F}(A_p, \mathfrak{F}_{p-1}(\mathbf{A}(\widehat{p}); B))$.

The sets $\mathfrak{F}(\mathbf{A}; B)$ will be called sets of iterated functions and elements of $\mathfrak{F}_p(\mathbf{A}; B)$ will be called *p-functions.*

Now if $\phi \in \mathfrak{F}_p(\mathbf{A}; B)$ then, in order to get an element of B from ϕ and $(a_1, \ldots, a_p) \in A_1 \times \cdots \times A_p$, one has to evaluate ϕ on a_p, evaluate the resulting function on a_{p-1} and so on. Instead of confusing with multiplicity of parentheses, I will write

$$\phi(a_p : a_{p-1} : \cdots : a_1) \text{ for } \phi(a_p)(a_{p-1}) \cdots (a_1).$$

The law of exponents in p variables is stated below.

Theorem 1.1.4 (The Law of Exponents).
For any $p \geq 1$ let $\mathbf{A} = (A_1, \ldots, A_p)$ be an ordered p-tuple of sets. Then for any set B there are natural bijections:

$$\begin{aligned} \Xi_p : \mathfrak{F}(\mathbf{A}; B) &\to \mathfrak{F}(\times \mathbf{A}, B) \\ \xi_p : \mathfrak{F}(\times \mathbf{A}, B) &\to \mathfrak{F}(\mathbf{A}; B) \end{aligned}$$

which are inverses of each other.

Proof. Simply define

$$\Xi_p : \mathfrak{F}(\mathbf{A}; B) \to \mathfrak{F}(\times A; B) \text{ by } \Xi(\phi)(a_1, \ldots, a_p) = \phi(a_p : \cdots : a_1) \text{ and}$$
$$\xi_p : \mathfrak{F}(\times A; B) \to \mathfrak{F}_p(\mathbf{A}; B) \text{ by } \xi_p(\phi)(a_p : \cdots : a_1) = \phi(a_1, \ldots, a_p)$$

Thanks to carefully chosen notation it is easily checked that ξ_p and Ξ_p are inverses of one another. □

Remark. The terms 'iterated functions" and "p-function" are not standard terms. They have been introduced to avoid repeated longwinded explanations.

1.2 Matrices

Matrices include the cases of (1×1) matrices or scalars and matrices having only one row or one column, that is vectors. Throughout, $\mathbf{k}$ will

$\mathbb{N}$	: the set of positive integers;
$\mathbb{Z}$	: the set of all integers;
$\mathbb{Q}$	: the field of rational numbers;
$\mathbb{Q}_\times$	: the multiplicative group of nonzero rational numbers;
$\mathbb{R}$	: the field of real numbers;
$\mathbb{R}_\times$	: the multiplicative group of nonzero real numbers;
$\mathbb{C}$	: the field of complex numbers
$\mathbb{C}_\times$	: the multiplicative group of nonzero complex numbers.
$\mathbf{k}^n$	: the set of column vectors with n entries of elements from $\mathbf{k}$;
$\mathbf{k}^{(n)}$	: the set of row vectors with n entries from $\mathbf{k}$;
$\mathbf{k}^n_\times$	: the nonzero vectors of $\mathbf{k}^n$.

TABLE 1.1. Frequently used symbols

denote an arbitrary field and $\mathbb{k}$ will denote either the real or complex numbers. The notations shown in Table 1.1 will be used throughout.

The set of $(m \times n)$ matrices whose entries belong to $\mathbf{k}$ will be denoted $M_{m,n}(\mathbf{k})$. Note that m is the number of rows and n is the number of columns. There are two exceptions:

1. The set of row vectors with p entries belonging to $\mathbf{k}$ will be denoted $\mathbf{k}^{(p)}$.

2. The set of column vectors with p entries belonging to $\mathbf{k}$ will be denoted $\mathbf{k}^p$.

Matrices will generally be enclosed in "brackets"; for example:

$$\begin{bmatrix} a & b \\ c & d \end{bmatrix}, \begin{bmatrix} a \\ b \\ c \end{bmatrix} \text{ or } \begin{bmatrix} a & b & c \end{bmatrix}.$$

However, column vectors will usually be represented by writing the entries as a comma-separated horizontal list enclosed in parentheses, as in $(a_1, \ldots, a_n)$. This device conserves paper and is consistent with the notation for elements of a Cartesian product. There is no ambiguity either since an element of $\mathbf{k}^{(n)}$ will be written as a $(1 \times n)$-matrix as in $[a_1, \ldots, a_n]$.

If A is a matrix, then its $(i, j)^{\text{th}}$ entry is the element situated on the ith row and j^{th} column; I will denote this by a_{ij}. The matrix whose $(i, j)^{\text{th}}$ entry is a_{ij} will be denoted briefly as $[a_{ij}]$.

$r_p(\mathrm{A})$ and $c_q(\mathrm{A})$ will denote the pth row and the qth column of A.

If A is an $(m \times n)$ matrix then its *transpose*, denoted A^t is the $(n \times m)$ matrix whose $(j, i)^{\text{th}}$ entry is a_{ij}.

If $A \in M_{p,m}(\mathbf{k})$ and $B \in M_{n,p}(\mathbf{k})$ then the product matrix $BA \in M_{n,m}$ is the matrix whose $(i,j)^{\text{th}}$ entry is $\sum_{k=1}^{p} b_{ik}a_{kj}$.

I will write $M_n(\mathbf{k})$ for $M_{n,n}(\mathbf{k})$. A square matrix, that is an element A of $M_n(\mathbf{k})$, (for some n) is said to be *invertible* or *nonsingular* if there is a matrix $A' \in M_n(\mathbf{k})$ such that $AA' = A'A = I_n$ where I_n (known as the identity matrix, of order n,) is the $(n \times n)$ matrix whose $(i,j)^{\text{th}}$ element is the "Kronecker delta symbol" δ_{ij} which is equal to 1 if $i = j$ and equal to 0 if $i \neq j$.

The nonsingular elements of $M_m(\mathbf{k})$ form a *group* (matrix multiplication being the group operation) called the *general linear group* (of $\mathbf{k}^n$), and denoted $GL(n; \mathbf{k})$.

We recall two important functions associated with a square matrix $A = [a_{ij}]$ of order n.

1. The *trace* of A is the sum of the diagonal elements of A. Thus, $\text{Trace}\,(A) = \sum_{i=1}^{n} a_{ii}$.

2. The *determinant* A, denoted det A, is defined by the equation

$$\det A = \sum_{\sigma} \varepsilon_{\sigma} a_{1\sigma(1)} \cdots a_{n\sigma(n)},$$

where the sum ranges over all permutations σ of $\{1, \ldots, n\}$ and ε_{σ} is the *sign of the permutation* σ. This is defined as

$$\varepsilon(\sigma) = \prod_{1 \leq j < i \leq n} \left(\frac{\sigma(i) - \sigma(j)}{i - j} \right).$$

An important property of the trace is that if A, B are square matrices of the same size then Trace AB = Trace BA.

The determinant is characterized by the following properties.

1. The Normalization Condition: The identity matrix has determinant 1: $\det I_n = 1$.

2. The Alternating Condition: if A' is obtained from A by interchanging two columns (or rows) then $(\det A') = -(\det A)$.

3. The Multilinearity Condition: Let A, A' and B be square matrices of the same size and suppose they are identical except for one column, say the αth column. If the matrix B is defined in terms of its columns by the conditions:

$$c_{\beta}(B) = \begin{cases} s \cdot c_{\alpha}(A) + s' \cdot c_{\alpha}(A') & \text{if } \beta = \alpha, \\ c_{\beta}(A) = c_{\beta}(A') & \text{if } \beta \neq \alpha. \end{cases}$$

Then $\det B = (s \cdot \det A + s' \cdot \det A')$.

One other property of the determinant is fundamental; however it can be deduced from the above.

4. THE MULTIPLICATIVITY PROPERTY: If A and B are two square matrices of the same size then $\det AB = (\det A) \cdot (\det B) = \det BA$.
This implies, in particular, that if A is nonsingular then $\det A \neq 0$.

1.3 Linear Algebra

It should be clear from the titles of this and the previous section that as far as this book is concerned, Linear Algebra is something different from the algebraic manipulation of vectors and matrices. Let me make a few points before getting down to business.

- To further emphasize my point of view, henceforth I will *not* use the term vector space but use the synonymous *linear space* instead.
- I will prove many results about linear transformations (and *hence* about matrices) but rarely will matrix manipulations be used in any essential manner.
- Although my main emphasis (and examples) in this chapter will be on linear spaces over $\mathbb{k}$, most of the linear algebra that I present here would be valid for linear spaces over an arbitrary field $\mathbf{k}$.

Let E be a linear space over a field $\mathbf{k}$, The *zero vector* of E will be denoted by 0_E, or simply by 0 if no confusion is likely to arise concerning the linear space to which it belongs.

Elements of a linear space, usually called *vectors*, will be denoted by italicized bold face letters such as $\boldsymbol{e}, \boldsymbol{f}, \boldsymbol{x}, \ldots$.

Let S be a subset of E. The following are fundamental notions:

1. S is said to be *linearly independent* if whenever a finite linear combination, $\sum_{i=1}^{m} c_i \boldsymbol{s}_i$ of elements of S is equal to 0_E, all the c_i's must be $0 \in \mathbf{k}$.
2. S is said to be *linearly dependent* if S is *not* linearly independent.
3. S is said to *span* E (or, S is called a *spanning set* for E,) if given any $\boldsymbol{x} \in E$ there are elements $\boldsymbol{s}_1, \ldots, \boldsymbol{s}_n \in S$ and $k_1, \ldots, k_n \in \mathbf{k}$ such that

$$\boldsymbol{x} = \sum_{i=1}^{n} k_i \boldsymbol{s}_i.$$

4. S is said to be a *basis* of E if S is linearly independent and also spans E. (The plural of "basis" is "bases"!)

It is easy to check that a spanning set S of E is a basis iff every vector of E has a *unique* expression as a linear combination of elements of S.

The next result, though fundamental, will not be proved since the proof would involve the use of some form of the Axiom of Choice.

Theorem 1.3.1 (The existence of Bases).
Let E be a linear space over $\mathbf{k}$.
(1) *If S is a linearly independent set of vectors in E, then S can be extended to a basis of E.*

(2) *If $S \subset E$ is a spanning set, then it contains a basis of E.*
In particular, since a singleton set $\{\boldsymbol{x}\}$ is linearly independent if $\boldsymbol{x} \neq 0$, every nonzero linear space has a basis. By convention, the singleton set consisting of the zero vector is regarded as a 0-dimensional vector space with the empty set as basis. □

I will say that a nonzero linear space E over $\mathbf{k}$ is *finite-dimensional* if it has a finite basis. In such a situation *all bases have the same cardinality* and this number, denoted $\dim E$, is called the *dimension* of E. The zero linear space is, by convention, regarded to have the empty set $\emptyset$, as a basis and hence regarded as a zero-dimensional linear space.

If E does not admit a finite basis, then E is said to be *infinite-dimensional.*

Remark 1.3.2. One word of caution concerning complex linear spaces (that is a linear space over $\mathbb{C}$) is in order at this point.

If E is a $\mathbb{C}$-linear space and $\mathcal{E} = \{\boldsymbol{e}_1, \ldots, \boldsymbol{e}_n\}$ is a basis of E over $\mathbb{C}$, then $\mathcal{E}$ is *not* a basis of E when E is regarded as a linear space over $\mathbb{R}$. For example, the complex number 1 is a basis of $\mathbb{C}$ considered as a linear space over $\mathbb{C}$ but obviously not every complex number is a *real* multiple of the complex number 1! In fact, if $\mathcal{E}$ is a basis of E over $\mathbb{C}$ then $\mathcal{E}_{\mathbb{R}} = \{\boldsymbol{e}_1, \ldots, \boldsymbol{e}_n, \boldsymbol{e}'_1, \ldots, \boldsymbol{e}'_n\}$ is an $\mathbb{R}$-basis of E where for each $1 \leq k \leq n$ the vector $\boldsymbol{e}'_k = \sqrt{-1} \cdot \boldsymbol{e}_k$. In particular, if E is a linear space over $\mathbb{C}$ its dimension as an $\mathbb{R}$-linear space is twice its dimension as a $\mathbb{C}$-linear space.

Let E and F be linear spaces over the field $\mathbf{k}$. A function $T : E \to F$ is a *linear transformation* (or more precisely, a $\mathbf{k}$-linear transformation) if for all $\boldsymbol{x}, \boldsymbol{y} \in E$ and $\alpha, \beta \in \mathbf{k}$, we have the relation:

$$T(\alpha \cdot \boldsymbol{x} + \beta \cdot \boldsymbol{y}) = \alpha \cdot T(\boldsymbol{x}) + \beta \cdot T(\boldsymbol{y}).$$

It should be noted that the role of $\mathbf{k}$ is not a trivial one. For example, $\mathbb{C}$ is obviously a linear space over $\mathbb{C}$ and since $\mathbb{R} \subset \mathbb{C}$ is a subfield, it is also a linear space over $\mathbb{R}$. The function $f(x + \sqrt{-1} \cdot y) = (x - \sqrt{-1} \cdot y)$ is an $\mathbb{R}$-linear transformation but not a $\mathbb{C}$-linear transformation.

If $T : E \to E$ is a linear transformation of a linear space into itself, then T is said to be an *endomorphism* of E.

Let us return to the discussion of bases of linear spaces. The following simple result is extremely useful.

Theorem 1.3.3 (The Basis Criterion).
Let $S = \{\boldsymbol{s}_\imath : i \in \boldsymbol{I}\}$ be a subset of a linear space, E, over $\mathbf{k}$. Then the following are equivalent:
1. *S is a basis of E.*

2. *If F is any linear space and $f : S \to F$ any function then there is a* unique *linear transformation $\widetilde{f} : E \to F$ such that $\widetilde{f}|S = f$.*

Proof. Suppose that S is a basis of E. If $\boldsymbol{x} \in E$ there is a *unique* expression of $\boldsymbol{x}$ as a finite linear combination $\boldsymbol{x} = \sum_\imath \alpha_\imath \boldsymbol{s}_\imath$. Define $\widetilde{f} : E \to F$ by setting:

$$\widetilde{f}(\boldsymbol{x}) = \textstyle\sum_\imath \alpha_\imath f(\boldsymbol{s}_\imath).$$

Clearly $\widetilde{f}|S = f$. Moreover, if $T : E \to F$ is a linear transformation such that $T|S = f$, then for any $\boldsymbol{y} = \sum_\imath \beta_\imath \boldsymbol{s}_\imath$, we will have:

$$\begin{aligned} T(\boldsymbol{y}) = T\left(\textstyle\sum_\imath \beta_\imath \boldsymbol{s}_\imath\right) &= \textstyle\sum_\imath \beta_\imath T(\boldsymbol{s}_\imath) \\ &= \textstyle\sum_\imath \beta_\imath f(\boldsymbol{s}_\imath) \\ &= \widetilde{f}(\boldsymbol{y}). \end{aligned}$$

This establishes the uniqueness of the extension $\widetilde{f}$.

Conversely suppose 2 holds. Then S must be linearly independent. For suppose there are elements $\boldsymbol{s}_1, \ldots, \boldsymbol{s}_n, \boldsymbol{s}_{n+1}$ and a linear dependence relation:
$\boldsymbol{s}_{n+1} = \sum_{\imath=1}^n \beta_\imath \boldsymbol{s}_\imath$. Then for $j = 1, \ldots, (n+1)$ choose $\boldsymbol{f}_\jmath \in F$ such that:

$$\sum_{j=1}^n \beta_\jmath \boldsymbol{f}_\jmath \neq \boldsymbol{f}_{n+1}.$$

Then no function $f : S \to F$ which, for $j = 1, \ldots, (n+1)$ takes $\boldsymbol{s}_j$ to $\boldsymbol{f}_j$ can be extended to a linear transformation of E to F, contradicting 2. This shows that 2 implies that S is linearly independent.

It remains then to show that 2 implies that S is a spanning set. So, suppose 2 holds but S is *not* a spanning set. We have just proved that S is linearly independent, so let $\mathcal{B} \supset S$ be a basis. (We are appealing to Theorem 1.3.1). Then given any $\boldsymbol{e} \in E$ there are unique elements $\{k_{\boldsymbol{b}}(\boldsymbol{e})\}$ such that

$$\boldsymbol{e} = \sum_{\boldsymbol{b} \in \mathcal{B}} k_{\boldsymbol{b}}(\boldsymbol{e})\boldsymbol{b}.$$

Define $L : E \to E$ by setting

$$L(\boldsymbol{e}) = \sum_{\boldsymbol{b} \in S} k_{\boldsymbol{b}}(\boldsymbol{e})\boldsymbol{b}.$$

Then $L : E \to E$ is a linear transformation. But this means there are *two* linear transformations $E \to E$ whose restrictions to S is the inclusion map $S \to E$: namely, L and the identity map of E. This contradicts the uniqueness part of condition 2, unless $\mathcal{B} = S$.

This completes the proof of the Basis Criterion. □

Let E, F be linear spaces over $\mathbf{k}$. Then the set of $\mathbf{k}$-linear transformations from E to F, denoted $L_{\mathbf{k}}(E, F)$ or $L(E, F)$, form a linear space (over $\mathbf{k}$) if we define the linear space operations as follows. For any choices of $T, T' \in L(E, F)$ and $\alpha, \alpha' \in \mathbf{k}$,

$$(\alpha T + \alpha' T')(\boldsymbol{x}) = \alpha T(\boldsymbol{x}) + \alpha' T'(\boldsymbol{x}),$$

for all $\boldsymbol{x} \in E$.

We look at a few examples of linear transformations.

Example 1.3.4a. For any linear space, E, the identity map $1_E : E \to E$ and any scalar multiple of 1_E are linear transformations.

Example 1.3.4b. Let σ be any permutation of $\{1, \ldots, n\}$. Then the functions $T_\sigma : \mathbf{k}^n \to \mathbf{k}^n$ defined by

$$T_\sigma(x_1, \ldots, x_n) = (x_{\sigma(1)}, \ldots, x_{\sigma(n)})$$

define (via the Basis Criterion) linear transformations from $\mathbf{k}^n \to \mathbf{k}^n$.

We now recall some elementary notions concerning linear transformations and discuss their relationship to matrices. Let $T : E \to F$ be a linear transformation of linear spaces. Then the *image* of T is the set : $\operatorname{Image} T = \{T(\boldsymbol{x}) \,:\, \boldsymbol{x} \in E\}$

and the *kernel* of T is the set : $\ker T = \{\boldsymbol{x} \in E : T(\boldsymbol{x}) = 0_F\}$.
It is clear that Image T (resp. $\ker T$) is a linear subspace of F (resp. E). T is injective iff $\ker T = \{0_E\}$ that is, no nonzero vector of E is taken to zero by T. The dimensions of Image T and $\ker T$ are known as the *rank* and *nullity* of T respectively and denoted $\rho(T)$ and $\nu(T)$ respectively. We have the following relationship between the rank and nullity:

(The rank-nullity identity): .... $\dim E = \rho(T) + \nu(T)$.

In particular, if $\dim E = \dim F < \infty$ then a linear transformation is surjective (resp. injective) iff it is injective (resp. surjective.) A linear transformation which is bijective is said to be an *isomorphism* and two linear spaces are said to be *isomorphic* if there exists an isomorphism between them. If E and F are isomorphic I will use the notation $E \approxeq F$ and the notation $T : E \xrightarrow{\approxeq} F$ will signify that the linear transformation $T : E \to F$ is an isomorphism. Clearly two finite-dimensional spaces are isomorphic iff they have the same dimension; if $\mathcal{B} \subset E$ and $\mathcal{B}' \subset E'$ are bases with the same cardinality, by Proposition 1.3.3 any bijection between these bases extends to an isomorphism between E and E'.

Matrices and Linear Transformations:

Suppose that $T : E \to F$ is a $\mathbf{k}$-linear transformation. If $\mathcal{E} = (\boldsymbol{e}_1, \ldots, \boldsymbol{e}_n)$ and $\mathcal{F} = (\boldsymbol{f}_1, \ldots, \boldsymbol{f}_m)$ are ordered tuples which are bases of E and F respectively. Then there exist unique elements $c_{ij} \in \mathbf{k}$ such that for each $1 \le i \le n$, we have relations:
$(\star_i)$ .... $T(\boldsymbol{e}_i) = \sum_{k=1}^{m} c_{ki} \boldsymbol{f}_k$.
The mn elements c_{ij}, occurring in the equations $(*_i)$ as i varies over $1, \ldots, n$ are usually written in a rectangular array as shown below:

$$\begin{bmatrix} c_{11} & c_{12} & \cdots & c_{1n} \\ c_{21} & c_{22} & \cdots & c_{2n} \\ \cdot & \cdot & \cdot & \cdot \\ \cdot & \cdot & \cdot & \cdot \\ \cdot & \cdot & \cdot & \cdot \\ c_{m1} & c_{m2} & \cdots & c_{mn} \end{bmatrix}.$$

This is the *matrix* of T with respect to the bases $\mathcal{E}$ and $\mathcal{F}$ and denoted $\mathrm{M}(T; \mathcal{E}, \mathcal{F})$. If $E = F$, then the matrix, $\mathrm{M}(T; \mathcal{E}, \mathcal{E})$ is said to be the matrix of T with respect to the basis $\mathcal{E}$ and denoted $\mathrm{M}(T; \mathcal{E})$.

There is a very natural way in which one can associate with any matrix, $\mathrm{A} \in \mathrm{M}_{m,n}(\mathbf{k})$, a $\mathbf{k}$-linear transformation, $T(\mathrm{A}) : \mathbf{k}^n \to \mathbf{k}^m$.

I have to first introduce the notion of "standard bases" of the linear spaces $\mathbf{k}^p$.

Definition 1.3.5 (Standard bases).
For all $p \in \mathbb{N}$ the *standard basis* of $\mathbf{k}^p$ will be the p column vectors:

$$\begin{aligned} \boldsymbol{e}_1 &= (1, 0, \dots, 0) \\ \boldsymbol{e}_2 &= (0, 1, \dots, 0) \\ &\vdots \\ \boldsymbol{e}_p &= (0, \dots, 0, 1). \end{aligned}$$

Here each column vector has p entries and the ith vector has 1 in the ith place and 0's everywhere else. *The notation is ambiguous* since $\boldsymbol{e}_i$ will represent column vectors of different sizes in $\mathbf{k}^2$ and $\mathbf{k}^3$ for example. Only when standard bases of spaces of tuples of different dimensions are being simultaneously considered will I bother to remove the ambiguity: for example by putting a superscript $\boldsymbol{e}_i^N$ or writing $\boldsymbol{e}_i(N)$ to indicate the dimension of the space. This will not cause serious confusion.

Let $\mathrm{A} \in \mathrm{M}_{m,n}(\mathbf{k})$. Define $T(\mathrm{A}) : \mathbf{k}^n \to \mathbf{k}^m$ to be the unique linear transformation which, for $1 \le j \le n$, takes the jth standard basis vector, $\boldsymbol{e}_j(n) \in \mathbf{k}^n$ to the jth column of A. (The existence and uniqueness of $T(\mathrm{A})$ follows from the Basis Criterion.)

Remark 1.3.6. It is important to keep in mind that what I have just done above is to set up a "two-way street" for travelling between linear transformations and matrices.

It is *not* a one-to-one correspondence but this failure does not diminish its importance or usefulness. Indeed once bases are fixed in E and F, then there is a bijective correspondence between $L(E, F)$ and $\mathrm{M}_{p,q}(\mathbf{k})$ where $q = \dim E$ and $p = \dim F$.

Before I give an example of the kind of insight one can derive from this correspondence, I recall for you the following fundamental result.

Proposition 1.3.7 (Composition and products).
Let $S : E \to F$ and $T : F \to G$ be $\mathbf{k}$-linear transformations of finite-dimensional linear spaces. Let $\mathcal{E}$, $\mathcal{F}$ and $\mathcal{G}$ be ordered bases of E, F and G respectively. If $\mathrm{A} = \mathrm{M}(S; \mathcal{E}, \mathcal{F})$ and $\mathrm{B} = \mathrm{M}(T; \mathcal{F}, \mathcal{G})$ then the matrix of TS with respect to $\mathcal{E}$ and $\mathcal{G}$ is the product matrix BA. □

This immediately leads to a criterion for the nonsingularity of square matrices.

Proposition 1.3.8. *Let* A *be an* $(n \times n)$ *matrix. Then* $\mathrm{A} \in \mathrm{GL}(n; \mathbf{k})$ *iff the columns of* A *form a basis of* $\mathbf{k}^n$. □

Proof. The columns of A will be a basis iff the function which, for $k = 1, \ldots, n$, takes $c_k(\mathrm{A})$ to $\boldsymbol{e}_k(n)$ extends via the Basis Criterion to a linear transformation $T' : \mathbf{k}^n \to \mathbf{k}^n$. Let A′ be the matrix of T' with respect to the standard basis of $\mathbf{k}^n$. From the previous proposition, it follows that $\mathrm{AA}' = \mathrm{A}'\mathrm{A} = \mathrm{I}_n$. □

Remark 1.3.9 (Bases and General Linear Groups)**.** Suppose E is an n-dimensional linear space over $\mathbf{k}$. Then given any basis $\mathcal{B} = \{\boldsymbol{x}_1, \ldots, \boldsymbol{x}_n\}$, we get an isomorphism $T_{\mathcal{B}} : E \to \mathbf{k}^n$: this is the unique linear map which takes $\boldsymbol{x}_i$ to the $\imath^{\text{th}}$ standard basis vector of $\mathbf{k}^n$. Once this has been done any basis of E corresponds to a unique invertible matrix $\mathrm{M} \in \mathrm{GL}(n; \mathbf{k})$. If $\mathcal{E}$ is another basis of E the corresponding element $\mathrm{M}(\mathcal{E}) \in \mathrm{GL}(n; \mathbf{k})$ is the matrix *with respect to the basis* $\mathcal{B}$ of the unique linear transformation which takes the basis $\mathcal{B}$ to $\mathcal{E}$.

It is as though the set of bases of E was the group $\mathrm{GL}(n; \mathbf{k})$ except for the fact that we can choose the identity element of the group to correspond to any basis we want. Thinking of bases as elements of a group is a very useful idea as we shall see in the next section.

The following fundamental result explains the relationship between the matrices, $\mathrm{M}(T; \mathcal{E})$, of a linear endomorphism $T : E \to E$ with respect to different bases of E.

Theorem 1.3.10 (Change of bases)**.**
Let $\mathcal{E} = \{\mathbf{e}_1, \ldots, \mathbf{e}_n\}$ *and* $\mathcal{E}' = \{\mathbf{e}'_1, \ldots, \mathbf{e}'_n\}$ *be two ordered bases for* E. *Suppose that* $\phi : E \to E$ *is the unique linear transformation such that* $\phi(\mathbf{e}_\imath) = \mathbf{e}'_\imath$. *If* $\mathrm{A} = \mathrm{M}(\phi : \mathcal{E})$ *then for every linear transformation* $T : E \to E$, *the matrices of* T *with respect to* $\mathcal{E}$ *and* $\mathcal{E}'$ *are related as follows:*

$$\mathrm{M}(T : \mathcal{E}') = \mathrm{AM}(T : \mathcal{E})\mathrm{A}^{-1}.$$ □

Since the columns and rows of an $(m \times n)$ matrix belong to the linear space $\mathbf{k}^m$ and $\mathbf{k}^{(n)}$ respectively one can talk of the linear independence of these rows and columns. The dimension of the subspace of $\mathbf{k}^{(n)}$ (resp. $\mathbf{k}^m$) spanned by the rows (resp. columns) of matrix A is defined to be the *row rank* (resp. *column rank*) of A. For any matrix both the row rank and column ranks are equal to the rank of the linear transformation $T(\mathrm{A})$.

Two square matrices M and N are said to be *similar* if there is an invertible matrix A such that $\mathrm{N} = \mathrm{A}^{-1}\mathrm{MA}$. If two matrices arise as representations of the same lineartransformation of a linear space $T : V \to V$ with respect to different bases of V then these matrices are similar. Conversely, suppose $\mathrm{N} = \mathrm{A}^{-1}\mathrm{MA}$ for some nonsingular matrix A. Consider the linear transformation $T(\mathrm{A})$, that is the linear transformation $\mathbf{k}^n \to \mathbf{k}^n$ which takes the $\imath^{\text{th}}$ standard basis vector to the $\imath^{\text{th}}$ column of A. The nonsingularity of A implies that $\mathcal{E}' = \{T(e_1), \ldots, T(e_n)\}$ form a basis of $\mathbf{k}^n$. Define a linear transformation $S : \mathbf{k}^n \to \mathbf{k}^n$ by setting $S(\boldsymbol{e}_\imath)$ equal to the $\imath^{\text{th}}$ column of M. Then the matrix of S with respect to the basis $\mathcal{E}'$ is N.

Proposition 1.3.11. *Let* A *and* B *be two similar matrices. Then,*

$$\begin{aligned} \mathrm{Trace\,A} &= \mathrm{Trace\,B} \\ \textit{and} \quad \det \mathrm{A} &= \det \mathrm{B}. \end{aligned}$$

Proof. For suppose $\mathrm{A} = \mathrm{M}^{-1}\mathrm{BM}$ for some nonsingular matrix M. Then using the property, $\mathrm{Trace\,XY} = \mathrm{Trace}\,YX$, we get:

$$\mathrm{Trace\,A} = \mathrm{Trace\,(M^{-1}BM)} = \mathrm{Trace\,(BMM^{-1})} = \mathrm{Trace\,B}.$$

Since $\det \mathrm{MM}^{-1} = \det \mathrm{I}_n = 1,$, the multiplicativity property of the determinant implies that $\det(\mathrm{M}^{-1}) = (\det \mathrm{M})^{-1}$. Using the multiplicativity property again we get $\det A = (\det \mathrm{M})^{-1}(\det \mathrm{B})(\det \mathrm{M}) = \det \mathrm{B}$. □

The significance of this result is that if $T : V \to V$ is an endomorphism of a finite-dimensional linear space then we can (and we do) define *unambiguously* $\det T$ and Trace T, by choosing a basis $\mathcal{B}$ of V and setting:

$$\begin{aligned} \mathrm{Trace}\,T &= \mathrm{Trace\,M}(T;\mathcal{B}) \\ \det T &= \det \mathrm{M}(T;\mathcal{B}). \end{aligned}$$

Exercises for § 1.3

1.3.1 Quotient Spaces and Direct Sums

Quotient Spaces Let E be a linear space over $\mathbf{k}$ and $F \subset E$ a subspace of E. Define a relation $\sim_F$ on E by setting:

$$\boldsymbol{x} \sim_F \boldsymbol{y} \text{ iff } \boldsymbol{x} - \boldsymbol{y} \in F.$$

Show that $\sim_F$ is an equivalence relation. Let E/F denote the equivalence classes and $q : E \to E/F$ the function which takes a vector $\boldsymbol{x} \in E$ to its equivalence class, $[\boldsymbol{x}]_F$.

1. Show that addition and scalar multiplication, defined by:
$$\begin{aligned} [\boldsymbol{x}]_F + [\boldsymbol{y}]_F &\doteq [\boldsymbol{x}+\boldsymbol{y}]_F \\ \alpha \cdot [\boldsymbol{x}]_F &\doteq [\alpha \cdot \boldsymbol{x}]_F \end{aligned}$$
are well-defined and yield a **k**-linear space structure on E/F. E/F is known as the quotient (linear) space, E modulo F. Show that q is a surjective linear transformation whose kernel is F.

2. Let $T : E \to V$ be any surjective **k**-linear transformation onto a linear space V such that $\ker T = F$. Show that the assignment:
$$\boldsymbol{v} \mapsto [\boldsymbol{x}]_F, \qquad \text{where } T(\boldsymbol{x}) = \boldsymbol{v}$$
defines an isomorphism from V to E/F.

3. Show that $\dim(E/F) = \dim E - \dim F$.

Direct Sums 1. Let $E_1, \ldots, E_n$ be linear spaces over **k**. Recall that the Cartesian product, $E_1 \times \cdots \times E_n$, has a linear space structure with addition and scalar multiplication defined coordinate wise; that is these operations are defined by:
$$\begin{aligned} (\boldsymbol{x}_1, \ldots, \boldsymbol{x}_n) + (\boldsymbol{y}_1, \ldots, \boldsymbol{y}_n) &\doteq (\boldsymbol{x}_1 + \boldsymbol{y}_1, \ldots, \boldsymbol{x}_n + \boldsymbol{y}_n) \\ \alpha(\boldsymbol{x}_1, \ldots, \boldsymbol{x}_n) &\doteq (\alpha\boldsymbol{x}_1, \ldots, \alpha\boldsymbol{x}_n). \end{aligned}$$
where the '+' on the left side is the addition of n-tuples, and the '+' signs on the right are operations in the linear spaces E_k and similar abuses of notation have been perpetrated with respect to scalar multiplication. This defines the (external) *direct sum* of the spaces $E_1, \ldots, E_n$.

2. Let E be a linear space and let $E_1, \ldots, E_p$ be subspaces of E. We say that these subspaces *span* E if any vector $\boldsymbol{e} \in E$ can be written in the form $\boldsymbol{e} = \boldsymbol{e}_1 + \cdots + \boldsymbol{e}_p$, where $\boldsymbol{e}_k \in E_k$.

   The linear subspaces $E_1, \ldots, E_p$ are said to be *linearly disjoint* if whenever $\boldsymbol{e}_k \in E_k$ and $\boldsymbol{e}_1 + \cdots + \boldsymbol{e}_p = 0_E$, then for each $k = 1, \ldots, p$, $\boldsymbol{e}_k = 0 \in E_k$.

   Suppose the subspaces $E_1, \ldots, E_p$ span E.

   Show that $E_1, \ldots, E_p$ is linearly disjoint iff each vector of E can be written in a *unique* fashion as a sum of vectors from E_k.

   Show that, in this situation, $E \approxeq E_1 \oplus \cdots \oplus E_p$. (This is sometimes referred to as the *internal* direct sum of the spaces $E_\imath$).

1.3.2 On the rank-relation

A. Let $R, S, T : E \to E$ be endomorphisms of a $\mathbf{k}$-linear space, E. Show that

$$\rho(ST) \le \min\big(\rho(S), \rho(T)\big)$$
$$\text{and} \quad \rho(RS) + \rho(ST) \le \rho(S) + \rho(RST).$$

B. The Euler-Poincaré Identity:
Let $V_0, V_1, \ldots, V_n$ be finite-dimensional linear spaces; $\dim V_k = n_k$.

Suppose for each $k = 1, \ldots, n$ there are linear transformations $\Delta_k : V_k \to V_{k-1}$ such that $\Delta_{k-1}\Delta_k = 0$. (You should assume that Δ_0 is the zero linear transformation taking all vectors to the zero vector of the trivial linear space $\{0\}$ containing only a zero vector.)

Such a family of linear spaces and linear transformations is called a *chain complex* and denoted V_*.

Consider the linear subspaces:

$$\text{For } 0 \le k \le n \quad \begin{cases} Z_k &= \ker \Delta_k \subset V_k, \\ B_k &= \operatorname{Image} \Delta_{k+1} \subset V_k, \\ B_n &= \{0\} \subset V_n. \end{cases}$$

These are known as the spaces of *k-cycles* and *k-boundaries* of the chain complex V_* respectively. Clearly $B_k \subset Z_k$ for each k. The quotient space Z_k/B_k is known as the kth homology of the complex V_* and denoted $H_k(V_*)$ or simply H_k. Suppose $\dim H_k = h_k$. Then establish the Euler–Poincaré identity:

$$\sum_{k=0}^{n}(-1)^k h_k = \sum_{k=0}^{n}(-1)^k n_k.$$

Hint: Apply the rank–nullity relation to the *surjective* linear transformations $q_k : Z_k \to H_k$ and $\Delta_{k+1} : V_{k+1} \to B_k$, where $V_{n+1} = \{0\}$.

1.4 Euclidean and Unitary spaces

In this section we will only consider linear spaces over $\Bbbk$. If E is a linear space over $\mathbb{R}$ then a function $g : E \times E \to \mathbb{R}$ is said to be an *inner product* on E if g satisfies the following properties.

For all $\boldsymbol{x}, \boldsymbol{y}, \boldsymbol{z} \in E$ and $s, t \in \mathbb{R}$ the following relations hold.

(Bilinearity-1)	$g(s\boldsymbol{x} + t\boldsymbol{y}, \boldsymbol{z}) = sg(\boldsymbol{x}, \boldsymbol{z}) + tg(\boldsymbol{y}, \boldsymbol{z})$
(Bilinearity-2)	$g(\boldsymbol{x}, s\boldsymbol{y} + t\boldsymbol{z}) = sg(\boldsymbol{x}, \boldsymbol{y}) + tg(\boldsymbol{x}, \boldsymbol{z})$
(Symmetry)	$g(\boldsymbol{x}, \boldsymbol{y}) = g(\boldsymbol{y}, \boldsymbol{x}),$
(Positivity)	$g(\boldsymbol{x}, \boldsymbol{x}) \geq 0$
(Nonsingularity)	$g(\boldsymbol{x}, \boldsymbol{x}) = 0$ iff $\boldsymbol{x} = 0.$

The inner product will usually be denoted $\langle \cdot, \cdot \rangle$ instead of $g(\cdot, \cdot)$. One refers to $\langle \boldsymbol{x}, \boldsymbol{y} \rangle$ as the inner product of $\boldsymbol{x}$ and $\boldsymbol{y}$.

The positive square root of the inner product of a vector with itself is called the *norm* or length of the vector. The norm of $\boldsymbol{x}$ is denoted $||\boldsymbol{x}||$. Thus, $||\boldsymbol{x}|| = \sqrt{\langle \boldsymbol{x}, \boldsymbol{x} \rangle}$,

A *Euclidean space* is a finite-dimensional $\mathbb{R}$-linear space equipped with an inner product.

The inner product of any two vectors in an inner product space, satisfy the Cauchy–Schwartz inequality:

Cauchy Schwartz inequality: $\langle \boldsymbol{x}, \boldsymbol{y} \rangle \leq ||\boldsymbol{x}|| \cdot ||\boldsymbol{y}||.$

The most important consequence of the Cauchy-Schwartz inequality is the following relation which holds for any two vectors.

Triangle Inequality: $\boldsymbol{x} + \boldsymbol{y}|| \leq ||\boldsymbol{x}|| + ||\boldsymbol{y}||$

Examples of Inner Product Spaces

Example 1.4.1a (The standard inner product on $\mathbb{R}^n$).
$\mathbb{R}^n$ has the so-called *standard inner product*:

The Standard $\mathbb{R}^n$: $\langle (x_1, \dots, x_n), (y_1, \dots, y_n) \rangle = \sum_{i=1}^{n} x_i y_i.$

Example 1.4.1b (The L_2 inner product on $\mathcal{C}(\mathbf{I})$).
Let $E = \mathcal{C}(\mathbf{I})$ be the set of continuous $\mathbb{R}$-valued functions defined on $\mathbf{I} = [0, 1]$. This is an infinite-dimensional linear space on which one can define an inner product by setting

$$\langle f, g \rangle = \int_0^1 f(t)g(t)dt.$$

The only point that is not entirely trivial to check is the positivity condition; we do this as follows.

Suppose $f \in E$ is a nonzero function. suppose that $f(p) \neq 0$, where $p \in (0,1)$. Since f is continuous, there is an interval $I = [p-\delta, p+\delta]$ such that if $x \in I$ then, $|f(x)| > \frac{|f(p)|}{2} = \kappa$, say. Then we can calculate :

$$||f|| = \int_0^1 |f(t)|^2 dt \geq 2\delta \cdot \kappa^2 > 0,$$

showing that the norm of any nonzero function is strictly positive.

Example 1.4.1c (A nonexample!).
Let $F = \mathcal{R}(\mathbf{I})$ be the set of Riemann integrable functions on $[0,1]$. The same formula as the L_2 inner product on the set of continuous functions, does not yield an inner product on F. The positivity condition will be violated by the characteristic function of every finite subset of $[0,1]$, since these are Riemann integrable nonzero functions with "norm" equal to zero.

I now recall the analogous notion for complex linear spaces. If E is a linear space over $\mathbb{C}$, a function $h : E \times E \to \mathbb{C}$ is called a *Hermitian product* or complex inner product on E if for all $\boldsymbol{x}, \boldsymbol{y}, \boldsymbol{z} \in E$ and $\alpha, \beta \in \mathbb{C}$, we have the relations:

(conjugate-linearity$_1$)

$$h(\alpha\boldsymbol{x} + \beta\boldsymbol{y}, \boldsymbol{z}) = \overline{\alpha}h(\boldsymbol{x}, \boldsymbol{z}) + \overline{\beta}h(\boldsymbol{y}, \boldsymbol{z})$$

(linearity$_2$) $\quad h(\boldsymbol{x}, \alpha\boldsymbol{y} + \beta\boldsymbol{z}) = \alpha h(\boldsymbol{x}, \boldsymbol{y}) + \beta h(\boldsymbol{x}, \boldsymbol{z})$

(Conjugate symmetry) $\quad h(\boldsymbol{y}, \boldsymbol{x}) = \overline{h(\boldsymbol{x}, \boldsymbol{y})}$

(Nonsingularity) $\quad h(\boldsymbol{x}, \boldsymbol{x}) \geq 0$, and equality holds iff $\boldsymbol{x} = 0$.

As with inner products, a Hermitian product is also written $\langle \cdot, \cdot \rangle$ instead of $h(\cdot, \cdot)$. A finite-dimensional complex linear space with a Hermitian product will be called a *Unitary space.*

There is a version of the Cauchy–Schwarz inequality for unitary spaces:

$$|\langle \boldsymbol{x}, \boldsymbol{y} \rangle| \leq ||\boldsymbol{x}|| \cdot ||\boldsymbol{y}||,$$

holds for any two vectors, $\boldsymbol{x}, \boldsymbol{y}$ belonging to a Unitary space.
(Here on the left hand side of the inequality we have the absolute value of a complex number and on the right hand side, $||\boldsymbol{v}||$ represents the "norm" or positive square root of the hermitian product of $\boldsymbol{v}$ with itself.)

There is also a corresponding triangle inequality, $||\boldsymbol{x}+\boldsymbol{y}|| \leq ||\boldsymbol{x}||+||\boldsymbol{y}||$.

Since Euclidean and Unitary spaces admit a notion of length of vectors (indeed in Euclidean spaces there is also a notion of the "angle" between two vectors,) I will employ geometrical terminology when discussing such spaces.

EXAMPLES OF UNITARY SPACES

Example 1.4.2a (The standard Hermitian product on $\mathbb{C}^n$).
$\mathbb{C}^n$ has the so-called standard Hermitian product:

$$\langle (z_1, \ldots, z_n), (\zeta_1, \ldots, \zeta_n) \rangle = \sum_{j=1}^{n} \bar{z}_j \zeta_j.$$

Example 1.4.2b (Frobenius product on $\mathrm{M}_n(\mathbb{C})$).
The set of $(n \times n)$ matrices, $\mathrm{M}_n(\mathbb{C})$, becomes a Unitary space if we set:

$$\langle \mathrm{A}, \mathrm{B} \rangle = \mathrm{Trace}\,(\mathrm{A}^*\mathrm{B})$$

where A^* (usually called the *adjoint* of A) is the matrix whose $(i,j)^{\mathrm{th}}$ entry is $\bar{a}_{ji}$. When restricted to $\mathrm{M}_n(\mathbb{R})$ this yields an inner product:

$$\langle \mathrm{A}, \mathrm{B} \rangle = \mathrm{Trace}\,(A^t B),$$

where A^t is the transpose of A.

Remark. If we regard the entries a_{ij} of a matrix $\mathrm{A} \in \mathrm{M}_n(\mathbb{k})$ (in any fixed ordering) as the components of a vector in $\mathbb{k}^{n^2}$, then the Frobenius inner product of two matrices is the same as the standard (real or complex) inner product ($\mathbb{k}^{n^2}$) of the corresponding vectors.

Example 1.4.2c (Inner products induced by Hermitian products).
If E is a Unitary space, then the underlying *real linear space* $E(\mathbb{R})$ is a Euclidean space if we set $\langle \boldsymbol{x}, \boldsymbol{y} \rangle_{E(\mathbb{R})} = \Re(\langle \boldsymbol{x}, \boldsymbol{y} \rangle_E)$, where $\Re(z)$ denotes the real part of the complex number, z. This is known as the inner product *induced by a Hermitian product.*

In a unitary or Euclidean space a set of vectors $X = \{\boldsymbol{x}_\alpha\}_{\alpha \in A}$ is said to be *orthogonal* if whenever α, β are distinct elements of A, the inner product $\langle \boldsymbol{x}_\alpha, \boldsymbol{x}_\beta \rangle = 0$. An orthogonal set is necessarily a linearly independent set. (Exercise!) An orthogonal set of vectors is said to be *orthonormal* if each vector has length 1. An orthonormal basis (or onb for short) in E is a basis which is also an orthonormal set.

Let $T : E \to F$ be a $\mathbb{R}$-linear transformation between Euclidean spaces. Then, T is said to be an *isometry* if any of the following equivalent conditions hold.

1. T preserves inner products; that is, $\langle \boldsymbol{x}, \boldsymbol{y} \rangle = \langle T(\boldsymbol{x}), T(\boldsymbol{y}) \rangle$ for all $\boldsymbol{x}, \boldsymbol{y} \in E$;

2. T preserves lengths of vectors; that is, $||\boldsymbol{x}|| = ||T(\boldsymbol{x})||$ for all $\boldsymbol{x} \in E$.

   If $E = F$ is an Euclidean space we can also give one more equivalent condition:

3. If A is the matrix of T with respect to an orthonormal basis of E then $\mathrm{A}^{-1} = \mathrm{A}^t$. Such matrices are called *orthogonal matrices.*

   For Unitary spaces a $\mathbb{C}$-linear endomorphism is an isometry if the same conditions hold with one modification: the matrix of such an isometry, with respect to an onb satisfies $\mathrm{A}^{-1} = \mathrm{A}^*$, the adjoint of A.

Observe that the second condition implies that an *isometry is always injective.* Also note that the orthogonal (resp. unitary) matrices of a given size form a group under matrix multiplication. The group of $(n \times n)$ orthogonal (resp. unitary) matrices is denoted $\mathrm{O}(n)$ (resp. $\mathrm{U}(n)$).

Proposition 1.3.8 admits the following refinement.

Proposition 1.4.3. *Let* $\mathrm{A} \in \mathrm{M}_n(\mathbb{k})$. *Then* $T(\mathrm{A}) : \mathbb{k}^n \to \mathbb{k}^n$ *is an isometry iff the columns of* A *form an orthonormal basis of* $\mathbb{k}^n$. □

Example 1.4.4 (Isometries of a space with $\mathbb{k}^n$).
Let E be any Euclidean or Unitary space and let $\mathcal{E} = \{\boldsymbol{e}_1, \dots, \boldsymbol{e}_n\}$ be an onb of E. Then, for any $\boldsymbol{x} \in E$ there are unique elements $\alpha_1, \dots, \alpha_n \in \mathbb{k}$ such that $\boldsymbol{x} = \sum_{i=1}^n \alpha_i \boldsymbol{e}_i$. The α_i are determined by the relations

$$\alpha_i = \langle \boldsymbol{e}_i, \boldsymbol{x} \rangle \quad \text{for all } 1 \leq i \leq n.$$

It can be easily checked that the function $\phi : E \to \mathbb{k}^n$ defined by

(*)...................... $\phi(\boldsymbol{x}) = (\alpha_1, \dots, \alpha_n)$

is an isometry from E to $\mathbb{k}^n$ (with its standard Euclidean or Hermitian product).

In fact every isometry from E to $\mathbb{k}^n$ arises in this way.

Remark 1.4.5 (Orthonormal bases and $\mathbb{U}(n; \mathbb{k})$).
Here $\mathbb{U}(n; \mathbb{k})$ denotes the unitary group if $\mathbb{k} = \mathbb{C}$ and the orthogonal group if $\mathbb{k} = \mathbb{R}$. If $\mathcal{U} = \{\boldsymbol{u}_1, \dots, \boldsymbol{u}_n\}$ is an orthonormal basis of E, any onb, $\mathcal{O}$ corresponds to an isometry (the linear transformation taking $\mathcal{U}$ to the onb $\mathcal{O}$) and *vice versa.*

This is the geometric analogue of Remark 1.3.9 which was purely *algebraic.* The additional comments made there concerning the set of **k**-bases of a **k**-linear space and the groups $\mathrm{GL}(n, \mathbf{k})$ apply (*mutatis mutandis*) also to the set of onb's of inner products spaces and the groups $\mathbb{U}(n;\mathbb{k})$.

The Gram–Schmidt orthonormalization procedure

I will make a small detour to point out the usefulness of the observations that the bases (resp. onb's) of a unitary space can be identified with $\mathrm{GL}(n;\mathbb{k})$ (resp. $\mathbb{U}(n;\mathbb{k})$) by reviewing the Gram–Schmidt orthonormalization procedure from a group-theoretic perspective.

Given any basis, $\mathcal{B} = \{\boldsymbol{b}_1, \dots, \boldsymbol{b}_n\}$, of an Euclidean or Unitary space E, the Gram–Schmidt procedure yields an orthonormal basis $\mathcal{E} = \{\boldsymbol{e}_1, \dots, \boldsymbol{e}_n\}$ with the property that for each $1 \le k \le n$, the space spanned by $\boldsymbol{b}_1, \dots, \boldsymbol{b}_k$ and $\boldsymbol{e}_1, \dots, \boldsymbol{e}_k$ are the same.

The new idea is to think of the basis $\mathcal{B}$ as a nonsingular matrix B whose i^{th} column is $\boldsymbol{b}_i$.

At the 1st stage of the G–S (Gram–Schmidt) procedure one divides $\boldsymbol{b}_1$ by its length; yielding a new basis $\mathcal{B}_1 = \{\boldsymbol{e}_1, \boldsymbol{b}_2, \dots, \boldsymbol{b}_n\}$ where $\boldsymbol{e}_1 = \frac{\boldsymbol{b}_1}{||\boldsymbol{b}_1||}$. Let B_1 be the matrix corresponding to $\mathcal{B}_1$. Now the transformation $\mathcal{B} \mapsto \mathcal{B}_1$ can also be achieved by *right* multiplying the nonsingular matrix B by the diagonal matrix $\mathrm{T}_{0,1}$ with $c_1 = \frac{1}{||\boldsymbol{b}_1||}$ as $(1,1)^{\text{th}}$ entry and all other diagonal entries equal to 1.

At the second stage of the G–S procedure one gets a new basis:

$$\begin{aligned} \mathcal{B}_2 &= \{\boldsymbol{e}_1, \boldsymbol{e}_2, \boldsymbol{b}_3, \dots, \boldsymbol{b}_n\}, \\ \text{where } \boldsymbol{e}_2 &= \frac{\boldsymbol{f}_2}{||\boldsymbol{f}_2||}, \\ \text{and } \boldsymbol{f}_2 &= \boldsymbol{e}_1 + c\boldsymbol{b}_2; \end{aligned}$$

the constant c being chosen so as to make $\boldsymbol{e}_1$ and $\boldsymbol{f}_2$ orthogonal. If B_2 is the nonsingular matrix corresponding to this basis, it is easily checked that

$$\mathrm{B}_2 = \mathrm{B}_1 \cdot \mathrm{T}_{1,2}$$

where $\mathrm{T}_{1,2}$ is the matrix

$$\mathrm{T}_{1,2} = \begin{bmatrix} 1 & c/||\boldsymbol{f}_1|| & 0 & 0 & 0 \\ 0 & 1/||\boldsymbol{f}_1|| & 0 & \cdots & 0 \\ 0 & 0 & 1 & 0 & \vdots \\ \vdots & \vdots & 0 & \ddots & 0 \\ 0 & 0 & 0 & \cdots & 1 \end{bmatrix}.$$

Elementary matrix algebra (recall `column operations`) shows that if $\mathcal{B}_j$ and B_j are the basis and the corresponding nonsingular matrix obtained after the jth stage of the G-S process, then

$$\mathrm{B}_m = \mathrm{B}_{m-1} \cdot \mathrm{T}_{m-1,m}$$

where the matrices $\mathrm{T}_{k,k+1}$ satisfy the following conditions:

- they are upper triangular matrices i. e., their $(i,j)^{\text{th}}$ entries are 0 whenever $i > j$;
- the diagonal elements are positive.

Now it is easy to verify that the set of matrices which satisfy these two conditions form *a group*, which I will denote $\mathfrak{T}^+{}_n(\mathbb{k})$.($\mathfrak{T}$ is uppercase "T" in Gothic.) The above observations, concerning the Gram-Schmidt procedure together with the fact that the set of $(n \times n)$ orthogonal (resp.unitary) matrices form a group, leads to the following theorem which is a simplified and partial version of the celebrated Iwasawa Decomposition Theorem,

Theorem 1.4.6 (Iwasawa Decomposition).
If $\mathrm{B} \in \mathrm{GL}(n;\mathbb{k})$ *then there are unique matrices* $\mathrm{T_B} \in \mathfrak{T}_n^+(\mathbb{k})$ *and* $\mathrm{U_B} \in \mathbb{U}(n,\mathbb{k})$ *such that* $\mathrm{B} = \mathrm{U_B} \cdot \mathrm{T_B}$.

Proof. We will continue to use the notation that was being employed in describing the Gram–Schmidt procedure above.

Let $\mathcal{B}$ be the ordered basis of $\mathbb{k}^n$ whose i^{th} element is the ith column of the matrix $\mathrm{B} \in \mathrm{GL}(n;\mathbb{k})$. we know that after n steps the Gram-Schmidt process will yield an ordered orthonormal basis, $\mathcal{E} = (\boldsymbol{e}_1, \ldots, \boldsymbol{e}_n)$, of $\mathbb{k}^n$. Let $\mathrm{U_B} \in \mathbb{U}(n;\mathbb{k})$ be the matrix whose i^{th} column is $\boldsymbol{e}_\imath$.

From our earlier discussion, we know that there are n matrices,

$$\mathrm{T}_{0,1}, \mathrm{T}_{1,2}, \ldots, \mathrm{T}_{n-1,n} \in \mathfrak{T}_n^+(\mathbb{k})$$

such that $\mathrm{U}_B = \mathrm{B} \cdot \mathrm{T}_{0,1} \cdot \mathrm{T}_{1,2} \cdot \cdots \cdot \mathrm{T}_{n-1,n}$.

Since each of the matrices, $\mathrm{T}_{0,1}, \ldots, \mathrm{T}_{n-1,n}$ belong to the group $\mathfrak{T}_n^+(\mathbb{k})$, their product will also be in $\mathfrak{T}_n^+(\mathbb{k})$.

Define $\mathrm{T_B} = \mathrm{T}_{0,1} \cdots \mathrm{T}_{n-1,n}$.

Clearly, $\mathrm{B} = \mathrm{U_B} \cdot \mathrm{T_B}^{-1}$. This shows that any nonsingular matrix over $\mathbb{k}$ is a product of a 'unitary' and upper triangular matrix as required by the theorem.

To see that such a decomposition must be unique, it is only necessary to observe that $\mathbb{U}(n;\mathbb{k}) \cap \mathfrak{T}_n^+(\mathbb{k}) = \{\mathrm{I}_n\}$. □

Remark 1.4.7. Thus the G–S procedure which seems like a rather laborious piece of mechanical manipulation of vectors, in fact contains within itself, information concerning the structure of the general linear groups. As I have already mentioned the simple result proved here has a celebrated profound generalization to the realm of `semisimple Lie groups`.

One last word: unitary and Euclidean spaces are special cases of **normed linear spaces**, which will be the kind of space for which we will develop the tools of Differential Calculus. It might be appropriate to simply define these spaces here.

Definition (Normed Linear Spaces).
Let E be a $\Bbbk$-linear space. A *norm* on E is a function $\nu : E \to \mathbb{R}$ satisfying the conditions:

1. **Positivity:** $\nu(\boldsymbol{x}) \geq 0$ for all $\boldsymbol{x} \in E$ and $\nu(\boldsymbol{x}) = 0$ iff $\boldsymbol{x} = 0$;

2. **Subadditivity:** $\nu(\boldsymbol{x} + \boldsymbol{y}) \leq \nu(\boldsymbol{x}) + \nu(\boldsymbol{y})$ for all $\boldsymbol{x}, \boldsymbol{y} \in E$;

3. **Homogeneity:** $\nu(\alpha \cdot \boldsymbol{x}) = |\alpha| \cdot \nu(\boldsymbol{x})$ for all $\alpha \in \Bbbk$ and $\boldsymbol{x} \in E$, where $|\alpha|$ is the absolute value of $\alpha \in \Bbbk$..

as in the case of Euclidean spaces, $\nu(x)$, the norm of $\boldsymbol{x}$ is usually denoted $||\boldsymbol{x}||$. A *normed linear space* is a $\Bbbk$-linear space equipped with a norm.

Normed spaces will be studied in some depth in Chapter 3.

Exercises for § 1.4

1.4.1 G–S Procedure in a linear space of functions

Let $\mathcal{P}_2$ be the linear space of real polynomials of degree ≤ 2 that is

$$\mathcal{P}_2 = \Big\{ \sum_{\imath=0}^{2} c_\imath t^\imath \,:\, c_\imath \in \mathbb{R} \Big\}$$

with the obvious rules of addition and scalar multiplication. Define an inner product on $\mathcal{P}_2$ by setting

$$\langle f(t), g(t) \rangle = \int_{-1}^{1} f(t)g(t)dt.$$

Apply the G–S procedure to the basis $\{1, t, t^2\}$ of $\mathcal{P}_2$ and obtain an orthonormal basis for $\mathcal{P}_2$. These are (upto sign) the first three *Legendre polynomials*: this sequence of polynomials (usually denoted $P_n(\mu)$) are very useful in Mathematical Physics.

1.4.2 Orthogonal Complements and Projections

1. Let E be a Euclidean or unitary space $V \subset E$ any subspace. The *orthogonal complement* of V is the set $V^\perp \doteq \{\boldsymbol{x} \in E : \langle \boldsymbol{x}, \boldsymbol{v} \rangle = 0 \text{ for every } \boldsymbol{v} \in V\}$. Show that $V, V^\perp$ are linearly disjoint and span E and hence $E \approxeq V \oplus V^\perp$.

If $\boldsymbol{x} \in E$ and $\boldsymbol{x} = \boldsymbol{v} + \boldsymbol{v'}$ where $\boldsymbol{v} \in V$ and $\boldsymbol{v'} \in V^\perp$, then the function $P_V : E \to E$ defined by $P_V(\boldsymbol{x}) = v$ is a linear transformation known as the *orthogonal projection of E along V*. Show that $P_V^2 = P_V$ and that $\operatorname{Trace}(P_V) = \dim V$.

2. More generally if E is any linear space over $\mathbb{k}$, an endomorphism $T : E \to E$ is said to be an idempotent or a projection if $T^2 = T$. Show that if T is a projection then $E \approxeq (\ker T) \oplus (\operatorname{Image} T)$ and that $\operatorname{Trace} T$ is equal to the rank of T.

1.4.3 On the isometries of $\mathbb{R}^n$

For the purposes of this exercise, which explicitly describes the group of isometries of $\mathbb{R}^n$, I am going to use a slightly different (but essentially equivalent) definition of isometry from the one given in the text. The advantage one derives is quite an easy route to a nontrivial result concerning the structure of the orthogonal groups and a decomposition of any orthogonal transformation into a product of "simple" transformations .

> An isometry of $\mathbb{R}^n$ is a bijection $f : \mathbb{R}^n \to \mathbb{R}^n$ such that f and f^{-1} preserves the distances between points. A 0-isometry is an isometry which keeps the origin of $\mathbb{R}^n$ fixed.

(a) Show that the 0-isometries of $\mathbb{R}^n$ form a group, $\mathfrak{I}_0(\mathbb{R}^n)$. We will see that $\mathfrak{I}_0(\mathbb{R}^n)$ is isomorphic to the orthogonal group, $\mathrm{O}(n)$. In fact, part (d) of this exercise points out an interesting "factorization" for elements of $\mathrm{O}(n)$.

(b) Let $V \subset \mathbb{R}^n$ be a linear subspace and suppose that $g : V \to V$ is a 0-isometry of V. Let $W \doteq V^\perp$ be the orthogonal complement of V in $\mathbb{R}^n$. Define $\bar{g} : \mathbb{R}^n \to \mathbb{R}^n$, by the formula:

$$\bar{g}(\boldsymbol{v}, \boldsymbol{w}) = (g(\boldsymbol{v}), \boldsymbol{w})$$

where on both sides of the equation, elements of $\mathbb{R}^n$ are being represented as elements of $V \oplus V^\perp$.

Show that $\bar{g}$ is a 0-isometry of $\mathbb{R}^n$.

Conversely show that if a 0-isometry, of $\mathbb{R}^n$, $h : \mathbb{R}^n \to \mathbb{R}^n$, leaves a subspace $V \subset \mathbb{R}^n$ pointwise fixed, then there is a 0-isometry g of $V^\perp$ such that $h = \bar{g}$.

(c) Let $\boldsymbol{e}$ be a unit vector in $\mathbb{R}^n$. Then the *reflection*, in the hyperplane orthogonal to $\boldsymbol{e}$, is defined by the formula:

$$\boldsymbol{x} \mapsto \big(\boldsymbol{x} - 2 \cdot \langle \boldsymbol{x}, \boldsymbol{e} \rangle \cdot \boldsymbol{e}\big).$$

This reflection is denoted $\rho_{\boldsymbol{e}} : \mathbb{R}^n \to \mathbb{R}^n$.

A reflection in a Euclidean space, E, is a linear transformation of the form $\rho_{\boldsymbol{e}}$ for some unit vector $\boldsymbol{e} \in E$.

Some important (but easily derived) properties of such reflections are listed below in the form of mini-exercises.

1. Show that $\rho_{\boldsymbol{e}}$ is a 0-isometry of $\mathbb{R}^n$.
2. Show that if $\boldsymbol{x}, \boldsymbol{y} \in \mathbb{R}^2$ are any two vectors of equal length then there is a reflection in $\mathbb{R}^2$ which interchanges them.
3. Show that with respect to some, and hence, any onb of $\mathbb{R}^n$, the matrix of a reflection is an orthogonal matrix. What is the determinant of a reflection?
4. Show that if $\rho : K \to K$ is a reflection, where $K \subset \mathbb{R}^n$, then so is the extended linear transformation $\bar{\rho} : \mathbb{R}^n \to \mathbb{R}^n$.
5. Suppose $r_\theta : \mathbb{R}^2 \to \mathbb{R}^2$ is a 0-isometry of $\mathbb{R}^2$ given by $r_\theta(\boldsymbol{e}_1) = (\cos\theta, \sin\theta)$ and $r_\theta(\boldsymbol{e}_2) = (-\sin\theta, \cos\theta)$, Show that r_θ is the composition/product of two reflections in $\mathbb{R}^2$.

(d) Now suppose that $\mathbf{I} : \mathbb{R}^n \to \mathbb{R}^n$ is a 0-isometry and that $\mathbf{I}(\boldsymbol{e}_n) = \boldsymbol{x}$, where $\boldsymbol{e}_n$ is the nth standard basis vector. Show that there is a reflection $\rho_{\mathbf{I}}$ of $\mathbb{R}^n$ such that $\rho_{\mathbf{I}}\big(\mathbf{I}(\boldsymbol{e}_n)\big) = \boldsymbol{e}_n$.

Using induction on n, or otherwise, show that any 0-isometry of $\mathbb{R}^n$ is a composition of at most n reflections.

(HINT: Consider the 2-dimensional subspace spanned by $\boldsymbol{x}$ and $\boldsymbol{e}_n$.)

(e) Finally suppose $\xi : \mathbb{R}^n \to \mathbb{R}^n$ is an isometry; i. e., a distance-preserving bijection. Show that for any $\boldsymbol{y} \in \mathbb{R}^n$, the translation, $\tau_{\boldsymbol{y}}(\boldsymbol{v}) \doteq \boldsymbol{v} + \boldsymbol{y}$, is an isometry. Hence show that any isometry i

of $\mathbb{R}^n$ is a function of the form $i(\boldsymbol{x}) = \mathrm{O}\boldsymbol{x} + \boldsymbol{y}$, where O is an orthogonal matrix. [1]

1.5 Eigenvalues and Semisimple Matrices

Let $T : E \to E$ be an endomorphism of a $\mathbf{k}$-linear space. An *eigenvalue* of T is an element $\lambda \in \mathbf{k}$ such that $\ker(T - \lambda \cdot 1_E) \neq \{0\}$. If λ is an eigenvalue of T, then any nonzero vector in $\ker(T - \lambda \cdot 1_E)$ is an *eigenvector* of T, corresponding to the eigenvalue λ.

If A is a square matrix in $\mathrm{M}_n(\mathbf{k})$ then $\boldsymbol{x} \in \mathbf{k}^n$ is said to be an *eigenvector* of A, corresponding to the eigenvalue λ, if $\boldsymbol{x} \neq 0$ and $\mathrm{A}\boldsymbol{x} = \lambda\cdot\boldsymbol{x}$, where on the left hand side $\boldsymbol{x}$ is being thought of as an $(n \times 1)$-matrix and left multiplied by A.

The well-known condition for the existence of nontrivial solutions for homogeneous linear equations implies that λ is an eigenvalue of A iff λ is a root of the polynomial equation:

$$\chi_{\mathrm{A}}(t) \doteq \det(T\cdot\mathrm{I}_n - \mathrm{A}) = 0.$$

The polynomial, χ_{A}, is known as the *characteristic polynomial* of A. It is easy to see that if $T : E \to E$ is a linear transformation of a finite-dimensional linear space then we can *unambiguously* define its characteristic polynomial as the characteristic polynomial of its matrix, $\mathrm{M}(T; \mathcal{E})$, with respect to any basis $\mathcal{E}$ of E.

Proposition 1.5.1. *Let* $\lambda_1, \dots, \lambda_m$ *be* distinct *eigenvalues of an endomorphism,* $T : E \to E$*. For each* $i = 1, \dots, m$ *let* $\boldsymbol{x}_i$ *be an eigenvector corresponding to* λ_i*. Then* $\{\boldsymbol{x}_1, \dots, \boldsymbol{x}_m\}$ *is a linearly independent set.*

Proof. If $\{\boldsymbol{x}_1, \dots, \boldsymbol{x}_m\}$ is a linearly dependent set, then we can choose a *minimal* subset of linearly dependent vectors. Since eigenvectors are always nonzero such a minimal subset must have at least two elements. Without loss of generality we may assume that $\{\boldsymbol{x}_1, \dots, \boldsymbol{x}_k\}$ is such a minimal set. Suppose that

(1)...................... $c_1\boldsymbol{x}_1 + \cdots + c_k\boldsymbol{x}_k = 0$

is a nontrivial linear dependence relation. Usually this would mean that not all the c_i's are 0. But, in fact, the minimality condition implies that *none* of the c_i's are zero. Applying the linear transformation T to both sides of the above equation we get

---

[1] This exercise is based on a lecture given by R. R. Simha to the students of a summer programme "Nurture" sponsored by the NBHM.

(2) .... $\lambda_1 c_1 \boldsymbol{x}_1 + \cdots + \lambda_k c_k \boldsymbol{x}_k = 0$

Multiplying (1) by λ_1 and subtracting from (2) we get

$$\sum_{i=2}^{k} (\lambda_i - \lambda_1) c_i \boldsymbol{x}_i = 0$$

which is a nontrivial linear dependence relation (since the c_i's are nonzero and the λ's are distinct) between $k-1$ of the $\boldsymbol{x}$'s, contradicting the choice of $\{\boldsymbol{x}_1, \ldots, \boldsymbol{x}_k\}$ as a minimal dependent set. □

In particular, we have

Theorem 1.5.2. *If* $\dim E = n$ *and* $T : E \to E$ *has* n distinct *eigenvalues, then there is a basis of* E *with respect to which the matrix of* T *is diagonal.*

Proof. Choose eigenvectors $\boldsymbol{x}_1, \ldots, \boldsymbol{x}_n$ corresponding to the distinct eigenvalues; these are linearly independent by the previous result and must be a basis since $\dim E = n$. □

A linear transformation $T : E \to E$ is said to be *semisimple* if one can choose a basis $\mathcal{E}$ of E such that $\mathrm{M}(T; \mathcal{E})$ is a diagonal matrix. A square matrix will be called *semisimple* if it is similar to a diagonal matrix. The term *diagonalizable* is sometimes used as a synonym for semisimple, in the context of matrices.

Remark. It is quite standard to use the term "semisimple" for linear transformations $T : E \to E$; but not quite so usual to apply it to square matrices as I have done. But I find "diagonalizable" an ugly adjective and "semisimple" is not half as hard to spell!

Clearly, the simplest possible nontrivial linear transformation $T : E \to E$ of a linear space into itself is multiplication by a nonzero element $\alpha \in \mathbf{k}$; the matrix of such a linear transformation with respect to any basis is $\alpha \cdot \mathrm{I}_n$. A semisimple transformation is only slightly more complicated: its matrix (in block diagonal form) has only nonzero blocks along the diagonal and each of these blocks are of the form $\beta\mathrm{I}$.

This may seem a very special kind of linear transformation, but we will see that they are abundant.

To this end I will discuss an ancient theorem in Algebra which dates back to Newton and show its relevance to the question of semisimplicity of matrices.

You may have, at some point, come across problems of the following genre.

PROBLEM 1. Let α and β be the roots of the equation

$$x^2 + bx + c = 0.$$

Express $\alpha^2 + \beta^2$ in terms of b and c.

ANSWER: $(b^2 - 2c)$.

PROBLEM 2. Let $\alpha_1, \ldots, \alpha_5$ be the roots of the equation:

$(*)\ldots\ldots\quad x^5 + a_1x^4 + a_2x^3 + a_3x^2 + a_4x + a_5 = 0.$

Express the sum of the squares of the α_i's in terms of $a_1, \ldots, a_5$.

ANSWER:

Now Problem 1 can be solved directly by first solving the quadratic equation and then working out the sum of the squares of the roots. In fact *any* function of α and β can be evaluated in terms of b and c.

But Problem 2 is different. The earlier method is no longer available because as you probably know there is no formula using the elementary operations (including the extraction of nth roots) which expresses the roots of an equation of degree ≥ 5 in terms of the coefficients involved.

On the other hand the answer to Problem 2 is easily found if we argue as follows. If $\alpha_1, \ldots, \alpha_5$ are the five roots then the left hand side of $(*)$ is the product: $\prod_{i=1}^{5}(x - \alpha_i)$ and comparing coefficients we see that:

$$\sum_{i=1}^{5} \alpha_i = -a_1,$$

and

$$\sum_{i \neq j} \alpha_i\alpha_j = a_2.$$

The answer follows from the identity:

$$\Big(\sum_{i=1}^{5} \alpha_i\Big)^2 = \sum_{i=1}^{5} \alpha_i^2 + 2 \cdot \sum_{i<j} \alpha_i\alpha_j.$$

I wish to discuss the general principle which is illustrated by these examples. I will begin with some elementary definitions (with which you are doubtless familiar) and some notation.

If S is a finite set then a bijection $f : S \to S$ is also called a *permutation* of S. In particular, a permutation of the set, $\underline{n} \doteq \{1, \ldots, n\}$, is called a permutation of n symbols. This is because if σ is such a permutation and $X = \{x_1, \ldots, x_n\}$ is any set with n elements, then σ can be associated to the permutation of X, which takes x_k to $x_{\sigma(k)}$. The group of permutations of $\underline{n}$ is the *symmetric group* of degree n, and is denoted $\mathfrak{S}_n$.

A *monomial* on $\mathbb{C}^n$ of degree d is a function $m : \mathbb{C}^n \to \mathbb{C}$ of the form

$$\boldsymbol{x} = (x_1, \ldots, x_n) \mapsto c \cdot x_1^{d_1} \ldots x_n^{d_n},$$

where $c \in \mathbb{C}$ and the d_i's are nonnegative integers with $\sum d_i = d$. A *polynomial* on $\mathbb{C}^n$ (or a polynomial in n variables) is a finite sum of monomials. Note that constant functions are monomials of degree 0. Consider a polynomial, $p = \sum_j m_j$ where the m_j's are monomials. The degree of p is the maximum of the degrees of m_j. The polynomial p is a *homogeneous* polynomial if it is a linear combination of monomials of the same degree. (I am implicitly assuming that each polynomial p has a unique expression as a linear combination of monomials. For $n = 1$ this is easy; the general case can be established by an induction argument which is not quite so easy.)

Observe that if $\sigma \in \mathfrak{S}_n$, then it transforms a polynomial p on $\mathbb{C}^n$ to another polynomial p^σ, where $p^\sigma(x_1, \ldots, x_n) = p(x_{\sigma(1)}, \ldots, x_{\sigma(n)})$. A polynomial is called a *symmetric polynomial* if $p = p^\sigma$ for every $\sigma \in \mathfrak{S}_n$. Let us discuss a few examples.

Examples of Polynomials

Example 1.5.3a. A linear transformation $\alpha : \mathbb{C}^n \to \mathbb{C}$ is a homogeneous polynomial of degree 1; such polynomials are often called *linear forms.*

Example 1.5.3b. The function $\boldsymbol{x} \mapsto (x_1^2 + \cdots + x_n^2)$ is a homogeneous polynomial of degree 2 (or a *quadratic* polynomial.)

We will now look at polynomials on the linear space, $V_n = \mathrm{M}_n(\mathbb{C})$, with the entries of a matrix being the n^2 variables.

Example 1.5.3c (Traces and Determinants). The trace function is a linear form on V_n. The function $\Delta : V_n \to \mathbb{C}$ defined by $\Delta(\mathrm{M}) = \det \mathrm{M}$ is a homogeneous polynomial of degree n.

Note that while the example 1.5.3b above is a symmetric polynomial the others are not.

To proceed further, we have to look at polynomials in an indeterminate T, whose coefficients are polynomials on $\mathbb{C}^n$. A simple minded way to think of these is to think of polynomials on $\mathbb{C}^{n+1}$ and replace the last coordinate by a symbol 'T'.

Let us consider the polynomial $\Sigma(T)$ in the indeterminate, T, defined by:

$$\begin{aligned}\Sigma(\boldsymbol{x};T)=\Sigma(x_1,\ldots,x_n;T) &= \prod_{j=1}^{n}(T-x_j)\\ &= T^n+\sum_{k=1}^{n}(-1)^k s_k(\boldsymbol{x})T^{n-k}.\end{aligned}$$

What we are doing in the second line is grouping together the terms involving T^m and writing the coefficient of T^m as $\big((-1)^{n-m}s_{n-m}(\boldsymbol{x})\big)$. The s_k are homogeneous polynomials of degree k in $(x_1,\ldots,x_n)$ and it is obvious that a permutation of the indices $1,\ldots,n$ will leave each s_k unchanged. The polynomial s_k is called the kth *elementary symmetric function* of the n variables $x_1,\ldots,x_n$.

For example the elementary symmetric functions in 3 variables are

$$\begin{aligned}s_1(x) &= x_1+x_2+x_3,\\ s_2(x) &= x_2x_3+x_3x_1+x_1x_2,\\ s_3(x) &= x_1x_2x_3.\end{aligned}$$

Notice that in this discussion of elementary symmetric functions there is some abuse of notation. The functions given above really should have been denoted s_k^3 for $k=1,2,3$. But avoiding this kind of ambiguity is difficult to achieve without making reading very difficult. Presumably you will be careful enough to remember how many components the vector x (which is the argument of $s_k(x)$) possesses; if a little bit of care is exercised, no problems will occur.

A *polynomial map* P from $\mathbb{C}^m$ to $\mathbb{C}^n$ is a function $P:\mathbb{C}^m\to\mathbb{C}^n$ of the form $P(x_1,\ldots,x_m)=\big(p_1(x_1,\ldots,x_m),\ldots,p_n(x_1,\ldots,x_m)\big)$ where each p_k is a polynomial on $\mathbb{C}^m$.

Example 1.5.4. For any $n\geq 1$ the function $\Sigma_n:\mathbb{C}^n\to\mathbb{C}^n$ given by

$$\Sigma_n(x_1,\ldots,x_n)=(s_1(\boldsymbol{x}),\ldots,s_n(\boldsymbol{x})).$$

is a polynomial map.

I can now state the classical result I wanted to present in this section.

Theorem 1.5.5 (Newton).

Let $p:\mathbb{C}^n\to\mathbb{C}$ be any symmetric polynomial. Then there is a unique

polynomial $\widetilde{p}: \mathbb{C}^n \to \mathbb{C}$ *such that* $p = \widetilde{p} \circ \Sigma_n$.
In other words, every symmetric polynomial is a polynomial in the elementary symmetric functions.

This is one of the major results which I use but do not prove. Some remarks on this result follow.

REMARKS 1.5.6 (ON NEWTON'S THEOREM)

Remark 1.5.6a. I am not certain if it was Newton who first established this result. He did enunciate the famous *Newton formulae* which express $x_1^r + \cdots + x_n^r$ in terms of the elementary symmetric functions of $x_1, \ldots, x_n$ and it seems unlikely that he would have missed the general principle.

Remark 1.5.6b. If you now look back at the second example I offered at the beginning of this section you will see that it was a special case of this theorem that I had used.

Remark 1.5.6c. I will omit the proof for the reason that although simple, it is long and the methods used are not useful in Calculus. In the next few paragraphs I will explain one consequence of this theorem which is relevant to the study of semisimple matrices; this consequence will be used several times in the rest of the book.

Consider the *characteristic polynomial* of a matrix A, defined as
$\chi_{\mathrm{A}}(T) = \det(T \cdot \mathrm{I}_n - \mathrm{A}) = T^n + c_1(\mathrm{A})T^{n-1} + \cdots + c_{n-1}(\mathrm{A})T + c_n(\mathrm{A})$.
The $c_k(\mathrm{A})$ are polynomials on V_n; in fact, c_1 and c_n are, modulo sign, the trace and determinant of A.

I will call the set of roots of this polynomial the *spectrum* of A and denote it as shown below:

$$\mathrm{Spectrum}(\mathrm{A}) \doteq \{\lambda_1(\mathrm{A}), \ldots, \lambda_n(\mathrm{A})\}.$$

These are the eigenvalues of A, each eigenvalue appearing as many times as its multiplicity as a root of $\chi_{\mathrm{A}}(T) = 0$. Since $\det(T \cdot \mathrm{I}_n - \mathrm{A}) = \prod_{j=1}^{n}(T - \lambda_j(\mathrm{A}))$, $c_1(\mathrm{A}), \ldots, c_n(\mathrm{A})$, are the elementary symmetric functions of the spectrum of A.

REMARKS 1.5.7 (ON THE SPECTRUM)

Remark 1.5.7a (Spectra of endomorphisms).
The identity:

$$\det(T \cdot \mathrm{I}_n - \mathrm{P}^{-1}\mathrm{AP}) = (\det \mathrm{P})^{-1}\big(\det(t \cdot \mathrm{I}_n - \mathrm{A})\big)(\det \mathrm{P})$$

implies that similar matrices have the same characteristic polynomial and hence, the same spectrum. In view of this one can speak of the characteristic polynomial (resp. eigenvalues, spectrum) of an endomorphism.

Remark 1.5.7b (The spetrum is an unordered n-tuple).
I have denoted the spectrum of a matrix by $\{\lambda_1(\mathrm{A}), \ldots, \lambda_n(\mathrm{A})\}$ rather than $\big(\lambda_1(\mathrm{A}), \ldots, \lambda_n(\mathrm{A})\big)$. This is because there is no natural way to order an arbitrary n-tuple of complex numbers.

Remark 1.5.7c. Note that the coefficients of the characteristic polynomial of a matrix, A, are not symmetric functions of the entries of A; they are (modulo sign) the elementary symmetric functions of the spectrum of A. Since the $c_i(\mathrm{A})$ are symmetric functions of the spectrum of A the fact that the λ_i's are not ordered does not matter.

Recall that if $T : E \to E$ is an endomorphism of a p-dimensional linear space (over $\mathbb{C}$) with distinct eigenvalues, then by Theorem 1.5.2, T is semisimple. Moreover, if $\mathcal{V} = \{\boldsymbol{v}_1, \ldots, \boldsymbol{v}_p\}$ are eigenvectors corresponding to the distinct eigenvalues then $\mathcal{V}$ is a basis of E.

Consider the polynomial mapping, $\Xi_n : \mathrm{M}(n;\mathbb{C}) \to \mathbb{C}^n$, defined by:

$$\Xi_n : \mathrm{A} \mapsto \big(c_1(\mathrm{A}), \ldots, c_n(\mathrm{A})\big),$$

and the function $\Delta(\mathrm{A}) = \prod_{i \neq j} \big(\lambda_i(\mathrm{A}) - \lambda_j(\mathrm{A})\big)^2$.

Obviously $\Delta(\mathrm{A})$ is a symmetric polynomial of the variables, $\lambda_1(\mathrm{A}), \ldots, \lambda_n(\mathrm{A})$ and hence by Theorem 1.5.5 there is a polynomial $\widetilde{\Delta} : \mathbb{C}^n \to \mathbb{C}$ such that $\Delta(\mathrm{A}) = \widetilde{\Delta}(c_1(\mathrm{A}), \ldots, c_n(\mathrm{A}))$. (Note that the $c_k(\mathrm{A})$'s are, modulo signs, the coefficients of the characteristic polynomial of A.)

What this means is that while $\Delta(\mathrm{A})$ is defined as a polynomial function in the variables $a_{11}, \ldots, a_{nn}$, we may also express it as a polynomial function (the polynomial $\widetilde{\Delta}$) of the eigenvalues $\lambda_1(\mathrm{A}), \ldots \lambda_n(\mathrm{A})$ of the matrix A.

Finally observe that $\Delta(\mathrm{A}) = 0$ iff the eigenvalues of A are *not* all distinct. We can now state (the proof has already been argued in the previous paragraph) the result which I have been after.

Theorem 1.5.8. *Let V_n be the set of $(n \times n)$ matrices over $\mathbb{C}$. There is a polynomial $\Delta : V_n \to \mathbb{C}$ such that if $\Delta(\mathrm{A}) \neq 0$ then* A *is semisimple.* □

Remark 1.5.9. You may be wondering why I have made such a long detour to introduce you to this result and why it is significant. The applications will follow in Chapter 3 but let me point out the geometric content of this theorem. From your exposure to coordinate geometry you will have noticed that the set of points at which a polynomial function *vanishes* is a "thin" subset of the domain: for example, the zero-set of $y^2 - 4x$ defined on $\mathbb{R}^2$ is a parabola; the zero-set of z defined on $\mathbb{R}^3$ is the (x, y)-plane. This is also true for polynomials in arbitrary number of variables. The point is that the set on which a non-constant polynomial *does not* vanish is almost all of the linear space; the precise sense in which I am using the phrase "almost all" will be clarified later. In view of this remark what Theorem 1.5.8 says is that "almost all" $(n \times n)$ matrices over $\mathbb{C}$ are semisimple. Since diagonal matrices are easy to deal with, the fact that almost any square matrix is similar to a diagonal matrix is very useful.

We end these preliminaries by explicitly setting out certain conventions concerning notation which you may have already grasped.

Conventions concerning Notation

Throughout this book an effort will be made to use different type-faces for different kinds of objects:

- English letters that appear in mathematical formulae, as symbols, will be, for the most part written in *italics* as in "$E = mc^2$."
- Linear transformations between linear spaces will be denoted by ordinary upper case Greek letters (like Φ, Ψ) or *italicized* upper case English letters (like S, T);
- For matrices I will use upright upper case letters (like $\mathrm{A}, \mathrm{B}, \ldots$);
- Vectors will be denoted by **bold** italicized lower case English letters (like $\boldsymbol{x}, \boldsymbol{y}, \ldots$),
- Bases of linear spaces will be denoted by "calligraphic" and similar "fancy" letters, such as $\mathcal{A}, \mathcal{B}, \mathcal{C}, \mathcal{E}, \ldots$ etc.

This convention will not be, indeed *cannot* be, applied with absolute consistency. Almost every mathematical object I will discuss belongs to some linear space or the other and hence is a "vector" and therefore should be denoted by the bold-italic typeface! but it is useful to know that A is almost certainly a matrix and that $\mathcal{A}$ or $\mathcal{B}$ is almost certainly a basis of some linear space.

In particular, special mention should be made of the following exception:

1. When I discuss Differential Calculus our functions will be defined on subsets of normed linear spaces. Suppose $U \subset E$ is a subset of a normed linear space and $f : U \to F$ is a function into another normed linear space. In this situation the argument of the

function is, of course, a vector. But unless the function is a linear transformation, the fact that the argument of the function is a vector is irrelevant. So when f is *not* a linear transformation, I will write $f(x)$ and not $f(\boldsymbol{x})$.

2. Similarly, if the set of vectors

$$\{\boldsymbol{x} + t\boldsymbol{y} \, : \, t \in [0,1]\}$$

is contained in U, then $x + ty$ will be vector $\boldsymbol{x} + t\boldsymbol{y}$ regarded as a point of U.

Reader is not expected to appreciate, at present, the need for making this distinction. It will become clear in Chapter 4.

2
Linear Spaces: New Spaces for Old

As suggested by the title, this chapter is concerned with the "creation" of new linear spaces. Why is it necessary to devote an entire chapter to such an enterprise in order to learn Differential Calculus? In this preamble I hope to convince you that, indeed, this is essential.

In Calculus, one is concerned with `differentiable` functions defined between `open subsets` of normed linear spaces: these open sets are a generalization, in the context of linear spaces, of the open intervals, (a, b). (Recall that in elementary calculus, differentiability of functions makes sense only for functions defined on open intervals: for functions defined on closed intervals, [a,b], one has to define the notion of *one-sided derivatives.*)

Let $U \subset E$ be an open set in a normed linear space E and suppose $f : U \to F$ is a function taking values in a normed linear space, F. For real-valued functions defined on an open interval, The derivative of a function at a point p is geometrically visualized as the slope of the line which best approximates the graph of the function near the point p.

In the more general context we will study functions $f : U \to F$, where U is an open subset of a normed linear space E. then the derivative of f at the point $x \in U$, (which will be denoted $\mathbf{D}f(x)$), will be the *linear transformation* from E to F which best approximates f near the point x. Observe that according to this way of thinking, for a real-valued function $f : (a, b) \to \mathbb{R}$ the derivative is a linear transformation

$\mathbf{D}f(x) : \mathbb{R} \to \mathbb{R}$ and not a real number! But elements of $L(\mathbb{R}, \mathbb{R})$ are completely determined by their value on 1 and the number $\mathbf{D}f(x)(1)$ is what you learnt earlier to call the `differential coefficient` of f and denote by the symbol df/dx or $f'(x)$.

To continue, let us consider the function $\mathbf{D}f : U \to L(E, F)$ which takes a point $x \in U$ to the derivative of f at the point x. We will see that if E and F are normed spaces then there is a natural way to define a norm on $L(E, F)$ and it may happen that $x \mapsto \mathbf{D}f(x)$ is also a differentiable function from U to $L(E, F)$. Then the derivative of this function at a point $x \in U$ will be called the second derivative of f at x and denoted $\mathbf{D}^2 f(x)$. This is an element of $L(E, L(E, F))$ and this is a completely different kind of mathematical entity from the first derivative: it is *not* a linear transformation from E to F. In the case when $E = F = \mathbb{R}$, one does not come across this phenomenon because $L(\mathbb{R}, \mathbb{R})$ is isomorphic to $\mathbb{R}$ via the function which takes a linear transformation $\alpha : \mathbb{R} \to \mathbb{R}$ to $\alpha(1)$.

The derivative of this second derivative, if it exists, will be the *third* derivative of f and is an element of $L(E, L(E, L(E, F)))$ and so on. So derivatives of higher orders are objects of different linear spaces and thus it is *essential to develop a set of tools to handle all these objects in a uniform fashion.* The aim of this chapter is to develop these tools.

The spaces $L(E, L(E, F)), L(E, L(E, L(E, F))), \ldots$ surely remind you of the sets $\mathfrak{F}_p(A; B)$ and p-functions that I introduced in Section §1.1. I now introduce an analogous notation for the analogous spaces of linear

will write $\mathfrak{L}_1(E; F)$ for the linear space $L(E, F)$
and $\mathfrak{L}_p(E, F)$ for the linear space $L(E, \mathfrak{L}_{p-1}(E; F))$.

The elements of $\mathfrak{L}_p(E; F)$ will be called "linear p-functions". (This is *ad hoc* terminology: don't look for it in other books!) By applying the Law of Exponents we will obtain a more easily understood description of the inhabitants of these frightening-looking linear spaces.

It turns out that the spaces $\mathfrak{L}_p(E; F)$ are isomorphic to the linear space of `multilinear functions` from the p-fold Cartesian product, $\times^p E$, of E to F. One can bring multilinear functions into the ambit of Linear Algebra using the notion of `tensor products` of linear spaces, but since it is not necessary to use tensor products in order to learn Differential Calculus, I will treat them in a somewhat informal fashion.

Let us now get down to business ...

You are already familiar with some ways in which new spaces can arise from a given collection of linear spaces and from linear transformations between them.

1. **Direct Sum:** If $E_1, \ldots, E_n$ are linear spaces over the *same* field $\mathbf{k}$ then their Cartesian product, denoted $\prod_{i=1}^n E_i$, has a natural $\mathbf{k}$-linear space structure given by component-wise addition and scalar multiplication. More precisely, if $\boldsymbol{v} = (\boldsymbol{v}_1, \ldots, \boldsymbol{v}_n)$ and $\boldsymbol{w} = (\boldsymbol{w}_1, \ldots, \boldsymbol{w}_n)$ are elements of $\prod_{i=1}^n E_i$, and $\alpha \in \mathbf{k}$, then addition and scalar multiplication can be defined by setting:
$$\begin{aligned} \boldsymbol{v} + \boldsymbol{w} &= (\boldsymbol{v}_1 + \boldsymbol{w}_1, \ldots, \boldsymbol{v}_n + \boldsymbol{w}_n) \\ \text{and} \quad \alpha \cdot \boldsymbol{v} &= (\alpha \boldsymbol{v}_1, \ldots, \alpha \boldsymbol{v}_n). \end{aligned}$$
The Cartesian product of $E_1, \ldots, E_n$, when equipped with this notion of addition and scalar multiplication, is known as the *direct sum* of $E_1, \ldots, E_n$ and will be denoted $E_1 \oplus \cdots \oplus E_n$.

2. **Kernels and Images:** If $T : E \to F$ is a linear transformation then the kernel of T and the image of T are "new" linear spaces.

We move on to discuss more substantial procedures for obtaining new spaces: the ones which are essential for the study of Differential Calculus.

2.1 Duality

Let E be a linear space over $\mathbf{k}$ and let E^* be the set of linear transformations from E to $\mathbf{k}$. If $\alpha, \beta \in E^*$ and $c \in \mathbf{k}$, then the function $(\alpha + \beta)$ which takes $\boldsymbol{v} \in E$ to $\alpha(\boldsymbol{v}) + \beta(\boldsymbol{v})$ is clearly a linear transformation $E \to \mathbf{k}$. The same is true of the function, $c \cdot \alpha$ which takes $\boldsymbol{v}$ to $c \cdot \alpha(\boldsymbol{v})$. So the set E^* is equipped with the structure of a linear space over $\mathbf{k}$. This is called the *dual space* of E. The elements of E^* are called *linear forms* on E.

If $\alpha \in E^*$ and $\boldsymbol{v} \in E$ we will denote $\alpha(\boldsymbol{v}) \in \mathbf{k}$ by $\langle \alpha, \boldsymbol{v} \rangle$. (This seemingly peculiar notation was invented by physicists but is so convenient that it is now almost universally used, with slight variations, by mathematicians as well.)

Proposition 2.1.1 (The Dual Basis).
Suppose E is finite-dimensional, with a basis $\mathcal{E} = \{\mathbf{e}_1, \ldots, \mathbf{e}_n\}$. Consider the set of elements, $\mathcal{E}^* = \{\epsilon_m : 1 \leq m \leq n\}$ of the dual space defined by the relations $\langle \epsilon_m, \mathbf{e}_k \rangle = \delta_{km}$.

Then $\mathcal{E}^*$ is a basis of E^*.

Hence $\dim E = \dim E^*$.

Proof. Since $\alpha \in E^*$ is a linear transformation on E it is completely determined by its values on the set $\mathcal{E}$. Clearly such $\alpha \in E^*$ may be written *uniquely* as $\alpha = \sum_{j=1}^{n} a_j \epsilon_j$, where $a_j = \langle \alpha, \boldsymbol{e}_j \rangle$. This implies that that $\mathcal{E}^* = \{\epsilon_1, \ldots, \epsilon_n\}$ is a basis of E^*. □

The collection $\mathcal{E}^* = \{\epsilon_1, \ldots, \epsilon_n\}$ is known as the *basis of E^* dual to* $\mathcal{E}$.

Example 2.1.2 (Duality in Euclidean Spaces).
Let E be a Euclidean space. Then, the bilinearity of the inner product implies:
1. For each $\boldsymbol{x} \in E$, the function $\mathcal{D}_E(\boldsymbol{x}) : E \to \mathbb{R}$ defined by $\mathcal{D}_E(\boldsymbol{x})(\boldsymbol{v}) = \langle \boldsymbol{x}, \boldsymbol{v} \rangle$ is a linear transformation from E to $\mathbb{R}$.
2 $\mathcal{D}_E : E \to E^*$ is a linear transformation.
3. The nonsingularity of the inner product implies that $\mathcal{D}_E$ is an isomorphism.

In fact, it is easily checked that a basis, $\mathcal{X} = \{\boldsymbol{x}, \ldots, \boldsymbol{x}_n\}$, is an onb *iff* $\Xi = \{\xi_1, \ldots, \xi_n\}$ is the basis dual to $\mathcal{X}$ where $\xi_k = \mathcal{D}_E(\boldsymbol{x}_k)$. So, the double role played by the "chevron pair" $\langle \cdot, \cdot \rangle$: first as the inner product of two vectors and then to denote the value of an element of E^* on an element of E simply amounts to omitting explicit mention of the isomorphism $\mathcal{D}_E$ in the chain of equalities:
$$\begin{aligned} \langle \xi, \boldsymbol{v} \rangle &= \mathcal{D}_E(\boldsymbol{x})(\boldsymbol{v}) \\ &= \langle \boldsymbol{x}, \boldsymbol{v} \rangle \end{aligned}$$
where on the top line we have evaluation of linear forms and on the second line we have the inner product of vectors.

So far we have discussed only the process by which a linear space gives rise to a new linear space — its dual space. You may feel that there is nothing really happening here: after all, if E is finite dimensional then E and E^* have the same dimension and are therefore isomorphic. But the next result hints that something more might be going on.

Proposition 2.1.3. *Let $T : E \to F$ be a linear transformation. Define a function, $T^* : F^* \to E^*$ by requiring that for all $\alpha \in F^*$ and all $\boldsymbol{x} \in E$, $\langle T^*(\alpha), \boldsymbol{x} \rangle = \langle \alpha, T(\boldsymbol{x}) \rangle$.*

T^*, thus defined, is a linear transformation, called the dual (or transpose) of T. *This procedure of dualizing linear transformations satisfies the following properties.*

1. *for every linear space V, $(1_V)^* = 1_{V^*}$.*

2. *If $S : F \to G$ is another linear transformation then $(ST)^* = T^*S^*$.*

Proof. Observe that the hypotheses define $T^*(\alpha)$ as the composition:

$$E \xrightarrow{T} F \xrightarrow{\alpha} \mathbf{k}.$$

So $T^*(\alpha)$ is certainly an element of E^*. The assignment, $T \mapsto T^*$, is a linear transformation since,

$$\begin{aligned} T^*(\alpha+\beta) &= (\alpha+\beta)\circ T, \\ &= \alpha\circ T + \beta\circ T, \\ &= T^*(\alpha) + T^*(\beta). \end{aligned}$$

The remaining parts are even easier: they are being left as exercises. □

The next few remarks should help you in acquiring more insight upon the fundamental operation of the transpose. (Note that the transpose of a linear transformation is generally referred to as its adjoint.)

REMARKS 2.1.3 (ON "FUNCTORS")

Remark 2.1.3a. The significance of Proposition 2.1.3 is that it shows that the process of dualization is not merely a way to produce new linear spaces from old; it converts linear transformations between linear spaces to linear transformations between their duals respecting identity maps and compositions.

Remark 2.1.3b (A new idea).
In fact, dualization is an example of a new kind of mathematical entity known as "`functor`": the "domain" of the dualization functor is the class of *all* **k**-linear spaces and **k**-linear transformations between them; it converts linear spaces and linear transformations to their duals in such a way that:

1. The identity, 1_V of V is taken to the identity $1_{V^*} \in V^*$.
2. A commutative diagram of linear spaces and linear transformations is taken to a commutative diagram of the dual spaces and dual linear transformations.
3. The direction in which the linear transformations proceed get reversed under dualization.

I will not dwell any further on this since functors and associated notions are not necessary for our objective of learning Calculus; however I will mention (when the occasion arises) that some new operation is actually a `functor` or that a certain property is "functorial". Though I will not formally define this notion I believe that you will soon get a feel for what a functorial property is and why it is a useful idea.

Remark 2.1.3c. In fact, because of the reversal of directions of the linear transformations, dualization is sometimes called a `cofunctor`. A functor does not change the direction of the "arrows". We will soon meet such an entity: that is a *true* functor rather than a "cofunctor". Usually the word functor is used to refer to both functors and cofunctors.

I will now illustrate the usefulness of the notion of dualization by using duality to give elegant proofs of familiar results of matrix theory. I begin with an interesting observation that is quite easily proved.

Proposition 2.1.4. *If $T : E \to F$ is an injective (resp. surjective) linear transformation, then $T^* : F^* \to E^*$ is surjective (resp. injective.)*

Proof. Suppose that $T : E \to F$ is surjective and let $\beta \in F^*$ be any linear form which is in the kernel of T^*. Then from the definition of T^* we see that for all $\boldsymbol{x} \in E$ we have $\langle \beta, T(x) \rangle = \langle T^*(\beta), \boldsymbol{x} \rangle = 0$. But since T is surjective this means that β is zero on every vector of F. hence $\beta = 0$ and T^* is injective.

Now suppose that T is injective and let $\xi \in E^*$. We have to find $\psi \in F^*$ such that $T^*(\psi) = \xi$; that is we want to find $\psi \in F^*$ such that for each $\boldsymbol{x} \in E$, the relation $\langle \psi, T(\boldsymbol{x}) \rangle = \langle \xi, \boldsymbol{x} \rangle$ holds.

Choose a basis $\mathcal{V} = \{\boldsymbol{v}_\imath : i \in I\}$ of E. Since T is injective, the set $\{T(\boldsymbol{v}_\imath) : i \in I\}$ is a linearly independent set in F. Extend this to a basis, $\mathfrak{F}$, of F. (See Theorem 1.3.1.)

Now define a linear form $\psi : E \to \mathbf{k}$ by defining the value of $\imath$ on $\mathfrak{F}$ as follows. For each $i \in I$ and $\boldsymbol{v}_\imath$ define $\psi\big(T(\boldsymbol{v}_\imath)\big) = \langle \xi, \boldsymbol{v}_\imath \rangle$ and define ψ to take the value 0 on all the remaining vectors of $\mathfrak{F}$. Clearly $T^*(\psi) = \xi$. □

Some comments are in order.

REMARKS 2.1.4 (ON THE TRANSPOSE)

Remark 2.1.4a (The Axiom of Choice).
Notice that while proving that the dual of an injective map is surjective, I have used the Axiom of Choice by invoking Theorem 1.3.1.

Remark 2.1.4b (A "functorial" proof of Proposition 2.1.4).
Notice that to say that a linear transformation $T : E \to F$ is injective is equivalent to saying that there is a linear transformation $S : F \to E$ such that $ST = 1_E$ or that T has a left inverse. From the properties of dualization we see that this means that, $T^*S^* = 1_{E^*}$. But this means that T^* has a right inverse, which is equivalent to saying that T^* is surjective.

This proves one half of Proposition 2.1.4; you should construct a similar proof for the other half. I remind you that the Axiom of choice assures us that a surjective function has a right inverse.

You should prove the other half by a similar functorial argument. □

We now prove a well-known result in matrix theory which generalizes (albeit in a finite-dimensional context) the assertions of Proposition 2.1.4.

Theorem 2.1.5 (Equality of row and column rank).
Let $T : E \to F$ be a linear transformation between finite-dimensional linear spaces. Then $\rho(T) = \rho(T^*)$.

Note that there is no matrix in sight!

Proof. Suppose that $\dim E = m$, $\dim F = n$ and that the rank, $\rho(T)$, of T is equal to d. Choose a basis $\{\boldsymbol{e}_{r+1}, \dots, \boldsymbol{e}_d\}$ of $\ker T$ and extend to a basis of E:

$$\mathcal{E} = \{\boldsymbol{e}_1, \dots, \boldsymbol{e}_r, \boldsymbol{e}_{r+1}, \dots, \boldsymbol{e}_m\}.$$

From the rank–nullity relation, we know that $\rho(T) = m - d = r$, (say).

If we set $\boldsymbol{f}_j = T(\boldsymbol{e}_{(d+j)})$ for $1 \le j \le r$, then $\{\boldsymbol{f}_1, \dots, \boldsymbol{f}_r\}$ is a linearly independent subset of F which obviously spans (Image T). Extend this to a basis of F: $\mathcal{F} = \{\boldsymbol{f}_1, \dots, \boldsymbol{f}_n\}$.

Let $\mathcal{E}^* = \{\epsilon_1, \dots, \epsilon_m\}$ be the dual basis of E^* and
$\mathcal{F}^* = \{\phi_1, \dots, \phi_n\}$ be the dual basis of F^*.

Then we have,

$$\langle T^*(\phi_j), \boldsymbol{e}_i\rangle = \langle \phi_j, T(\boldsymbol{e}_i)\rangle = \begin{cases} 0 & \text{if } i \le d, \\ \delta(j, (i-d)) & \text{if } i > d. \end{cases}$$

(Here we have used $\delta(\alpha, \kappa)$ for $\delta_{\alpha\kappa}$ for the sake of clarity.)

Clearly $\rho(T^*) = m - d = r$. □

The next result justifies the somewhat mysterious title of Theorem 2.1.5.

Proposition 2.1.6. *Let E, F be finite-dimensional linear spaces with ordered bases $\mathcal{E} = \{\boldsymbol{e}_1, \dots, \boldsymbol{e}_m\}$ and $\mathcal{F} = \{\boldsymbol{f}_1, \dots, \boldsymbol{f}_n\}$ respectively. Suppose $T : E \to F$ is a linear transformation. If $\mathrm{A} = (a_{ij})$ is the matrix of T with respect to the bases, $\mathcal{E}$ and $\mathcal{F}$, then $\mathrm{M}(T^*; \mathcal{E}^*, \mathcal{F}^*) = \mathrm{A}^{\mathbf{t}}$.*

Proof. Let $\mathcal{E}^* = \{\epsilon_1, \dots, \epsilon_m\}$ and $\mathcal{F}^* = \{\phi_1, \dots, \phi_n\}$ be the dual bases of E^* and F^*. Then from the definitions of $\mathrm{M}(T; \mathcal{E}, \mathcal{F})$ and $\mathrm{M}(T^*; \mathcal{F}^*, \mathcal{E}^*)$ we see that

$$(*) \;\ldots\ldots\ldots\ldots\ldots\ldots\ldots \qquad \begin{aligned} a_{ij} &= \langle \phi_i, T(\mathbf{e}_j)\rangle \\ &= \langle T^*(\phi_i), \mathbf{e}_j\rangle \end{aligned}$$

and the result follows. □

An immediate corollary of the last two results is:
If A *is an* $(m \times n)$ *matrix over a field* $\mathbf{k}$*, then* the row rank of A is equal to its column rank.

Remark 2.1.7 (Transpose of a linear transformation).
Proposition 2.1.6 is highly significant. It shows that the operation of taking transposes of matrices is not just an operation on rectangular arrays of elements of $\mathbf{k}$, but is a well-defined operation on the set of *linear transformations between two linear spaces.*

A counter-example may help in driving home this point. Let V be the set of linear transformations from $\mathbf{k}^2$ to $\mathbf{k}^2$. Let us define an operation on (2×2) matrices $\mathrm{A} \mapsto \phi(\mathrm{A})$ by declaring that the $(i,j)^{\text{th}}$ element of $\phi(\mathrm{A})$ is the cube of the $(i,j)^{\text{th}}$ element of A. This does not determine an operation on the set of linear transformations from $\mathbf{k}^2$ to $\mathbf{k}^2$.

We leave this as an exercise with the hint that you should check whether or not, given two similar matrices, A, B, the matrices $\phi(\mathrm{A})$ and $\phi(\mathrm{B})$ are similar.

Now that we have seen that given a linear space E we can associate to this space a new space E^* one is naturally led to try to iterate the procedure and study the dual space of E^*.

The dual of E^* will be denoted E^{**}.

Theorem 2.1.8 (Reflexivity).
Let E be a linear space and define a function $\kappa_E : E \to E^{**}$ by setting

$$\langle \kappa_E(\boldsymbol{x}), \alpha\rangle = \langle \alpha, \boldsymbol{x}\rangle \textit{ for all } \alpha \in E^*.$$

Then κ_E is an injective linear transformation. If E is finite-dimensional, then $\kappa_E : E \to E^{**}$ is an isomorphism. This fact is known as the reflexivity of finite-dimensional spaces under duality.

Proof. For the sake of brevity, I will denote $\kappa_E(\boldsymbol{x})$ by $\boldsymbol{x}^{**}$. If $s, t \in \mathbf{k}$ and $\boldsymbol{x}, \boldsymbol{y} \in E$ then for any $\alpha \in E^*$, we have:

$$\begin{aligned} \langle (s\boldsymbol{x} + t\boldsymbol{y})^{**}, \alpha\rangle &= \langle \alpha, s\boldsymbol{x} + t\boldsymbol{y}\rangle && \text{(by the definition of } \kappa_E), \\ &= s\langle \alpha, \boldsymbol{x}\rangle + t\langle \alpha, \boldsymbol{y}\rangle && \text{(since } \alpha \text{ is linear)} \\ &= s\langle x^{**}, \alpha\rangle + t\langle y^{**}, \alpha\rangle \\ &= \langle sx^{**} + ty^{**}, \alpha\rangle \end{aligned}$$

showing that κ_E is a linear transformation from E to E^{**}.

To see that κ_E is injective suppose that $\boldsymbol{x}^{**} = 0$. Then since

$$\langle \alpha, \boldsymbol{x} \rangle = \langle \boldsymbol{x}^{**}, \alpha \rangle = 0,$$

every $\alpha \in E^*$ takes the value 0 on $\boldsymbol{x}$. This implies that $\boldsymbol{x} = 0$ for if $\boldsymbol{x}$ was nonzero, we could extend it to a basis and the dual basis would contain a linear form which does not vanish on $\boldsymbol{x}$. Hence κ_E is injective.

Since taking duals of finite-dimensional spaces leaves dimension unaltered, if E is finite-dimensional then $\dim E^{**} = \dim E$. Since κ_E is injective, from the rank-nullity relation we see that $\rho(\kappa_E) = \dim E = \dim E^{**}$ which means that κ_E is surjective. finite-dimensional. □

REMARKS 2.1.9 (THE NATURALITY OF κ)

The proof of Proposition 2.1.6 suggested that the dual basis of a dual basis might be the original one. So the previous result should not have come as a complete surprise.

Remark 2.1.9a (On Naturality).
You may ask what is so special about the previous result. After all, in the finite-dimensional context, all it does is to define an isomorphism between two linear spaces which are known to be of equal dimensions. (Not only am I proclaiming it as a THEOREM but even exalting it with a name!)

What is remarkable about κ_E is that it is "naturally" defined. I did not have to choose a basis of E or E^{**}, I did not need to know anything about the nature of the elements of E. It is as though the existence of E automatically brings forth, not only, $E^*, E^{**} \ldots$ etc., but also the injections $\kappa_E : E \to E^*$, $\kappa_{E^*} : E^* \to E^{**} \ldots$ etc.

By contrast, you can try as hard as you can, but you will not succeed in constructing an isomorphism between E and E^* unless you have some other data like a basis of E or an inner product on E. (See Example 2.1.2.)

In fact, even in the restricted context of finite-dimensional spaces, *there is no naturally defined inverse for* κ_E.

Remark 2.1.9b (Reflexivity in infinite-dimensional spaces).
We shall soon see that κ_E fails to be an isomorphism if E is infinite-dimensional. But later in Chapter 3, we will see that a slightly more refined, but restrictive, notion of linear transformation (more appropriate for infinite-dimensional spaces) restores the bijectivity of κ for an important class of linear spaces.

Remark 2.1.9c (Functoriality of the $**$ procedure).
Note that double dualization is a functor.

If $T : E \to F$ is a linear transformation and we define $T^{**} : E^{**} \to F^{**}$ to be the dual of $T^* : F^* \to E^*$, then clearly this procedure takes the identity map, 1_V, of any linear space to $1_{V^{**}}$ and commutative diagrams of linear spaces and linear transformations to commutative diagrams *without reversing arrows.*

Observe also that double dualization of linear transformations preserves both surjectivity and injectivity of linear maps, whereas dualization interchanges these properties.

As an example of what I mean by the naturality of the linear transformation κ I will now prove a result which tells us that if E, F are two linear spaces then κ_E and κ_F behave coherently with respect to the linear transformations between E and F and their double dualizations.

Proposition 2.1.10 (The naturality of κ).
Let $T : E \to F$ be a linear transformation between linear spaces. Then there is a commutative diagram:

$$\begin{array}{ccc} E & \xrightarrow{\ T\ } & F \\ {\scriptstyle \kappa_E}\downarrow & & \downarrow{\scriptstyle \kappa_F} \\ E^{**} & \xrightarrow{\ T^{**}\ } & F^{**} \end{array}$$

That is, for all $\boldsymbol{x} \in E$, we have: $\kappa_F\big(T(\boldsymbol{x})\big) = T^{**}\big(\kappa_E(\boldsymbol{x})\big)$.

Proof. We must show that for every $\alpha \in E^*$,

$$\langle T^{**}\big(\kappa_E(\boldsymbol{x})\big), \alpha\rangle = \langle \kappa_F\big(T(\boldsymbol{x})\big), \alpha\rangle.$$

By definition, $\langle \kappa_F\big(T(\boldsymbol{x})\big), \alpha\rangle = \langle \alpha, T(\boldsymbol{x})\rangle$.

On the other hand, since T^{**} is the dual of T^*, we get:

$$\langle T^{**}\big(\kappa_E(\boldsymbol{x})\big), \alpha\rangle = \langle \kappa_E(\boldsymbol{x}), T^*(\alpha)\rangle = \langle T^*(\alpha), \boldsymbol{x}\rangle = \langle \alpha, T(\boldsymbol{x})\rangle.$$

This proves the result. □

A couple of final observations concerning $\kappa_E : E \to E^{**}$ are necessary.

Remark 2.1.11 (Failure of Reflexivity).
If E is not finite-dimensional, then κ_E *cannot be surjective.*

Proof. Suppose E is infinite-dimensional and let $\{\boldsymbol{e}_i : i \in I\}$ be a basis of E. Let $\epsilon_i \in E^*$ be the "duals" of these basis vectors, that is $\langle \epsilon_i, \boldsymbol{e}_j\rangle =$

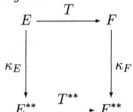

δ_{ij}. Then $\{\epsilon_i : i \in I\}$ is clearly a linearly independent set. Extend this, if necessary, to a basis $\mathcal{B}^*$ of E^*. It is easy to see that the element $\epsilon^{**} \in E^{**}$ which takes every ϵ_i to 1 and the remaining elements (if any) of $\mathcal{B}^*$ to 0 will not be in the image of $\kappa_E : E \to E^{**}$. □

Remark 2.1.12 (Reflexivity in Euclidean Spaces).
Let E be a Euclidean space. Then, via the inner product one can identify E^{**} with E itself. Recall (see Example 2.1.2) the isomorphism, $\mathcal{D}_E : E \to E^*$ obtained by setting $\langle \mathcal{D}_E(\boldsymbol{x}), \boldsymbol{y}\rangle = \langle \boldsymbol{x}, \boldsymbol{y}\rangle_E$ where $\langle\ ,\ \rangle_E$ indicates the Euclidean inner product.

From this and the definition of κ_E, we get:

$$\langle \kappa_E(\boldsymbol{x}), \mathcal{D}_E(\boldsymbol{y})\rangle = \langle \mathcal{D}_E(\boldsymbol{y}), \boldsymbol{x}\rangle = \langle \boldsymbol{x}, \boldsymbol{y}\rangle_E.$$

Thus we may (and in future, *will*) assume that if E is a Euclidean space then $E^{**} \equiv E$ and that $\kappa_E = 1_E$.

Exercises for § 2.1

We begin with the formal definition of block decomposition of matrices. This is something which is standard material in Linear Algebra; it is being set as an exercise on the notion of direct sums primarily so that I don't have to interrupt my text at an inconvenient point to explain the notation.

2.1.1 Direct Sums and Block Decompositions

For $i = 1, \ldots, n$ and $j = 1, \ldots, m$ let

- $(E_1, \ldots, E_m)$ and $(F_1, \ldots, F_n)$ be finite-dimensional $\mathbf{k}$-linear spaces;
- $\mathcal{E}_i = \{\boldsymbol{e}_1^{(i)}, \ldots, \boldsymbol{e}_{k_i}^{(i)}\}$ a basis of E_i;
- $\mathcal{F}_j = \{\boldsymbol{f}_1^{(j)}, \ldots, \boldsymbol{f}_{l_j}^{(j)}\}$ a basis of F_j;
- E and F the direct sums of these two families; that is,
$$E = \bigoplus_{i=1}^{m} E_i \text{ and } F = \bigoplus_{j=1}^{n} F_j.$$
- $\sigma_i : E_i \to E$ be the natural inclusion: that is, the linear transformation $\boldsymbol{v} \mapsto (\delta_{i1}\boldsymbol{v}, \ldots, \delta_{im}\boldsymbol{v})$ and
- $\pi_j : F \to F_j$ be the jth projection: that is, the linear transformation:
$$(\boldsymbol{f}_1, \ldots, \boldsymbol{f}_n) \mapsto \boldsymbol{f}_j.$$

For any linear transformation $T : E \to F$ define $T_{ij} : E_i \to F_j$ by $T_{ij} = \pi_j T \sigma_i$.

1. Using the Basis Criterion or otherwise show that the sets
$$\mathcal{E} = \{\mathbf{e}_1^{(1)}, \dots, \mathbf{e}_{k_1}^{(1)}; \mathbf{e}_1^{(2)} \dots, \mathbf{e}_{k_2}^{(2)}; \cdots ; \mathbf{e}_1^{(m)}, \cdots, \mathbf{e}_{k_m}^{(m)}\}$$
and
$$\mathcal{F} = \{\boldsymbol{f}_1^{(1)}, \dots, \boldsymbol{f}_{l_1}^{(1)}; \boldsymbol{f}_1^{(2)}, \dots; \boldsymbol{f}_{l_2}^{(2)}; \cdots ; \boldsymbol{f}_1^{(n)}, \dots, \boldsymbol{f}_{l_n}^{(n)}\}$$
are bases for E and F respectively.

2. Show that the matrix of T with respect to the bases, $\mathcal{E}$ and $\mathcal{F}$ is given by:
$$\text{(Block Decomposition)} \quad \mathrm{M}(T; \mathcal{E}, \mathcal{F}) = \begin{bmatrix} \mathrm{M}_{11} & \mathrm{M}_{12} & \dots & \mathrm{M}_{1m} \\ \mathrm{M}_{21} & \mathrm{M}_{22} & \vdots & \mathrm{M}_{2m} \\ \vdots & \vdots & & \vdots \\ \mathrm{M}_{n1} & \mathrm{M}_{n2} & \dots & \mathrm{M}_{nm} \end{bmatrix}$$
where M_{ji} is the matrix of the linear transformation $T_{ij} : E_i \to F_j$ with respect to the bases $\mathcal{E}_i$ of E_i and $\mathcal{F}_j$ of F_j.

The right-hand side of the last equation is called a *block decomposition* of the matrix M. Replacing the M_{ij}'s by the T_{ij}'s one gets a rectangular array of linear transformations which is referred to as the *block decomposition* of the linear transformation T with respect to the direct sum decompositions of E and F.

2.1.2 Computations involving duals and adjoints

1. Let $\mathcal{P}_n$ be the space of polynomials of degree $\leq n$ with real coefficients, with addition and scalar multiplication defined in the obvious manner.

   Choose bases $\{\phi_0, \dots, \phi_n\}$ of $\mathcal{P}_n$ for each n and let $\{\phi_1^*, \dots, \dots, \phi_n^*\}$ be the corresponding dual basis of $\mathcal{P}_n^*$.

   Define $\alpha_1 : \mathcal{P}_n \to \mathbb{R}$ to be the function $\alpha_1(f) = \left.\frac{df}{dt}\right|_{t=1/2}$. Write α_1 as a linear combination of the elements of the dual basis you have constructed. Express α_k in terms of your dual basis where α_k is the kth derivative evaluated at $t = 1/2$.

   Do the same for the linear transformations $\beta_k : \mathcal{P}_n \to \mathbb{R}$ where β_k evaluates the kth derivative of the polynomial at $t = 0$.

2. Using the same notations as above let $T_m : \mathcal{P}_n \to \mathcal{P}_{n+m}$ be the linear transformation which multiplies a polynomial by t^m. Determine the matrix of T_m with respect to the bases you had chosen. Do the same for the transformation $\frac{d}{dt} : \mathcal{P}_n \to \mathcal{P}_{n-1}$.

   Write down a formula for $\left(\frac{d}{dt}\right)^*(\phi_m)$.

2.1.3 On linear spaces over $\mathbb{C}$

(1) Let E be an n-dimensional linear space over $\mathbb{C}$ and suppose we are given an ordered $\mathbb{C}$-basis of E: $\mathcal{B} = \{\boldsymbol{b}_1, \ldots, \boldsymbol{b}_n\}$.

Let $E_{\mathbb{R}} \subset E$ be the $\mathbb{R}$-linear space spanned by $\mathcal{B}$. Let us denote $\sqrt{-1}$ by $\boldsymbol{i}$ for brevity. Show that given $\boldsymbol{z} \in E$, there are unique vectors $x(\boldsymbol{z}), y(\boldsymbol{z}) \in E_{\mathbb{R}}$ such that $\boldsymbol{z} = x(\boldsymbol{z}) + \boldsymbol{i} \cdot y(\boldsymbol{z})$.

Thus, E is isomorphic (*qua* $\mathbb{R}$-linear space) to $E_{\mathbb{R}} \oplus \boldsymbol{i} \cdot E_{\mathbb{R}}$.

(2) Suppose that V is a $\mathbb{R}$-linear space on which there is defined an endomorphism $J : E \to E$ such that $J^2 = -1_V$.

What can you say about the spectrum of J?

Let $E_{\pm}$ be the image of $(1_V \pm J)$. Show that there is a direct sum decomposition $E \cong E_+ \oplus E_-$. Show that E can be given the structure of a $\mathbb{C}$-linear space in such a way that multiplication by $\sqrt{-1}$ is the same as applying the endomorphism J.

(HINT: Observe that $J(E_{\pm}) = E_{\mp}$ and note that for any $\boldsymbol{v} \in E$, the vectors $(\boldsymbol{v} \pm J(\boldsymbol{v}))$ can be thought of as the usual basis of a complex plane contained in E.)

Definition An endomorphism, J, of a $\mathbb{k}$-linear space, V, is called a *complex structure* if $J^2 = -1_V$.

(3) Define a complex structure on $\mathbb{C}^2$, that is *distinct* from the usual complex structure.

2.1.4 On Annihilators

Let E be a linear space and $S \subset E$ any subset of E. The *annihilator* of S is the set: $S^0 = \{\alpha \in E^* : \langle \alpha, s \rangle = 0 \text{ for all } s \in S\}$.

Show that S^0 is a subspace of E^* and that $\kappa_E(S) \subset (S^0)^0$. When does equality hold?

Let $A, B \subset E$ be finite-dimensional subspaces of a linear space E. Show that

(a) $$(A \cap B)^0 = A^0 + B^0 \text{ and}$$

(b) $$(A + B)^0 = A^0 \cap B^0.$$

Would these results hold if A and B were not finite-dimensional?

2.1.5 Direct Sums and Annihilators

Let V_1, V_2 be subspaces of a linear space E such that $V = V_1 \oplus V_2$. Show that the annihilator of V_i is isomorphic to V_j^* where $i, j = 1, 2$ and $i \neq j$. Show that E^* is isomorphic to the direct sum of these annihilators.

2.1.6 On duals of infinite dimensional spaces

We have seen that if E is an infinite-dimensional linear space with a basis $\{\boldsymbol{x}_i : i \in I\}$ then the dual vectors $\{\xi_i^*\}_{i \in I}$ defined by $\langle \xi_i^*, \boldsymbol{x}_j \rangle = \delta_{ij}$ *do not* form a basis of the dual space. This *suggests* that E and E^* are not isomorphic. Using the hints provided, or otherwise, prove the following result.

Proposition. *Let E be a linear space over* any *field with a basis $\mathcal{B}$ which is* countably infinite, *then E cannot be isomorphic to the dual space of any linear space.*

Hence if E is a linear space with a *countably infinite* basis (for example the free vector space over $\mathbf{k}$ generated by the set $\mathbb{N}$,) then E and E^* cannot be isomorphic.

Hints:

1. Show that if E is infinite-dimensional then E^* contains a linearly independent set } Σ^* having the same cardinality as $\mathbb{R}$. For this, proceed as in 2 and 3 below.

2. Let $\mathcal{B}$ be a basis of E, indexed by the rational numbers; that is, suppose $\mathcal{B} = \{\boldsymbol{v}_r\}_{r \in \mathbb{Q}}$ is a basis of E. For each real number x let $\mathbf{s}(x)$ be a sequence of rational numbers converging to x and let

$$\Sigma(x) = \{\boldsymbol{v}_q \in \mathcal{B} : q \text{ occurs in } \mathbf{s}(x)\}.$$

Show that if $x \neq y$ then $\Sigma(x) \cap \Sigma(y)$ is a finite set.

3. Let $\sigma^*(x) : E \to \mathbf{k}$ be the unique linear map such that $\sigma^*(x) \mid \mathcal{B} = \chi(\Sigma(x))$, the characteristic or indicator function of $\Sigma(x)$. Prove that

$$\Sigma^* = \{\sigma^*(x) : x \in \mathbb{R}\}$$

is a linearly independent set in E^*.

(I would like to thank Anindya Sen for showing me this elegant argument.)

2.2 Multilinear Functions

We have spent a long time studying the simplest example of the space of linear transformations between two linear spaces: the situation where the range is 1-dimensional. (The case when the domain is 1-dimensional is trivial. As was pointed out in the preamble of this chapter; $L(\mathbf{k}, E)$ being isomorphic to E via $T \mapsto T(1)$.) However, our time has not been wasted: dual spaces are truly fundamental, indeed ubiquitous, and we will encounter them again and again in the remainder of this book.

Our main objective now is to study the spaces $\mathfrak{L}_p(E; F)$ where the "**higher derivatives**" of a function live. Again, we begin with the simplest case, when $p = 1$; then by definition, $\mathfrak{L}_1(E; F) = L(E, F)$.

Let E and F be finite-dimensional linear spaces and $\mathcal{E} = \{\mathbf{e}_1, \ldots, \mathbf{e}_n\}$ and $\mathcal{F} = \{\boldsymbol{f}_1, \ldots, \boldsymbol{f}_m\}$ be bases of of E and F respectively. Then you already know that $L(E, F)$ can be identified with the set of $(m \times n)$ matrices. Since my aim is to avoid matrices as far as possible, I want to restate this as follows.

Proposition 2.2.1. *Let E and F be finite-dimensional linear spaces and $\mathcal{E}, \mathcal{F}$ be bases for these spaces as described above. Then the linear transformations $\epsilon_{ij} : E \to F$ defined by $\epsilon_{ij}(\mathbf{e}_k) = \delta_{ik}\boldsymbol{f}_j$ form a basis of $L(E, F)$.* (Here the indices, i, j vary over the integers $\{1, \ldots, m\}$ and $1 \leq k \leq n$.)

Hence $\dim L(E, F) = (\dim E) \cdot (\dim F)$.

(Note that the matrix of ϵ_{ij} with respect to the bases $\mathcal{E}, \mathcal{F}$ has 1 at the $(j, i)^{\text{th}}$ spot and 0's everywhere else.) □

For the present we content ourselves with this rather trivial result concerning $L(E, F)$.

Let us recall the conventions which we introduced in § 1.1 (see page 5) for p-functions. For each linear space E and $p > 1$, $E[p]$ will denote the p-tuple $(E, \ldots, E)$ and $E^{[p]}$ the p-fold Cartesian product of E.

I will not think of $E^{[p]}$ as a linear space; this set with componentwise addition and scalar multiplication is the p-fold direct sum: $E \oplus \cdots \oplus E$, with p summands.

Since $\mathfrak{L}_p(E;F) \subset \mathfrak{F}_p(E[p];F)$, the bijection

$$\Xi_p : \mathfrak{F}_p(\mathbf{E}[p];F) \to \mathfrak{F}(E^{[p]},F)$$

restricts to an injective function (also denoted Ξ_p) on $\mathfrak{L}_p(E;F)$. We will see that the image of $\Xi_p : \mathfrak{L}_p(E,F) \to \mathfrak{F}(E^{[p]},F)$ consists of familiar entities.

We start with the case $p = 2$. Suppose that $G \in \mathfrak{L}_2(E;F)$. Then

$$\begin{aligned}\Xi_2(G)(\boldsymbol{e}_1,\boldsymbol{e}_2) &= G(\boldsymbol{e}_2 : \boldsymbol{e}_1)\\ &= G(\boldsymbol{e}_2)(\boldsymbol{e}_1)\end{aligned}$$

and it is clear from the second line of the right-hand side that if we hold one of the variables in $\underline{G} \doteq \Xi_2(G)$ fixed, $\underline{G}$ is a linear function, in the other variable. In fact, for all $a, b \in \mathbf{k}$ and $\boldsymbol{x}, \boldsymbol{y} \in E$ we have,

$$\begin{aligned}\underline{G}(\boldsymbol{e}_1, a{\cdot}\boldsymbol{x} + b{\cdot}\boldsymbol{y}) &= a{\cdot}\underline{G}(\boldsymbol{e}_1,\boldsymbol{x}) + b{\cdot}\underline{G}(\boldsymbol{e}_1,\boldsymbol{y})\\ \underline{G}(a{\cdot}\boldsymbol{x} + b{\cdot}\boldsymbol{y}, \boldsymbol{e}_2) &= a{\cdot}\underline{G}(\boldsymbol{x},\boldsymbol{e}_2) + b{\cdot}\underline{G}(\boldsymbol{y},\boldsymbol{e}_2)\end{aligned}$$

Functions from $E \times E$ to any linear space satisfying these conditions are called *bilinear*. Similarly, you can easily check that if $G \in \mathfrak{L}_p(E[p];F)$ then $\underline{G} \doteq \Xi_p(G) : E^{[p]} \to F$ is a p-linear transformation. That is, for any $1 \leq j \leq p$ and for all $a, b \in \mathbf{k}$ and $\boldsymbol{x}, \boldsymbol{y} \in E$, we have:

$$\left.\begin{array}{l}\underline{G}(\boldsymbol{e}_1,\ldots,\boldsymbol{e}_{j-1}, a{\cdot}\boldsymbol{x} + b{\cdot}\boldsymbol{y}, \boldsymbol{e}_{j+1},\ldots,\boldsymbol{e}_p)\\ a{\cdot}\underline{G}(\boldsymbol{e}_1,\ldots,\boldsymbol{e}_{j-1},\boldsymbol{x},\boldsymbol{e}_{j+1},\ldots,\boldsymbol{e}_p)\\ +{\cdot}b\underline{G}(\boldsymbol{e}_1,\ldots,\boldsymbol{e}_{j-1},\boldsymbol{y},\boldsymbol{e}_{j+1},\ldots,\boldsymbol{e}_p)\end{array}\right\}\ (p\text{-linearity})$$

Such functions from $E^{[p]}$ to F will be called *p-linear functions* from E to F (abuse of language!). A *multilinear* function from E to F is a function which is p-linear for some p. 2-linear functions are called bilinear functions.

You have seen p-linear (or multilinear) functions before, but, rather like Molière's bourgeois gentleman (who was astonished when he found out that he had been speaking "prose" all his life!) you did not know that you were dealing with multilinear functions.

Examples of Multilinear Functions

Example 2.2.2a. The function $\mathbb{R}^n \to \mathbb{R}$ which takes $(x_1,\ldots,x_n)$ to the product of the coordinates is a n-linear transformation from $\mathbb{R}^n \to \mathbb{R}$.

Example 2.2.2b (The "cross product" in $\mathbb{R}^3$).
On $\mathbb{R}^3$ the vector cross-product, which you may have come across in

Mechanics or Physics, is defined by the formula:

$$\begin{bmatrix} x_1 \\ x_2 \\ x_3 \end{bmatrix} \times \begin{bmatrix} y_1 \\ y_2 \\ y_3 \end{bmatrix} = \begin{bmatrix} x_2y_3 - x_3y_2 \\ x_1y_3 - x_3y_1 \\ x_1y_2 - x_2y_1 \end{bmatrix}.$$

is an *antisymmetric* bilinear function. The antisymmetry mentioned alludes to the fact that for any two $\boldsymbol{x}, \boldsymbol{y} \in \mathbb{R}^3$, $(\boldsymbol{x} \times \boldsymbol{y}) = -(\boldsymbol{y} \times \boldsymbol{x})$. Such a bilinear function from $\mathbb{R}^n$ to $\mathbb{R}^n$ does not exist for $n \neq 3$!

Example 2.2.2c. If V is a linear space over $\mathbb{R}$, an inner product on V is a bilinear function from V to $\mathbb{R}$.

Example 2.2.2d (The determinant).
An $(n \times n)$ matrix A may be regarded as an element of, $\times^n \mathbf{k}^n$, the n-fold Cartesian product of $\mathbf{k}^n$ simply by regarding its columns as elements of $\mathbf{k}^n$. Then the function $\mathrm{A} \mapsto \det(\mathrm{A})$ is an n-linear function from $\mathbf{k}^n$ to $\mathbf{k}$.

I hope that I have reassured you that although the higher derivatives are "inhabitants" of complicated spaces such as $\mathfrak{L}_p(E;F)$ they themselves can be identified with multilinear functions and *these are familiar objects.*

I am now going to introduce a further level of complexity!

Recall that the sets $\mathfrak{F}_p(A[p];B)$ (of which the linear spaces $\mathfrak{L}_p(E;F)$ were special subsets), were themselves special cases of the sets of type $\mathfrak{F}(A_1,\dots,A_p;B)$ and we should look at the linear analogues of these sets. So let us consider an ordered p-tuple of linear spaces $\mathbf{E} = (E_1,\dots,E_p)$ and another linear space F. As before, we inductively define

$$\mathfrak{L}_p(\mathbf{E};F) \doteq \begin{cases} L(E_1,F), & \text{if } p=1 \\ L\big(E_p, \mathfrak{L}_{p-1}\big((E_1,\dots,E_{p-1});F\big)\big), & \text{if } p>1. \end{cases}$$

Once again by applying the bijection Ξ_p we map $\mathfrak{L}_p\big((E_1,\dots,E_p);F\big)$ into a subset of $\mathfrak{F}(E_1 \times \cdots \times E_p;F)$.

Again, it is easy check that

$$\mathcal{M}(E_1 \times \cdots \times E_p, F) \doteq \Xi_p\Big(\mathfrak{L}_p\big((E_1,\dots,E_p);F\big)\Big)$$

consists of functions, $g : E_1 \times \cdots \times E_p \to F$ that are *separately* linear in each variable. This means that for any $1 \le j \le p$ and for all $a, b \in \mathbf{k}$ and all $\boldsymbol{x}, \boldsymbol{y} \in E_j$, we have:

$$\left.\begin{aligned} &g(\boldsymbol{e}_1,\dots,\boldsymbol{e}_{j-1}, a\cdot\boldsymbol{x} + b\cdot\boldsymbol{y}, \boldsymbol{e}_{j+1},\dots,\boldsymbol{e}_p) \\ &= a\cdot g(\boldsymbol{e}_1,\dots,\boldsymbol{e}_{j-1},\boldsymbol{x},\boldsymbol{e}_{j+1},\dots,\boldsymbol{e}_p) \\ &\quad + b\cdot g(\boldsymbol{e}_1,\dots,\boldsymbol{e}_{j-1},\boldsymbol{y},\boldsymbol{e}_{j+1},\dots,\boldsymbol{e}_p) \end{aligned}\right\} \quad \left(\begin{matrix}\text{Multilinearity} \\ \text{condition.}\end{matrix}\right)$$

Obviously such functions should, and will be, also called p-linear functions. Some important and familiar examples follow.

EXAMPLES OF MULTILINEAR FUNCTIONS (CONTD.)

Example 2.2.2e.
If V is a linear space over $\mathbf{k}$ then "scalar multiplication" of vectors by elements of $\mathbf{k}$, is a bilinear function, $\mu : \mathbf{k} \times V \to V$. This is part of the axioms that define a linear space.

Example 2.2.2f (Evaluation functions).
If E, F are linear spaces then the evaluation function $\boldsymbol{ev} : L(E,F) \times E \to F$ defined by $\boldsymbol{ev}(T, \boldsymbol{e}) = T(\boldsymbol{e})$ is a bilinear function.

Example 2.2.2g (Composition of linear transformations).
If E, F, G are linear spaces, then the "composition":

$$\Gamma : L(E,F) \times L(F,G) \to L(E,G)$$

defined by $\Gamma(S,T) = T \circ S$ is a bilinear function.

I am sure you will agree that the slight additional level of complication I introduced is worthwhile since many extremely useful and important operations are included in our augmented notion of multilinear functions. Further interesting examples pertaining to linear transformations will follow.

It is obvious that the set of multilinear functions $\mathcal{M}(E_1 \times \cdots \times E_p, V)$ is a linear space if we define the addition and scalar multiplication of functions in a point-wise fashion, using the linear space structure of V.

The relation between p-linear functions and linear p-functions, i.e. the relation between the spaces $\mathcal{M}(E_1 \times \cdots \times E_p, F)$ and $\mathfrak{L}_p\big((E_1, \ldots, E_p); F\big)$ will be elucidated by the "Law of Exponents" which follows. The proof is trivial. Once one observes that the bijection Ξ_p takes the elements of $\mathfrak{L}_p\big((E_1, \ldots, E_p); F\big)$ to multilinear functions, there is nothing more to do!

Theorem 2.2.3 (Law of Exponents II).
With the linear space structures as defined above, the spaces of multilinear functions are isomorphic to the appropriate space of linear p-functions, via the exponential law. More precisely, for any p-tuple, $\mathbf{E} = (E_1, \ldots, E_p)$, *of linear spaces and any linear space* F, *over the same field* $\mathbf{k}$, *the bijections*

$$\Xi_p : \mathfrak{F}(\mathbf{E}; F) \to \mathfrak{F}(\times\mathbf{E}, F)$$
$$\xi_p : \mathfrak{F}(\times\mathbf{E}, F) \to \mathfrak{F}(\mathbf{E}; F)$$

of the Law of Exponents restrict to mutually inverse isomorphisms:

$$\Xi_p : \mathfrak{L}_p(\mathbf{E}; F) \ast\!\!\to \mathcal{M}(\times\mathbf{E}, F)$$
$$\xi_p : \mathcal{M}(\times\mathbf{E}, F) \ast\!\!\to \mathfrak{L}_p(\mathbf{E}; F). \quad \square$$

(The symbol, '$\times\mathbf{E}$', as explained on page 5, represents the Cartesian product of the elements of the p-tuple, $\mathbf{E}$.)

This is the second *avatar* of the Law of Exponents; a third (and final!) *avatar* will appear in the next chapter; however, we will use the same symbols to denote these isomorphisms in their different manifestations.

Remark 2.2.4. Note that although $E_1 \times \ldots E_p$ and $E_1 \oplus \cdots \oplus E_p$ are identical as sets we will always write the Cartesian product (which, *a priori*, does not have a linear structure) for the domain of a multilinear function. This is because the same linear space may have different direct sum decompositions and functions which are multilinear on $E_1 \times \cdots \times E_p$ need not be multilinear on $E_1' \times \cdots \times E_p'$ even if $E_1 \oplus \cdots \oplus E_p \approx E_1' \oplus \cdots \oplus E_p'$.

I will write $\mathcal{M}^*(\times\mathbf{E})$ for $\mathcal{M}(\times\mathbf{E}, \mathbf{k})$. Elements of $\mathcal{M}^*(\times\mathbf{E})$ are known as *multilinear forms* on $\times\mathbf{E}$. (For example, recall that $\det : \mathrm{M}_n(\mathbf{k}) \to \mathbf{k}$ can be thought of as an n-linear form on $\mathbf{k}^n$.)

If $\mathbf{E}, \mathbf{F}$ are p-tuples of $\mathbf{k}$-linear spaces, then a *linear morphism* $\mathbf{T} : \mathbf{E} \Rightarrow \mathbf{F}$ is a p-tuple, $(T_1, \ldots, T_p)$ where for each i, $T_i : E_i \to F_i$ is a linear transformation. (This bit of nonstandard nomenclature will save a lot of words.)

The next result is trivially proved but suggests the possibility of reducing the study of multilinear functions to the study of linear transformations.

Proposition 2.2.5 (Induced multilinear functions).
(1) *Let $\mathbf{E}, \mathbf{F}$ be n-tuples of linear spaces. Suppose that $\mathbf{T} : \mathbf{E} \Rightarrow \mathbf{F}$ is a linear morphism. If V is any linear space, then $\mathbf{T}$ induces a linear transformation*

$$\mathbf{T}^\star : \mathcal{M}_n(\times\mathbf{F}, V) \to \mathcal{M}_n(\times\mathbf{E}, V) \text{ via the formula:}$$
$$\mathbf{T}^\star(\phi)(\boldsymbol{f}_1, \ldots, \boldsymbol{f}_n) = \phi(T_1(\boldsymbol{e}_1), \ldots, T_n(\boldsymbol{e}_n))$$

for every $\phi \in \mathcal{M}_n(\times\mathbf{F}, V)$.

(2) *Any linear transformation, $L : V \to W$ induces a linear transformation $L_\star : \mathcal{M}_n(\times\mathbf{E}, V) \to \mathcal{M}_n(\times\mathbf{E}, W)$ defined as $L_\star(\phi) = L \circ \phi$.* $\square$

Part (2) shows that certain multilinear functions with range W can be represented as a composition of a multilinear function with range V composed with linear transformations from V to W. This possibility motivates the next definition.

Definition 2.2.6 (Universal multilinear functions).
Let $\mathbf{E} = (E_1, \ldots, E_n)$ be an n-tuple of $\mathbf{k}$-linear spaces. A multilinear function, $U_{\mathbf{E}} : \times\mathbf{E} \to V_{\mathbf{E}}$, is said to be *universal* if given any multilinear function, $\phi : \times\mathbf{E} \to F$, there exists a unique linear transformation $\widetilde{\phi} : V_{\mathbf{E}} \to F$ such that the diagram:

$$\begin{array}{ccc} \times\mathbf{E} & \xrightarrow{\;U_{\mathbf{E}}\;} & V_{\mathbf{E}} \\ & {\scriptstyle\phi}\searrow \quad \swarrow{\scriptstyle\widetilde{\phi}} & \\ & F & \end{array}$$

commutes. Such a universal multilinear function, if it exists, will obviously reduce the study of multilinear functions with domain $\times\mathbf{E}$ to the study of linear transformations with domain $V_{\mathbf{E}}$. The linear transformation $\widetilde{\phi}$ is said to be induced by the *universal mapping property* of $U_{\mathbf{E}}$.

It might seem a rather tall order to produce universal multilinear functions for arbitrary tuples of linear spaces, but this can be done. More interestingly (for our purpose) it is not too difficult to construct such universal multilinear functions as long as one requires that all the linear spaces involved are Euclidean spaces.

But before going further we look at a special case (indeed the only case) where universal bilinear functions are easily constructed.

Example 2.2.7 (Trivial examples).
Let E be any $\mathbf{k}$-linear space. Then scalar multiplication, $\mu : \mathbf{k} \times E \to E$, (which, in part, defines the linear structure of E) is a universal bilinear function.

Proof. Let $\phi : \mathbf{k} \times E \to F$ be any bilinear function. Then from the definition of multilinearity, $\widetilde{\phi} : E \to F$ defined by $\widetilde{\phi}(\boldsymbol{e}) = \phi(1, \boldsymbol{e})$ is a linear transformation.

Clearly, $\phi(\alpha, \boldsymbol{e}) = \alpha{\cdot}\phi(1, \boldsymbol{e}) = \phi(1, \alpha{\cdot}\boldsymbol{e}) = \widetilde{\phi}\big(\mu(\alpha, \boldsymbol{e})\big)$. This shows that μ is universal. □

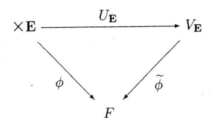

The next result asserts that any n-tuple of linear spaces admits an (essentially) unique universal multilinear function.

Proposition 2.2.8 (The uniqueness of tensor products).
Let $\mathbf{E} = (E_1, \ldots, E_n)$ *be an ordered* n*-tuple of linear spaces and suppose that* $\phi : \times\mathbf{E} \to V$, *and* $\Psi : \times\mathbf{E} \to W$ *are universal multilinear functions.*
Then V *and* W *are* "naturally isomorphic".

Proof. Consider the commutative diagram:

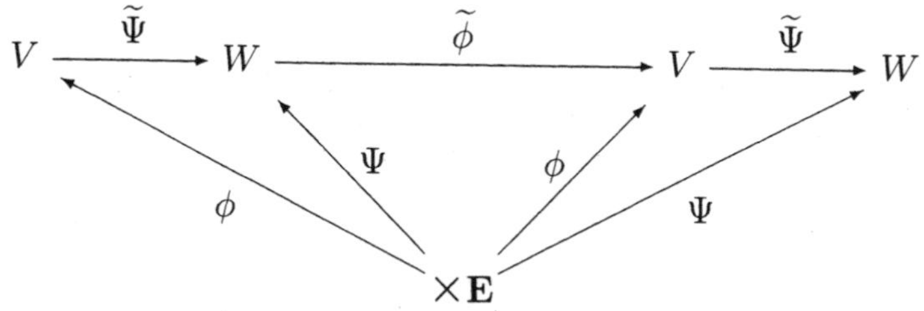

where $\widetilde{\phi}$ and $\widetilde{\Psi}$ are the linear transformations induced by the universal functions ϕ and Ψ respectively.

From the diagram it is clear that $\phi = (\widetilde{\phi} \circ \widetilde{\Psi}) \circ \phi$. So the uniqueness of the linear transformation induced by universality implies that $(\widetilde{\phi} \circ \widetilde{\Psi}) = 1_V$. Interchanging the roles of V, W and ϕ, Ψ we see that $(\widetilde{\Psi} \circ \widetilde{\phi}) = 1_W$. thus V and W are isomorphic via linear transformations which arise out of the data given by the universality conditions of ϕ and Ψ. This is why we say that V and W are *naturally isomorphic.* □

Remark. Notice that this proof of the uniqueness of universal multilinear functions did not involve the field of scalars or finite dimensionality of the spaces involved. Though I will construct universal multilinear functions or "tensor products", only for Euclidean spaces, you should keep in mind that they can be defined for linear spaces over arbitrary fields, without any finite-dimensionality conditions.

I now formally define tensor products; henceforth I will use standard terminology. The phrase "universal multilinear functions" was used to motivate what is a completely novel way of defining a mathematical object: as an entity satisfying a universal mapping property.

Definition 2.2.9 (Tensor Products).
If $\mathbf{E} = (E_1, \ldots, E_n)$ is an n-tuple of linear spaces, the *tensor product* of $\mathbf{E}$ is a universal multilinear function on $\times\mathbf{E}$. *Both the function and the range together constitute the tensor product.* The usual notation for the

tensor product is: $\otimes : \times\mathbf{E} = (E_1 \times \cdots \times E_n) \to E_1 \otimes \cdots \otimes E_n.$

The multilinear function '$\otimes$' is called the *canonical function* of the tensor product and the range is, by abuse of language, called the tensor product of the n-tuple $(E_1, \ldots, E_n)$. If several tensor products are being considered at the same time then I will be more specific about the canonical functions. For example if two ordered pairs (E, F) and (V, W) are being considered at the same time, I will write,

$$\begin{aligned} \otimes_{E,F} : E \times F &\to E \otimes F, \\ \otimes_{V,W} : V \times W &\to V \otimes W. \end{aligned}$$

REMARKS 2.2.9 (ON TENSOR PRODUCTS)

Remark 2.2.9a. In most books you will find that the definition of the tensor product employs the indefinite article. For example the definition might start: "A tensor product of E and F is a ...". In such texts, the first result proved is the uniqueness of the tensor product. I have first proved the uniqueness and so in the above definition speak of *the* tensor product of $\mathbf{E}$.

Remark 2.2.9b (A caveat).
I will not be constructing *true* tensor products. What I will do is to construct, for any n-tuple, $\mathbf{E}$, of Euclidean spaces, a multilinear function which has the universal mapping property for multilinear functions whose domain is $\times\mathbf{E}$ and whose range is also a *Euclidean space.* I will sometimes call this a *Euclidean tensor product.*

You should keep in mind the shortcomings of the construction of universal multilinear functions that follows.

The great advantage of defining objects in terms of universal mapping properties is that many properties of the object so defined can be ascertained without knowing "what the object looks like". Here is a small example.

Proposition 2.2.10 (The dimension of a tensor product).
Let $\mathbf{E} = (E_1, \ldots, E_n)$ be an n-tuple of finite dimensional linear spaces. Then,

$$\dim(E_1 \otimes \cdots \otimes E_n) = \dim(\otimes\mathbf{E}) == \prod_{k=1}^{n} (\dim E_k).$$

Proof. From the hypotheses of the universal mapping property which defines a tensor product, it is clear that the space of multilinear forms on $\times\mathbf{E}$ and the dual space of the tensor product are isomorphic. (Any $\alpha \in (E_1 \otimes \cdots \otimes E_n)^*$ yields a multilinear form, $(\alpha \circ \otimes)$, on $\times\mathbf{E}$ and vice versa.)

So, using the Law of exponents we get the relations:

$$\dim(\otimes\mathbf{E}) = \dim(\otimes\mathbf{E})^* = \dim \mathcal{M}^*(\times\mathbf{E}) = \dim\big(\mathfrak{L}_n(\mathbf{E};\mathbf{k})\big).$$

To compute the dimension of $\mathfrak{L}_n(\mathbf{E}:\mathbf{k})$ it is only necessary to repeatedly use Proposition 2.2.1 and the result follows. □

As yet we know nothing about what the space $\otimes\mathbf{E}$ might look like, except that it has the same dimension as the space of multilinear forms on $\times\mathbf{E}$. As a first guess we could try to define a multilinear function from $\times\mathbf{E}$ to $\mathcal{M}^*(\times\mathbf{E})$ which behaved like a canonical function of the tensor product. But it is not clear how one can associate with each n-tuple, $(\boldsymbol{e}_1, \ldots, \boldsymbol{e}_n)$, a multilinear form and so even defining any kind of multilinear function seems difficult.

On the other hand, given a multilinear form, μ, the evaluation function:

$$\begin{aligned} \boldsymbol{ev} : (\times\mathbf{E}) \times \big(\mathcal{M}^*(\times\mathbf{E})\big)^* &\to \mathbf{k}, \text{ defined by} \\ \big((\boldsymbol{e}_1, \ldots, \boldsymbol{e}_n), \mu\big) &\mapsto \mu(\boldsymbol{e}_1, \ldots, \boldsymbol{e}_n) \end{aligned}$$

is a function of $(n+1)$ variables, which for fixed μ, is a multilinear function of the first n variables and for a fixed choice, $(\boldsymbol{e}_1, \ldots, \boldsymbol{e}_n)$, dependent linearly on μ. So this gives us a multilinear function:

$$\phi_{\mathbf{E}} : \times\mathbf{E} \to \big(\mathcal{M}^*(\times\mathbf{E})\big)^*.$$

If the spaces E_k are Euclidean spaces and $\mathbf{k} = \mathbb{R}$, this function is a Euclidean tensor product.

Theorem 2.2.11 (Euclidean Tensor Products).
Let $\mathbf{E} = (E_1, \ldots, E_n)$ be an n-tuple of Euclidean spaces. Define the function, $\phi : \times\mathbf{E} \to \big(\mathcal{M}^*(\mathbf{E})\big)^*$ by setting

$$\phi(\boldsymbol{e}_1, \ldots, \boldsymbol{e}_n)(\mu) = \mu(\boldsymbol{e}_1, \ldots, \boldsymbol{e}_n)$$

for every n-linear form, μ on $(E_1 \times \cdots \times E_n)$.

Then ϕ defines the Euclidean tensor product of $(E_1, \ldots, E_n)$.

Proof. Let V be a Euclidean space. I will show that given any multilinear function $T : \times\mathbf{E} \to V$, there is a unique linear transformation

$$\widetilde{T} : \big(\mathcal{M}^*(\times\mathbf{E})\big)^* \to V^{**}$$

such that there is a commutative diagram:

$$(\dagger)\ \cdots\cdots\cdots\cdots \qquad \begin{array}{ccc} \times\mathbf{E} & \xrightarrow{\ T\ } & V \\ \Big\downarrow{\scriptstyle \phi} & & \Big\downarrow{\scriptstyle \kappa_V} \\ (\mathcal{M}^*(\times\mathbf{E}))^* & \xrightarrow{\ \widetilde{T}\ } & V^{**} \end{array}$$

So modulo κ_V, which for the Euclidean space, V, is equal to the identity, 1_V, the function, ϕ is a universal multilinear function on $\times\mathbf{E}$.

The definition of $\widetilde{T}$:

(I will use the second part of the innocuous-looking Proposition 2.2.5.) Let $T : \times\mathbf{E} \to V$ be any multilinear function. For each $\alpha \in V^*$ consider the composite function

$$\begin{array}{rcl} \alpha \circ T : \times\mathbf{E} & \to & \mathbb{R} \text{ defined by:} \\ (\mathbf{e}_1, \ldots, \mathbf{e}_n) & \mapsto & \big\langle \alpha, T(\mathbf{e}_1, \ldots, \mathbf{e}_n)\big\rangle. \end{array}$$

Now $\big\langle \alpha, T(\mathbf{e}_1, \ldots, \mathbf{e}_n)\big\rangle$ depends linearly on α and for each fixed $\alpha \in V^*$ is a multilinear form on $\times\mathbf{E}$. So $\alpha \mapsto \alpha \circ T$ is a linear transformation $\widehat{T} : V^* \to \mathcal{M}(\times\mathbf{E})^*$. Define the linear transformation, $\widetilde{T}$, induced by the multilinear function T to be the transpose of $\widehat{T}$.

Commutativity of $\dagger$:

Let $(\mathbf{e}_1, \ldots, \mathbf{e}_n) \in \times\mathbf{E}$ and $\alpha \in V^*$. Since, by definition, $\widetilde{T}$ is the transpose of $\widehat{T}$, we see that

$$\begin{aligned} \big\langle \widetilde{T}\big(\phi(\mathbf{e}_1, \ldots, \mathbf{e}_n)\big), \alpha\big\rangle &= \langle \alpha, \widehat{T}\big(\phi(\mathbf{e}_1, \ldots, \mathbf{e}_n)\big)\rangle \\ &= \big\langle T^*(\alpha), \phi(\mathbf{e}_1, \ldots, \mathbf{e}_n)\big\rangle \quad (\text{since } T^*(\alpha) = \alpha \circ T) \\ &= \langle \alpha, T(\mathbf{e}_1, \ldots, \mathbf{e}_n)\rangle \\ &= \langle \kappa_V\big(T(\mathbf{e}_1, \ldots, \mathbf{e}_n)\big), \alpha\rangle, \end{aligned}$$

proving that $(\dagger)$ commutes.

We leave it as an exercise for the reader to check the uniqueness of $\widetilde{T}$. □

Since universal multilinear functions (tensor products) are *essentially unique*, (this is the content of Proposition 2.2.8), when dealing with Euclidean spaces, I will take the liberty of using the evaluation (multilinear) function:

$$\phi : E_1 \times \cdots \times E_n \to \big(\mathcal{M}^*(E_1 \times \cdots \times E_n)\big)^*$$

as a tensor product and so refer to ϕ as the *canonical function* of the tensor product and use the standard notation:

$$\otimes \doteq \phi : E_1 \times \cdots \times E_n \to E_1 \otimes \cdots \otimes E_n.$$

Remark. This practice is common although there are eloquent denunciations of this "abusive double dualization" in the literature! (See, for example, the Preface in [Lan65].)

2.3 More about tensor products

In view of the importance of tensor products in mathematics and physics, I briefly discuss some of their important properties. Although our construction of tensor products is valid only in the context of Euclidean spaces, the results in this section are stated and proved for all **k**-linear spaces. (To simplify the exposition I will deal, for the most part, with the tensor product of two linear spaces.)

Theorem 2.3.1 (Properties of the canonical function).

1. *If* $\mathcal{E} = \{\boldsymbol{e}_1, \ldots, \boldsymbol{e}_m\}$ *and* $\mathcal{F} = \{\boldsymbol{f}_1, \ldots, \boldsymbol{f}_n\}$ *are bases for the* **k**-*linear spaces* E *and* F *respectively, then:*

$$\mathcal{E} \otimes \mathcal{F} \doteq \{\boldsymbol{e}_i \otimes \boldsymbol{f}_j \doteq \otimes(\boldsymbol{e}_i, \boldsymbol{f}_j) : 1 \leq i \leq m, 1 \leq j \leq n\}$$

   form a basis of $E \otimes F$.

2. $\otimes : E \times F \to E \otimes F$ *is* never *injective.*

3. $\otimes : E \times F \to E \otimes F$ *is* generally not *surjective.*

Proof. 1. Let G be any linear space and $\mu : E \times F \to G$ a bilinear function. From the bilinearity of μ it follows that

$$\mu\Big(\sum_{i=1}^{m} c_i \boldsymbol{e}_i, \sum_{j=1}^{n} d_j \boldsymbol{f}_j\Big) = \sum_{\substack{1 \leq i \leq m \\ 1 \leq j \leq n}} c_i d_j \mu(\boldsymbol{e}_i, \boldsymbol{f}_j).$$

This shows that any bilinear function on $E \times F$ is completely determined by the values it takes on the mn ordered pairs:

$$\{(\boldsymbol{e}_i, \boldsymbol{f}_j) : 1 \leq i \leq m, 1 \leq j \leq n\}.$$

Now observe that if $L : E \otimes F \to G$ is a linear transformation, then the composition $E \times F \xrightarrow{\otimes} E \otimes F \xrightarrow{L} G$ is a bilinear function.

Thus given any elements $\{g_{ij} \in G : 1 \leq i \leq m, 1 \leq j \leq n\}$ there is a unique linear transformation taking $\boldsymbol{e}_i \otimes \boldsymbol{f}_j$ to g_{ij}. The result follows from the Basis Criterion.

2. Indeed no nontrivial multilinear function M can be injective because:
$$M(c\cdot\boldsymbol{x}, \boldsymbol{y}) = c\cdot M(\boldsymbol{x}, \boldsymbol{y}) = M(\boldsymbol{x}, c\cdot\boldsymbol{y}).$$

3. We elucidate the reason for the failure of surjectivity of $\otimes$ by looking at the situation when both E and F have dimension 2. Using the notation for bases used in the first part, we see that:
$$\otimes\Big(\sum_{i=1}^{2} c_i\boldsymbol{e}_i, \sum_{j=1}^{2} d_j\boldsymbol{f}_j\Big) = \sum_{\substack{i=1,2\\ j=1,2}} c_i d_j(\boldsymbol{e}_i \otimes \boldsymbol{f}_j).$$
Hence if
$\boldsymbol{v} = w\cdot(\boldsymbol{e}_1 \otimes \boldsymbol{f}_1) + x\cdot(\boldsymbol{e}_1 \otimes \boldsymbol{f}_2) + y\cdot(\boldsymbol{e}_2 \otimes \boldsymbol{f}_1) + z\cdot(\boldsymbol{e}_2 \otimes \boldsymbol{f}_2) \in E \otimes F$
is in the image of $\otimes : E \times F \to E \otimes F$ then w, x, y, z must satisfy the condition $w\cdot z - x\cdot y = 0$. This is also clearly a sufficient condition. Similar, but more complicated, conditions determine the image of the canonical function when the dimensions of E and F are greater than 2.

In any case, if $\mathcal{E}, \mathcal{F}$ are, as in the first part, bases of E, F respectively then the vector $\boldsymbol{v}$ will be in the image of $\otimes : E \otimes F \to E \otimes F$ iff $wz = xy$. □

Remark 2.3.2. Parts (2) and (3) of the above result imply that in order to define a linear transformation $L : E \otimes F \to V$ from the tensor product of two spaces to another linear space,

- it suffices to define L on vectors of the form $\boldsymbol{e} \otimes \boldsymbol{f}$, however:
- since any symbol like $\boldsymbol{e} \otimes \boldsymbol{f}$ can be represented by many pairs of elements $(\boldsymbol{e}', \boldsymbol{f}') \in E \times F$ there remains a non-trivial task of checking that $L(\boldsymbol{e} \otimes \boldsymbol{f})$ is well defined.

In practice, the only way one defines a linear transformation on a tensor product is by defining a *bilinear function* $\lambda : E \times F \to V$ and then looking at the induced linear transformation from $\widetilde{\lambda} : E \otimes F \to V$. Obvious analogues of this procedure apply to the situation of defining linear transformations on a p-fold tensor product, $E_1 \otimes$

$\cdots \otimes E_p$. Some explicit examples of such procedures will follow soon.

EXAMPLES OF DEFINING LINEAR FUNCTIONS ON TENSOR PRODUCTS

Example 2.3.3a. Let E and F be finite-dimensional. Consider the bilinear function, $\ell_{E,F} : E^* \times F \to L(E, F)$ defined by: $\ell_{E,F}(\alpha, \boldsymbol{e})(\boldsymbol{f}) = \langle \alpha, \boldsymbol{e}\rangle \cdot \boldsymbol{f}$. The induced linear transformation $\widetilde{\ell}_{E,F} : E^* \otimes F \to L(E, F)$ is an isomorphism and will be denoted $\lambda_{E,F}$ in future.

Proof. As before, let $\mathcal{E}, \mathcal{F}$ be bases for E, F and let $\mathcal{E}^* = \{\epsilon_1, \ldots, \epsilon_m\}$ be the basis dual to $\mathcal{E}$. It is clear that the matrix of $\ell_{E,F}(\epsilon_i, \boldsymbol{f}_j)$ with respect to the bases $\mathcal{E}$ and $\mathcal{F}$ is the matrix E_{ij} which has 1 in the $(j, i)^{\text{th}}$ spot and 0's elsewhere. Thus the basis $\mathcal{E}^* \otimes \mathcal{F}$ is mapped onto the basis of $L(E, F)$ introduced in Proposition 2.2.1 on page 51. Hence $\lambda_{E,F}$ is an isomorphism. □

As a continuation of the previous example, here is a basis-free definition of the trace of an endomorphism $T : E \to E$.

Example 2.3.3b (Definition of Trace).
Consider the bilinear function $\boldsymbol{ev} : E^* \times E \to \mathbb{R}$ given by $\boldsymbol{ev}(\alpha, \boldsymbol{v}) = \langle \alpha, \boldsymbol{v}\rangle$. If $T \in L(E, E)$ then $\widetilde{\boldsymbol{ev}}(\lambda_{E,E}^{-1}(T)) = \operatorname{Trace} T$.

Proof. Both $\widetilde{\boldsymbol{ev}} \circ \lambda^{-1}$ and Trace are linear transformations from $L(E, E)$ to $\mathbf{k}$. It is easily checked that they agree on the basis $\{E_{ij}\}$. □

We now consider an example with a slightly different flavour.

Example 2.3.3c (The "flip" function).
Let E, F be linear spaces. Consider the function $T : E \times F \to F \times E$ given by $T(\boldsymbol{e}, \boldsymbol{f}) = (\boldsymbol{f}, \boldsymbol{e})$. If $B : F \times E \to V$ is any bilinear function, then $(B \circ T)$ is a bilinear function on $E \times F$. In particular, choose B to be the canonical function, $\otimes_{F,E} : F \times E \to F \otimes E$. Denote the induced linear transformation $\widetilde{B \circ T} : E \otimes F \to F \otimes E$ by $T_\#$. In the same way one can define a linear transformation $T'_\# : F \otimes E \to E \otimes F$. It is easy to check that these two linear transformations are inverses of one another.

This establishes an important property of tensor products:

Proposition 2.3.4 (Commutativity of the Tensor Product).
If E, F are linear spaces, then $E \otimes F$ is naturally isomorphic to $F \otimes E$.

The adjective "natural" is used since the isomorphisms $T_{\#}, T'_{\#}$ do not depend on any special choices; they arise from the canonical functions and the *flip* operation. □

We continue with examples of linear transformations defined on tensor products.

Example 2.3.3d (The tensor product of linear transformations).
Let $S : E \to V$ and $T : F \to W$ be linear transformations. Regard the ordered pair (S, T) as a linear morphism, $\phi : (E, F) \Rightarrow (V, W)$. (Recall Proposition 2.2.5 on page 55.)

The induced bilinear function $S \otimes T \doteq \phi^\star(\otimes_{V,W}) : E \otimes F \to V \otimes W$, is known as the tensor product of the linear transformations S and T. Thus, for all $\boldsymbol{e} \in E$ and $\boldsymbol{f} \in F$, $(S \otimes T)(\boldsymbol{e} \otimes \boldsymbol{f}) = (S(\boldsymbol{e})) \otimes_{V,W} (T(\boldsymbol{f}))$.

Remark 2.3.5. The notion of tensor products first entered mathematics in the guise of the "Kronecker product" of matrices which is a matrix version of the tensor product of linear transformations. (This will be dealt with in the Exercises which follow.)

Here are some important properties of this new operation.

Proposition 2.3.6 (Functorial properties).
Suppose that E, F, V, W, V', W' *are linear spaces and suppose that*

$$\begin{aligned} S, S' : E &\to V, \\ T, T' : F &\to W, \\ L : V &\to V', \\ \textit{and } M : W &\to W' \textit{ linear transformations.} \end{aligned}$$

Then:

1. $1_E \otimes 1_F = 1_{E \otimes F}$;

2. $(L \otimes M) \circ (S \otimes T) = (LS) \otimes (MT)$;

3. $$\begin{aligned} (S + S') \otimes T &= S \otimes T + S' \otimes T, \\ \textit{and } S \otimes (T + T') &= S \otimes T + S \otimes T'. \end{aligned}$$

Thus one can think of the tensor product as a "bilinear functor of two variables"!

Proof. The proof is best done using commutative diagrams. By definition, $(S \otimes T) : E \otimes F \to V \otimes W$ is the unique linear transformation which makes the following diagram commute (Note that in this diagram the vertical arrows are bilinear functions, the top horizontal "arrow" represents a linear morphism; only the lower horizontal arrow is a linear transformation!)

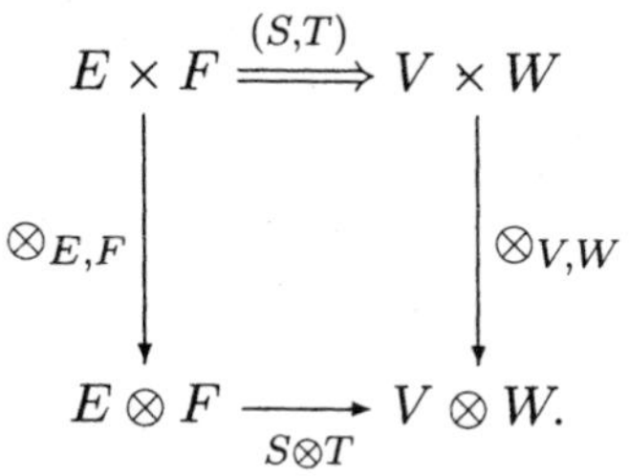

1. In the above diagram if we replace V by E, W by F, S by 1_E and T by 1_F, it is obvious that the identity of $E \otimes F$ as the lower horizontal arrow will make the diagram commute. It follows that $(1_E) \otimes (1_F) = 1_{E \otimes F}$.

2. Consider the diagram

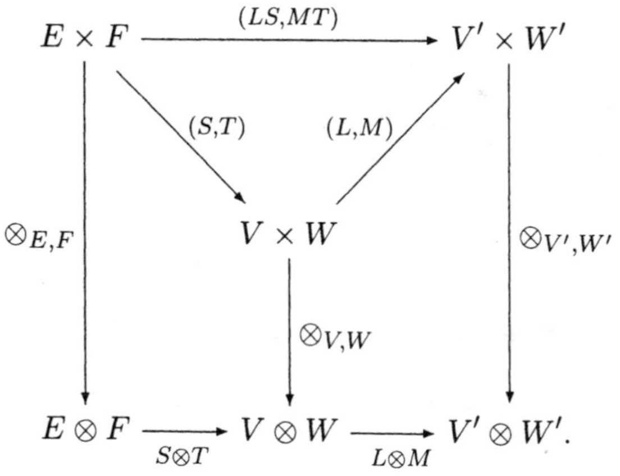

Since $(L, M) \circ (S, T)$ is obviously equal to $\big((LS), (MT)\big)$ it follows that the top triangle commutes; the "trapezoidal" sub-diagrams commute by definition of $S \otimes T$ and $L \otimes M$. Hence the whole diagram commutes and we can conclude that $(L \otimes M) \circ (S \otimes T) = (LS) \otimes (MT)$.

3. This is being left as an exercise for the reader.

□

Exercises on Tensor Products

These exercises for the most part depend only on the universality of the canonical function $\otimes$. Although I have established the universal canonical function in the context of Euclidean spaces, these exercises are valid for arbitrary finite-dimensional linear spaces and are being offered in this wider context.

2.3.1 The associativity of the tensor product

(a) Let E, F, G be linear spaces. Show that $(E \otimes F) \otimes G$ and $E \otimes (F \otimes G)$ are naturally isomorphic.

(Hint: Show that both can be the target space of a universal 3-linear function on $E \times F \times G$.)

(b) Show that the associative property is true for tensor products of linear transformations.

2.3.2 The Kronecker product of matrices

Let E, F, V, W be finite-dimensional linear spaces and suppose that

$$\begin{aligned} \mathcal{E} &= \{\boldsymbol{e}_1, \ldots, \boldsymbol{e}_n\} \\ \mathcal{F} &= \{\boldsymbol{f}_1, \ldots, \boldsymbol{f}_q\} \\ \mathcal{V} &= \{\boldsymbol{v}_1, \ldots, \boldsymbol{v}_m\} \\ \mathcal{W} &= \{\boldsymbol{w}_1, \ldots, \boldsymbol{w}_p\} \end{aligned}$$

are bases for these spaces. Let $S : E \to V$ and $T : F \to W$ be linear transformations and $\mathrm{A} = \mathrm{M}(S; \mathcal{E}, \mathcal{V})$ and $\mathrm{B} = \mathrm{M}(T; \mathcal{F}, \mathcal{W})$ the matrices of the linear transformations with respect to these bases.

Endow the basis $\mathcal{E} \otimes \mathcal{F} = \{\boldsymbol{e}_i \otimes \boldsymbol{f}_j \; : \; 1 \leq i \leq n, 1 \leq j \leq q\}$ with the lexicographic ordering; that is, $\boldsymbol{e}_k \otimes \boldsymbol{f}_l < \boldsymbol{e}_{k'} \otimes \boldsymbol{f}_{l'}$ if $k < k'$ and if $k = k'$ then if $l < l'$. Give a similar ordering to the basis $\mathcal{V} \otimes \mathcal{W}$.

Show that with respect to the bases $\mathcal{E} \otimes \mathcal{F}$ and $\mathcal{V} \otimes \mathcal{W}$ the matrix of $S \otimes T$ has a block decomposition:

$$\begin{bmatrix} a_{11}\mathrm{B} & a_{12}\mathrm{B} & \ldots & a_{1n}\mathrm{B} \\ a_{21}\mathrm{B} & a_{22}\mathrm{B} & \ldots & a_{2n}\mathrm{B} \\ \vdots & \vdots & \ldots & \vdots \\ a_{m1}\mathrm{B} & a_{m2}\mathrm{B} & \ldots & a_{mn}\mathrm{B} \end{bmatrix}.$$

(a) Can there be an ordering of the bases $\mathcal{E} \otimes \mathcal{F}$ and $\mathcal{V} \otimes \mathcal{W}$ such that the matrix of $S \otimes T$ has a block decomposition whose $(k, l)^{\text{th}}$ element (where $1 \leq k \leq q$ and $1 \leq l \leq p$) is $b_{kl}\mathrm{A}$?

(b) Show that $\rho(\mathrm{A} \otimes \mathrm{B}) = \rho(\mathrm{A}) \cdot \rho(\mathrm{B})$, where, of course ρ is the rank of a matrix.

2.3.3 Tensor Products, composition of linear transformations

Let E, F, G be finite-dimensional linear spaces and $\lambda_{E,F} : E^* \otimes F \to L(E, F)$ and $\lambda_{F,G} : F^* \otimes G \to L(F, G)$ the isomorphisms defined in Example 2.3.3a.

Show that the following diagram commutes:

$$\begin{array}{ccc}
L(E,F) \otimes L(F,G) & \xrightarrow{\lambda_{E,F}^{-1} \otimes \lambda_{F,G}^{-1}} & (E^* \otimes F) \otimes (F^* \otimes G) \\
 & & \downarrow T_F \\
\circ \downarrow & & E^* \otimes F^* \otimes F \otimes G \\
 & & \downarrow 1_{E^*} \otimes \widetilde{\boldsymbol{ev}} \otimes 1_G \\
L(E,G) & \xrightarrow{\lambda_{E,G}^{-1}} & E^* \otimes G,
\end{array}$$

where T_F is the multilinear function induced by the "flip" function which interchanges the second and third factors in the Cartesian product, $E^* \times F \times F^* \times G$, $\widetilde{\boldsymbol{ev}}$ the bilinear function which evaluates linear forms on F and $\circ$ the bilinear function induced by the composition of linear transformations.

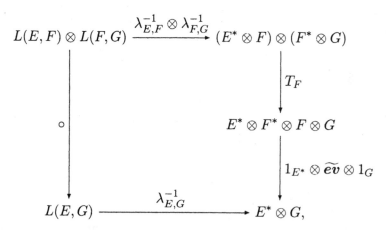

3
Normed Linear Spaces, Metric Spaces

In the remainder of the book we will deal only with linear spaces over $\mathbb{R}$ or $\mathbb{C}$.

Now that I have built up a suitably flexible algebraic framework for dealing with the linear spaces and transformations on them which arise in the study of `differentiable functions` between subsets of linear spaces of dimension greater than one, it is time to do some more spadework and develop the basic tools of analysis in linear spaces.

The simplest class of linear spaces for which it is possible to do `Differential Calculus` is the class of normed linear spaces, which were defined on page 25.

In Section 1, we study the basic properties of `convergent` sequences and `continuous` functions in normed linear spaces. Although both these notions depend on a specific norm, we shall prove that in the case of finite-dimensional spaces a change of norm does not affect either the convergence of sequences or the continuity of functions.

The other important result established in this section is a necessary and sufficient condition for a linear transformation between normed linear spaces to be a continuous function.

Section 2 introduces the important notion of `compact` sets. Compact sets generalize the closed and bounded intervals of $\mathbb{R}$ and share with their more familiar cousins, important properties such as:

- any sequence of points in a compact set contains a subsequence which converges to a point belonging to the set;
- any continuous $\mathbb{R}$-valued function on a compact set is bounded and attains its sup and inf at points belonging to the set.

Many of the concepts introduced (resp. results proved) in these two sections are useful (resp. valid) in the wider context of `metric spaces` and after defining metric spaces, I have made it a point to work in this wider context, whenever possible. This involves no extra effort in learning the material but gives the reader the advantage of acquiring a wider perspective.

In the preamble to Chapter 2, I had mentioned that if E, F are normed linear spaces, then the space of linear maps $L(E, F)$ has a naturally defined norm. This was somewhat inaccurate: in Section 3 we show how to define a norm on the set of *continuous* linear transformations from E to F. Naturally the continuous analogues of the spaces such as E^*, E^{**} or $\mathfrak{L}_p(\mathbf{E}; F)$ also have natural norms. Since the last of these are spaces that appear when we deal with differentiable functions and their higher derivatives, we study these spaces at some length.

Section 4 is devoted to a discussion of the convergence of infinite series in normed linear spaces. There is only one result here that will be new to you: the Dominated Convergence Theorem or DCT, for short. Many important examples of power series are discussed here.

The final section describes two useful techniques which can help decide if two continuous functions are, in fact, identical.

3.1 Convergence and Continuity

We recall that if E is a linear space over $\Bbbk$, then a *norm* on E is a function $\nu : E \to \mathbb{R}$ satisfying the following conditions for all $c \in \Bbbk$ and $\boldsymbol{x}, \boldsymbol{y} \in E$:

(Positivity)	$\nu(\boldsymbol{x}) \geq 0$ and $\nu(\boldsymbol{x}) = 0$ iff $\boldsymbol{x} = 0$
(Homogeneity)	$\nu(c\cdot\boldsymbol{x}) = \lvert c\rvert\cdot\nu(\boldsymbol{x})$
(Subadditivity)	$\nu(\boldsymbol{x}+\boldsymbol{y}) \leq \nu(\boldsymbol{x}) + \nu(\boldsymbol{y})$

where $|c|$ is the absolute value of the real or complex number c.

EXAMPLES OF NORMED LINEAR SPACES

Example 3.1.1a (Real and Complex Numbers).
The usual absolute values on $\mathbb{R}$ and $\mathbb{C}$ are norms. So the Real and Complex numbers are normed linear spaces.

Example 3.1.1b (Some commonly used norms on $\mathbb{R}^n$).
The following functions define norms on $\mathbb{R}^n$ with the names shown

$$\underline{\ell_\infty\text{-norm}} \qquad \ell_\infty(x_1, \ldots, x_n) = \max\{|x_1|, \ldots, |x_n|\}$$

$$\underline{\ell_1\text{-norm}} \qquad \ell_1(x_1, \ldots, x_n) = \sum_{i=1}^{n} |x_i|$$

$$\underline{\ell_2\text{-norm}} \qquad \ell_2(x_1, \ldots, x_n) = +\sqrt{x_1^2 + \cdots + x_n^2}$$

Example 3.1.1c. If E is a Euclidean space then
$$||v|| = +\sqrt{\langle v, v\rangle}$$
is a norm. If $E \approxeq \mathbb{R}^n$ with the standard inner product then this norm coincides with the ℓ_2-norm.

Example 3.1.1d (Norms on Unitary Spaces).
This is a slight generalization of the previous example. Let E be a unitary space. Then there are two ways of producing a norm on E. First, observe that $\langle \boldsymbol{v}, \boldsymbol{w}\rangle_{\mathbb{R}} = \Re(\langle \boldsymbol{v}, \boldsymbol{w}\rangle)$ (where $\Re(z)$ is the real part of the complex number z) is an inner-product on the underlying real linear space $E_{\mathbb{R}}$. Hence this yields a norm as in (c) above. Otherwise one could simply define a norm on the space by setting $||v|| = \sqrt{\langle v, \boldsymbol{v}\rangle}$. The subadditivity property of these norms follows from the Cauchy–Schwartz inequality.

In fact, most of the norms discussed before have complex analogues.

Example 3.1.1e (The ℓ_p-norms on $\mathbb{k}^n$).
In general, if $1 \le p < \infty$, then the function,

$$\ell_p : (x_1, \ldots, x_n) \mapsto \{|x_1|^p + \cdots + |x_n|^p\}^{1/p}$$

is a norm on $\mathbb{k}^n$. The ℓ_p-norms will usually be denoted $|| \cdot |||_p$; that is I will write $||\boldsymbol{x}||_p$ for $\ell_p(\boldsymbol{x})$ for $\boldsymbol{x} \in \mathbb{k}^n$.

Remark (On the notation for norms).
The "double absolute value sign", $|| \cdot ||$, will be used to denote a norm (which is unspecified) on a general linear space E. If several norms on

the same linear space are being considered simultaneously, I will either use the norm functions with sub- or super-scripts (as in, ν, ν' or ν_i) or apply the double absolute value with subscripts or superscripts (as in, $|| \, ||'$ or $|| \, ||_i$.) If $E, F, \cdots$ are normed linear spaces which I need to consider at the same time I will denote the norms $|| \, ||_E, || \, ||_F, \cdots$.

Now for norms of a somewhat different flavour.

Example 3.1.1f (Norms on a space of functions).
Let $\mathcal{C}(\mathbf{I})$ be the linear space of continuous real-valued functions defined on $\mathbf{I} = [0, 1]$, with pointwise addition and multiplication by real numbers as the linear space operations.
(1) The sup-norm on $\mathcal{C}(\mathbf{I})$

The function
$$f \mapsto \sup\{|f(t)| \, : \, t \in \mathbf{I}\}$$
is a norm on $\mathcal{C}(\mathbf{I})$ usually called the *sup-norm* and denoted $|| \, ||_{\text{sup}}$.

(2) The L^1-norm on $\mathcal{C}(\mathbf{I})$
This norm is defined by the function:
$$f \mapsto ||f||_{L^1} \doteq \int_0^1 |f(t)|dt.$$

It is customary to use $|| \cdot ||_1$ instead of the subscript L^1. This is a norm on $\mathcal{C}(\mathbf{I})$. In checking the norm properties the only thing which is not entirely trivial is to show that $||f||_1 = 0$ implies that f is identically 0. To see this recall that if $f(t_0) \neq 0$ for some $t_0 \in \mathbf{I}$ then there is a closed interval containing t_0 where f is nonzero and hence on this interval $|f(x)| \geq C > 0$; but then $||f||_{L^1} > 0$. This norm will be called the L^1-norm on $\mathcal{C}(\mathbf{I})$ and this *normed linear space* will be denoted $\mathcal{C}_{L^1}(\mathbf{I})$.

Example 3.1.1g (A "non-example").
Let $E = \mathcal{R}[0, 1]$ be the set of Riemann integrable functions on $[0, 1]$. Then $N(f) = \int_0^1 |f(t)|dt$ is certainly subadditive and homogeneous. But it obviously does not define a norm on E since the characteristic function of any finite set $F \subset [0, 1]$ is a nonzero Riemann integrable function such that $N(f) = 0$.

Example 3.1.1h. If $E_1, \ldots, E_n$ are normed linear spaces one can define norms, $||\cdot||_{\ell_p}$, on $E = (E_1 \oplus \cdots \oplus E_n)$, by setting

$$||(\boldsymbol{e}_1, \ldots, \boldsymbol{e}_p)||_{\ell_p} = \left\{ \sum_{j=1}^{n} (\nu_j(\boldsymbol{e}_j))^p \right\}^{\frac{1}{p}}$$

where ν_j is the norm on E_j. These are called the *ℓ_p-norms on the direct sum.*

A norm on E enables us to define a notion of "distance" between two points of E. If (E, ν) is a normed linear space, for any $\boldsymbol{x}, \boldsymbol{y} \in E$, define the distance between the points x and y to be equal to $d_\nu(x, y) \doteq \nu(\boldsymbol{x} - \boldsymbol{y})$.

(Notice that on the left hand side of the previous equation, x, y are simply points in E; hence no bold italic font for them, but to define their difference we have to use the linear space structure, hence $\boldsymbol{x} - \boldsymbol{y}$ is used on the right hand side!)

The function d_ν is known as the *metric* induced by the norm ν and satisfies the following properties for all $x, y, z \in E$:

(Positivity)	$d_\nu(x, y) \geq 0$ and equality holds iff $x = y$;
(Symmetry)	$d_\nu(x, y) = d_\nu(y, x)$;
(Triangle Inequality)	$d_\nu(x, y) \leq d_\nu(x, z) + d_\nu(z, y)$

These properties are immediate consequences of the three defining properties of the norm ν.

These three properties are sufficient to develop a theory of convergence and continuous functions; so we introduce a new concept simply by declaring that these conditions define a new mathematical entity.

Definition 3.1.2 (Metrics and Metric Spaces).
Let S be any set. A function $d : S \times S \to \mathbb{R}$ satisfying the conditions of positivity, symmetry and the triangle inequality is called a *metric* on S. *A metric space is a set S together with a metric $d : S \times S \to \mathbb{R}$.*

Some remarks are in order at this point.

Remark 3.1.2a. A metric space should be thought of as a set on which there is a notion of "distance between two points" ($d(x, y)$ representing the distance between x and y.) However do not place *too great a trust on intuition* when dealing with metric spaces; see for example, the exercise on "STRANGE METRICS" at the end of this section.

Remark 3.1.2b. When dealing with the metric on a normed linear space I will usually *omit the subscript* ν of d_ν, since if a different norm was used it would be a *different* normed linear space.

The metric d_ν induced by a norm ν satisfies two other properties which are extremely useful.

Proposition 3.1.3. *Let* (E, ν) *be a normed linear space. Then the induced metric,* $d \doteq d_\nu$*, satisfies the following properties.*
1. *The metric is* translation-invariant, *that is for any* $\boldsymbol{x}, \boldsymbol{y}, \boldsymbol{t} \in E$ *we have the relation* $d(x+t, y+t) = d(x, y)$.
2. *For any real* $r \in \mathbb{R}$, $d(r\boldsymbol{x}, r\boldsymbol{y}) = |r| \cdot d(x, y)$. □

Remark. In the next two sections we will be studying convergence-related properties of subsets of normed linear spaces. These are, of course, metric spaces. Whenever an assertion and its proof does not involve the use of the two properties mentioned in Proposition 3.1.3 (or some notion which is specifically related to the linear space structure, such as dimension) the assertion will be true for all metric spaces and such statements will be made for the more general context.

In a metric space one can introduce some geometric nomenclature in describing certain important subsets. This allows us to use language which is consonant with our intuitive view of the world around us. However do not forget the warning in Remark 3.1.2a. While these terms do not replace proofs, their use do clarify the essential ideas underlying the analytical "ϵ-δ" type arguments one uses in analysis.

We begin by extending a familiar notion to normed linear spaces: a set $S \subset E$ of a normed linear space E is said to be *bounded* if there exists an $R > 0$ such that $x \in S$ implies that $||\boldsymbol{x}|| < R$.

If X is a metric space with metric d and $x \in X$, then for each real number $r > 0$ we define the following sets which will play an important role in the sequel. The *open* ball, *closed* ball and *sphere* of radius r centred at x, are defined as follows:

$$\mathbb{B}_r(x) \doteq \{y \in E : d(x,y) < r\} \quad \text{(the open ball of radius } r \text{ centred at } x);$$
$$\overline{\mathbb{B}}_r(x) \doteq \{y \in E : d(x,y) \leq r\} \quad \text{(the closed ball of radius } r \text{ centred at } x);$$
$$\mathbb{S}_r(x) \doteq \{y \in E : d(x,y) = r\} \quad \text{(the sphere of radius } r \text{ centred at } x).$$

The terminology of balls and spheres centred around points is most useful when applied to normed linear spaces (and their subsets) since these are metric spaces with the metric induced by the norm.

If E is a normed linear space, then the sets $\mathbb{B}_1(0_E)$, $\overline{\mathbb{B}}_1(0_E)$ and $\mathbb{S}_1(0_E)$ are usually referred to as the unit open ball, the closed unit ball and the unit sphere of E respectively. Generally spheres play an important role only in normed linear spaces.

Remark 3.1.3c (A caveat).
Note that in some books (for example [Rud61], [Sim63]) what I have called the ball of radius r centred at x is described as the *neighbourhood* or *sphere* of radius r centred at x!

EXAMPLES OF BALLS AND SPHERES IN NORMED LINEAR SPACES

Example 3.1.4a (Balls in the ℓ_m-norm, $m = 1, 2, \infty$).
Figure 3.1.1 illustrates the balls of $\mathbb{R}^2$ with respect to the norms (a) ℓ_1 (b) ℓ_2 and (c) ℓ_∞.

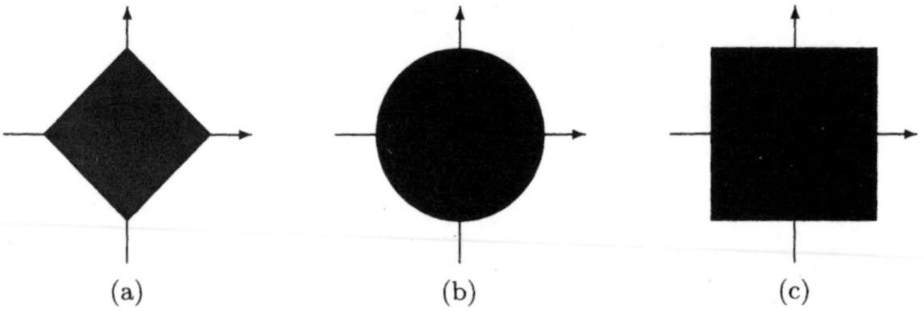

FIGURE 3.1.1. The unit balls in (a) $(\mathbb{R}^2, \ell_1)$, (b) $(\mathbb{R}^2, \ell_2)$, (c) $(\mathbb{R}^2, \ell_\infty)$

The "diamond" in (a) has $(\pm 1, 0), (0, \pm 1)$ as vertices, the "box" in (c) has $(\pm 1, \pm 1)$ and $(\mp 1, \pm 1)$ as vertices.

Let us look at an example of a ball which is not so easily visualized.

Example 3.1.4b (A ball in an infinite-dimensional space).
Let $E = \mathcal{C}(\mathbf{I})$ be the continuous functions on $[0, 1]$ with the sup-norm. Then the open ball in E, of radius r centred at the origin consists of continuous $\mathbb{R}$-valued functions defined on $[0, 1]$ whose graphs are contained in the rectangle:
whose graphs $$\{(x, y) \in \mathbb{R}^2 : x \in [0, 1] \text{ and } -r < y < r\}.$$

Since the norm in a linear space has properties analogous to the absolute value of real or complex numbers, we can define the notion of convergence of sequences and series in an arbitrary normed linear space, simply by "replacing the absolute value sign by $|| \cdot ||$".

Thus we will say $\boldsymbol{x}_n$ tends to $\boldsymbol{x}$ if $||\boldsymbol{x} - \boldsymbol{x}_n||$ tends to 0 as $n \to \infty$. Almost all the theorems you learnt in your earlier courses concerning convergence of series and sequences continue to hold for *arbitrary* norms on any finite-dimensional space over $\mathbb{k}$. This will be proved shortly, for the ℓ_∞ norm on $\mathbb{R}^n$. That these results continue to hold for arbitrary norms on $\mathbb{k}^n$ is more difficult and will take some time to establish.

At this point a word concerning our notation for sequences is in order.

Recall that if S is a set, then a sequence of points of S (or a sequence in S) is a function $f : \mathbb{N} \to S$; the values, $f(n)$ of f are referred to as the elements of the sequence and one says that $\{f(n)\}$ is a sequence in S. In what follows, I will be looking mainly at subsets of normed linear spaces and hence the elements of these subsets are, of course, vectors. But when I do not use their "vectorial" aspect (for example, use linear space operations on these elements) I will use the standard $x, y, z, \ldots$ to represent them instead of the "vectorial font" $\boldsymbol{x}, \boldsymbol{y}, \boldsymbol{z} \ldots$. In particular, if S is a subset of a normed linear space and $\mathbf{x}$ is a sequence of points belonging to S, then its elements will be denoted x_n *not* $\boldsymbol{x}_n$.

I will denote a sequence of points $\{x_n\}$ by the same letter (in this case, the letter "x",) in bold-face unitalicized type: for example, $\mathbf{x} = \{x_n\}$.

Let E be a normed linear space and $\mathbf{x} = \{x_n\}_{n \in \mathbb{N}}$ a sequence of points of E.

The following are familiar but fundamental notions.

1. The sequence, $\mathbf{x}$, is a *bounded* sequence if there is a real number R such that $||x_n|| < R$ for all $n \in \mathbb{N}$.

2. The sequence, $\mathbf{x}$, is a *convergent* sequence if there is a point, $\mathbf{x}^* \in E$, such that $\lim_{n \to \infty} d(x_n, \mathbf{x}^*) = 0$. This point $\mathbf{x}^*$ is called the *limit* of the sequence $\mathbf{x}$. It is easy to verify that the limit of a sequence, if it exists, is unique.

3. The sequence, $\mathbf{x}$, is a *Cauchy* sequence if the real numbers, $a_{mn} \doteq d(x_m, x_n)$ have the property that for any $\varepsilon > 0$, $a_{mn} < \varepsilon$ provided m, n are large enough.

Remark (Convergence in Metric Spaces).

The notion of convergent sequence and Cauchy sequence, as defined above can be transported to metric spaces easily. If X is a metric space and $\mathbf{z} = \{z_n\}$ a sequence of points in X, then it is a Cauchy sequence if

$d(z_m, z_n)$ is a Cauchy sequence converging to 0 and $\mathbf{z}$ converges if there is a $\mathbf{z}^* \in X$ such that $d(\mathbf{z}^*, z_n)$ tends to 0 as $n \to \infty$. Boundedness of sequences is not a useful notion in the context of metric spaces.

Notice that these properties can be expressed in geometric language by saying that:

1′. $\mathbf{x}$ is a *bounded* sequence in a normed linear space, E, iff $\mathbf{x} \subset \mathbb{B}_R(0)$ for some $R > 0$;

2′. $\mathbf{x}$ converges to $\mathbf{x}^*$ if for every $\varepsilon > 0$, all but finitely many elements of $\mathbf{x}$ lie in $\mathbb{B}_\varepsilon(\mathbf{x}^*)$.

3′. In a normed linear space, $\mathbf{x}$ is a Cauchy sequence if given any $\varepsilon > 0$, there exists $N = N(\varepsilon)$ such that if $m, n > N$, then $(\boldsymbol{x}_m - \boldsymbol{x}_n) \in \mathbb{B}_\varepsilon(0)$.

Notice that condition 1′ is a particular instance of the more general notion of a *bounded set.*

Definition. A subset A of a normed linear space is said to be *bounded* if $A \subset \mathbb{B}_R(0)$ for some $R > 0$.

Because of its importance we point out an obvious implication of (2′).

Remark 3.1.5 (Balls and convergent sequences).
Whether or not a sequence, $\mathbf{x} = \{x_n\}$, is convergent with respect to a particular metric depends only on the balls of *small radius.* Further, in view of Proposition 3.1.3, if we are looking at sequences in a normed linear space, convergence is determined by the balls of small radius *centred at the origin.*

The basic properties of these three kinds of sequences are described in the next result.

Theorem 3.1.6. *Let E be a normed linear space and $\mathbf{x}$ a sequence of points in E. If $\mathbf{x}$ is a Cauchy sequence then $\mathbf{x}$ is a bounded sequence. If $\mathbf{x}$ is a convergent sequence then it is also a Cauchy sequence.*

Proof. Suppose that $\mathbf{x}$ is a Cauchy sequence in a normed linear space. Choose an integer N so that if $m, n > N$, then $||\boldsymbol{x}_m - \boldsymbol{x}_n|| < 1$.

Let $R = \max\{||\boldsymbol{x}_k|| : k \leq N\}$. Then clearly, $\mathbf{x} \subset \mathbb{B}_{(R+1)}(0)$, showing that Cauchy sequences are bounded sequences.

Now suppose that $\mathbf{x}$ is a convergent sequence having limit $\mathbf{x}^*$. Then if $\varepsilon > 0$ is any preassigned positive real number, we can find a positive integer N such that $q > N$ implies that $||x_q - \mathbf{x}^*|| < \varepsilon/2$. Now if if $m, n > N$, then $|a_{mn}| = ||x_m - x_n|| \leq ||x_m - \mathbf{x}^*|| + ||\mathbf{x}^* - x_n|| < \varepsilon$. So convergent sequences are necessarily Cauchy. □

Perhaps you had expected me to prove that *in any normed linear space E a sequence converges if and only if it is a Cauchy sequence.*

UNFORTUNATELY THIS IS FALSE!

Example 3.1.7 (Cauchy sequences of two kinds).

Let E and F be the linear space of continuous functions from $[0,1]$ to $\mathbb{R}$, equipped with the sup-norm and the L^1-norm respectively.

If $\{f_n\}$ is a Cauchy sequence in E then

$$\sup_{0 \leq x \leq 1} |f_n(x) - f_m(x)| \to 0 \quad \text{as } m, n \to \infty$$

which means that $\{f_n\}$ is a *uniformly* convergent sequence of continuous functions on [0,1]. Hence the f_n's converge to a continuous function. So in E, Cauchy sequences are convergent sequences.

The following example illustrates that this is not so for the space, F.

For $n > 0$, consider the following sequence of functions $f_n : [0,1] \to \mathbb{R}$, whose graph is sketched alongside.

$$f_n(t) = \begin{cases} 0 & \text{if } t < \frac{n-1}{2n}, \\ 1 & \text{if } t > \frac{n+1}{2n} \\ n(t - \frac{n-1}{2n}) & \text{otherwise.} \end{cases}$$

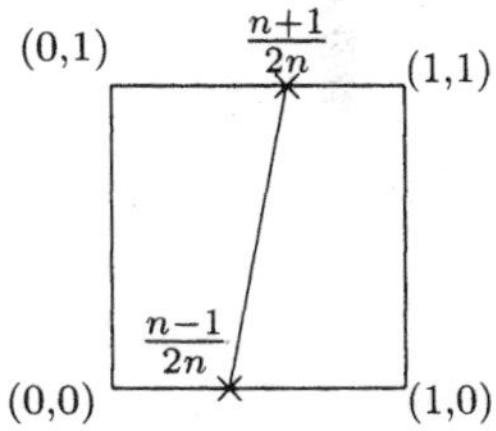

Clearly if $0 \leq t < \frac{1}{2}$, then $f_n(t) \to 0$ as $n \to \infty$ and if $\frac{1}{2} < t \leq 1$ then $f_n(t) \to 1$ as $n \to \infty$. Hence if f_n converges to g with respect to the L^1-norm, then g will have to have a discontinuity at $t = \frac{1}{2}$. But then $g \notin F$! Thus F can contain *non-convergent* Cauchy sequences.

The example above indicates that the problem is that a normed linear space may have some "holes" in the same sense that the rational numbers have some holes (these are the irrational real numbers) and this is what prevents sequences that "ought to converge" (i.e. Cauchy sequences) from converging. This example also indicate that the problem is not one relating to the underlying linear space structure, but lies in *the norm*; since it is this, that determines which sequences are Cauchy.

I will say that a norm on a linear space is *complete* if every Cauchy sequence (with respect to this norm) converges. A linear space with a complete norm is known as a *Banach space.*

Proposition 3.1.8 (Completeness of the ℓ_∞-norms on $\mathbb{k}^n$).
The ℓ_∞-norm on $\mathbb{k}^n$ is complete. In other words, if $\mathbf{x}$ is a sequence in the normed linear space $(\mathbb{k}^n, ||.||_\infty)$, then it is convergent if and only if it is a Cauchy sequence.

Proof. Suppose that $\mathbf{x} = \{\boldsymbol{x}_m : m \in \mathbb{N}\}$ is a Cauchy sequence in the normed space $(\mathbb{k}^n, || \cdot ||_\infty)$. Then if $\boldsymbol{x}_m = (x_1^m, \ldots, x_n^m) \in \mathbb{k}^n$, we have:

$$\max_i |x_i^p - x_i^q| \to 0 \quad \text{as } p, q \to \infty.$$

So for $i = 1, \ldots, n$, there are elements, $x_i^* \in \mathbb{k}$, where $x_i^* = \lim_{p\to\infty} x_i^p$.
Obviously $\mathbf{x}$ converges to the vector $\mathbf{x}^* = (x_1^*, \ldots, x_n^*)$. □

My next objective is to prove a theorem which asserts that on a finite-dimensional linear space, the notion of convergence is independent of the norm. More precisely, I wish to establish that

> *if E is a finite-dimensional $\mathbb{k}$-linear space, ν_1 and ν_2 two norms on E and $\mathbf{x} = \{x_n\}$ a sequence in E then $\mathbf{x}$ is a Cauchy sequence with respect to ν_1 iff it is a Cauchy sequence with respect to ν_2. Further if $\mathbf{x}$ is a Cauchy sequence, then its limit with respect to both the norms exist and are the same.*

To this end I introduce a notion of *equivalence* of norms on a linear space. Let E be any linear space, ν_1 and ν_2 two norms on E. We say ν_1 and ν_2 are *equivalent* if there are constants, k, K such that for every $\boldsymbol{x} \in E$, we have the relations:

$$k \cdot \nu_2(\boldsymbol{x}) \leq \nu_1(\boldsymbol{x}) \leq K \cdot \nu_2(\boldsymbol{x}).$$

Remark 3.1.9 (Geometric formulation of Equivalence of Norms).
In terms of open balls one can reformulate the condition expressed by the above inequalities as follows:

ν_1 and ν_2 are equivalent iff every ν_1-open ball centred at a point $x \in E$ contains a ν_2-open ball also centred at x and vice versa.

From this formulation, one sees quite painlessly, that the equivalence of norms is an equivalence relation on the set of norms on the relevant linear space.

This immediately implies the next result.

Proposition 3.1.10. *Let E be a linear space and suppose that ν_1 and ν_2 are equivalent norms on E. Then a sequence $\mathbf{x} = \{x_n\}$ is Cauchy (resp. bounded) with respect to ν_1 iff it is Cauchy (resp. bounded) with respect to ν_2.* □

I now present examples of equivalent and nonequivalent norms.

Example 3.1.11a (Norms on $\Bbbk^n$).
On $\Bbbk^n$ the ℓ_1-norm and ℓ_∞-norm are equivalent. Simply observe that, for any $\boldsymbol{x} \in \Bbbk^n$, we have the relations:

$$\tfrac{1}{n}||x||_1 \leq ||\boldsymbol{x}||_\infty \leq ||\boldsymbol{x}||_1.$$

We will prove, as the first major result of this section, that every norm on $\Bbbk^n$ is equivalent to the $||\cdot||_\infty$-norm.

Example 3.1.11b (Nonequivalent norms).
Consider the linear space, $E = \mathcal{C}(\mathbf{I})$ of continuous functions from $[0,1]$ to $\mathbb{R}$.

In the sup-norm, the closed unit ball consists of functions, $f : [0,1] \to \mathbb{R}$ such that $|f(t)| \leq 1$ for all $t \in [0,1]$.

However, if we equip E with the L^1-norm, the situation is quite different. For each $n \in \mathbb{N}$, consider the function,

$$f_n(t) = \begin{cases} 0 & \text{if } 0 \leq t \leq \frac{1}{2} - \frac{1}{2n}, \\ 4n(\frac{1}{2} - t) & \text{if } \frac{1}{2} - \frac{1}{2n} \leq t \leq \frac{1}{2}, \\ f_n(1-t) & \text{if } \frac{1}{2} \leq t \leq 1. \end{cases}$$

Clearly $\int_0^1 f_n(t)dt = 1$ and so this is a sequence in the closed unit ball of E with respect to the L^1-norm. But obviously these functions do not constitute a bounded set with respect to the sup-norm.

It is clear that the sup-norm and L^1-norms on $\mathcal{C}(\mathbf{I})$ are not equivalent.

My next objective is to show that *every* norm on $\Bbbk^n$ is equivalent to the ℓ_∞-norm. I will work with $\mathbb{R}^n$ for the sake of simplicity of exposition; the complex case follows, almost immediately.

One half of the equivalence is easy.

Proposition 3.1.12. *Let ν be any norm on $\mathbb{R}^n$. Then there is a constant, K_ν such that $\nu(\boldsymbol{x}) \leq K_\nu ||\boldsymbol{x}||_\infty$ for all $\boldsymbol{x} \in \mathbb{R}^n$.*

Proof. Let $\boldsymbol{e}_\imath(n)$ be the $\imath^{\text{th}}$ standard basis vector of $\mathbb{R}^n$. If $\boldsymbol{x} \in \mathbb{R}^n$ then

$$\begin{aligned} \nu(\boldsymbol{x}) &= \nu\Big(\sum_{\imath=1}^{n} x_i \boldsymbol{e}_\imath(n)\Big) \\ &\leq \sum_{\imath=1}^{n} |x_\imath| \cdot \nu\big(\boldsymbol{e}_\imath(n)\big) \\ &\leq K_\nu ||x||_\infty, \end{aligned}$$

where $K = \sum_{\imath=1}^{n} \nu\big(\boldsymbol{e}_\imath(n)\big)$. □

To completely prove that every norm on $\mathbb{R}^n$ is equivalent to $||\cdot||_\infty$, I have to extend to the present setting of normed linear spaces some of the elementary results concerning continuous real-valued functions defined on subsets of $\mathbb{R}$.

I begin with the definition of continuous functions or maps between metric spaces.

Definition 3.1.13 (Maps between metric spaces).
If X, Y are metric spaces, a function $f : X \to Y$ is *continuous* at $x \in X$ if for any $\varepsilon > 0$ there is a $\delta > 0$, possibly depending on x and ε such that $d_X(x, x') < \delta$ implies $d_Y\big(f(x), f(x')\big) < \varepsilon$. Moreover, $f : X \to Y$ will be called a continuous function or *map* if f is continuous at each point of X.

Proposition 3.1.14 (Geometric formulation of continuity). *Let $f : X \to Y$ be a function between metric spaces. Then f is continuous at $x \in X$ iff for any $\varepsilon > 0$, there is a $\delta > 0$ such that $f\big(\mathbb{B}_\delta(x)\big) \subset \mathbb{B}_\varepsilon\big(f(x)\big)$.* □

Remark. Thus it follows that continuity of a function $f : X \to Y$ is determined by the images of balls of *small radii* in X and the balls of Y having small radii. In particular, if $f : E \to F$ is a function between normed linear spaces, the continuity of f (regarded as a map of metric spaces) will not be affected if the norms are replaced by equivalent norms.

Examples of Maps

Example 3.1.15a (Trivial examples).
1. For arbitrary metric spaces X and Y, a constant function (i.e. one whose image is a single point of Y) is continuous.
2. For any metric space X, the identity function of the space, $1_X : X \to X$, is continuous.

These follow immediately from the geometric formulation of continuity. □

Example 3.1.15b (The distance function).
If X is a metric space and $p \in X$, then $d_p : X \to \mathbb{R}$ defined by $d_p(x) = d(p,x)$ is a map.

Example 3.1.15c (Linear space operations).
If E is a normed linear space over $\Bbbk$ then the operations of scalar multiplication and linear space addition can be regarded as functions:

$$\begin{aligned} \Bbbk \times E &\to E & (\alpha, \boldsymbol{x}) &\mapsto \alpha \cdot \boldsymbol{x} \\ E \times E &\to E & (\boldsymbol{x}, \boldsymbol{y}) &\mapsto \boldsymbol{x} + \boldsymbol{y} \end{aligned}$$

respectively. Now $\Bbbk \times E$ and $E \times E$ are the underlying sets of the normed linear spaces $\Bbbk \oplus E$ and $E \oplus E$ respectively and hence are normed linear spaces with norms:

$$\begin{aligned} ||(\alpha, \boldsymbol{x})|| &= \sup(|\alpha|, ||\boldsymbol{x}||_E), \text{ and} \\ ||(\boldsymbol{x}, \boldsymbol{y})|| &= \sup(||\boldsymbol{x}||_E, ||\boldsymbol{y}||_E) \end{aligned}$$

respectively. The homogeneity and subadditivity properties of $|| \quad ||_E$ imply that these functions are continuous.

Example 3.1.15d (Norms on $\mathbb{R}^n$ and the ℓ_∞-metric).
If $\nu : \mathbb{R}^n \to$ is a norm then $\nu : (\mathbb{R}^n, \ell_\infty) \to \mathbb{R}$ is a continuous function. This is an immediate consequence of Proposition 3.1.12.

Example 3.1.15e (Linear transformations defined on $(\mathbb{R}^n, ||\cdot||_\infty)$).
Any linear transformation $S : (\mathbb{R}^n, \ell_\infty) \to E$ taking values in a normed linear space is continuous.

Proof. Let $S : \mathbb{R}^n \to E$ be a linear transformation. If $S = 0$ then it is a constant function, hence continuous. So let us suppose that $S \neq 0$.

Since $S(\boldsymbol{x}+\boldsymbol{y}) - S(\boldsymbol{x}) = S(\boldsymbol{y})$, a linear transformation is continuous *if and only if it is continuous at the origin.* To check continuity at 0, note that if $\boldsymbol{e}_\imath(n)$ is the $\imath^{\text{th}}$ standard basis vector of $\mathbb{R}^n$ and $||S(\boldsymbol{e}_\imath(n))||_E = s_\imath$, then for the vector $\boldsymbol{x} = (x_1, \ldots, x_n) \in \mathbb{R}^n$ we have:

$$||S(\boldsymbol{x})||_E = ||\textstyle\sum_\imath x_\imath S(\boldsymbol{e}_\imath(n))||_E \le ||x||_\infty \sum_\imath s_\imath.$$

Since $S \neq 0$, not all the $s_\imath$'s are zero. Hence $||S(\boldsymbol{x})||_E < \varepsilon$ if $||\boldsymbol{x}||_\infty < \varepsilon/s$ where $s = \sum_\imath s_\imath$. This shows that S is continuous at 0 and hence that S is a map. □

We now establish a simple criterion for the continuity of linear transformations between normed linear spaces.

Theorem 3.1.16 (Continuity of Linear Transformations).
Let E, F be normed linear spaces and $T : E \to F$ a linear transformation. Then T is a continuous function iff $T\big(\mathbb{S}_1(0_E)\big)$ is a bounded subset of F.

Proof. We have already observed that a linear transformation is *a map* iff it is continuous at the origin of the domain.

So we only have to show that T will be continuous at 0_E iff the unit sphere of E is mapped by T onto a bounded subset of F.

So suppose that $T\big(\mathbb{S}_1(0_E)\big) \subset \mathbb{B}_R(0_F)$, that is $||T(\boldsymbol{x})||_F < R$ whenever $||\boldsymbol{x}||_E = 1$.

Let $\varepsilon > 0$ be arbitrarily assigned. If $||\boldsymbol{x}||_E < \frac{\varepsilon}{R}$, then clearly from the homogeneity of the norm it will follow that $||T(\boldsymbol{x})||_f < \varepsilon$ proving the continuity of T at 0_E.

Conversely suppose T is continuous at 0_E. Then there exists an $r > 0$ such that $||T(\boldsymbol{x})||_F \leq 1$ if $||\boldsymbol{x}||_E \leq r$. Then, if $\boldsymbol{e} \in \mathbb{S}_1(0_E)$, we have the relations:

$$||T(\boldsymbol{e})||_F = \tfrac{1}{r} \cdot ||T(r \cdot \boldsymbol{e})||_F \leq \tfrac{1}{r},$$

showing that T takes the unit sphere of E to a bounded subset of F. □

Remark (Bounded Linear Operators).
Because of the criterion for continuity in the previous theorem, continuous linear transformations between normed linear spaces are often referred to as *bounded linear operators.*

Linear transformations with infinite dimensional normed linear spaces as domains, need not be continuous.

Example 3.1.17 (A *discontinuous* linear transformation).
Consider the space F, of continuous functions from $[0, 1]$ to $\mathbb{R}$, equipped with the L^1-norm and the linear transformation, $\epsilon : F \to \mathbb{R}$ defined by $\epsilon(\phi) = \phi(\frac{1}{2})$. We have seen in Example 3.1.11b on page 79 that the L^1-unit ball is not a bounded set with respect to the sup-norm. This implies that even if $\phi \in \mathbb{S}_1(F)$, $|\epsilon(\phi)|$ can be arbitrarily large. Thus ϵ is a discontinuous linear form on F. □

Having familiarized ourselves with some important examples relating to continuity, we return to the business of proving the equivalence of all norms on a finite-dimensional linear space.

Let X be a metric space. A subset $A \subset X$ is said to be a *closed* set if, whenever $\mathbf{x} = \{x_n\}$ is a convergent sequence of points of A, the limit, $\mathbf{x}^*$, of the sequence is also a point of A.

Note that if E is a normed linear space , the property of a subset $A \subset E$ being closed or bounded is clearly dependent on the norm. But in view of Lemma 3.1.10, if two norms on a space are equivalent, then the notions of closed (resp. bounded) sets in the two normed linear spaces arising from these two norms coincide.

Also observe that in view of Proposition 3.1.12 if ν is a norm on $\mathbb{R}^n$ and $\mathbf{x} = \{x_n\}$ a Cauchy sequence with respect to ℓ_∞ then it is also a Cauchy sequence with respect to the norm ν. Hence we have:

Proposition 3.1.18. *Let ν be any norm on $\mathbb{R}^n$ and $A \subset \mathbb{R}^n$ a set which is closed with respect to the ℓ_∞-norm. Then A is a ν-closed set.* □

I have not yet given examples of closed sets. The following shows how to describe *all* closed sets.

Example 3.1.19 (Closed sets).
From the definition it is immediate that if $f : X \to Y$ is a map of metric spaces and y any point of Y which is in the image of f, then the *inverse image* of y, that is, the set $f^{-1}(y) = \{x \in X \,:\, f(x) = y\}$ is a closed subset of X.

In fact, *every closed set* $A \subset X$ is of this particular form! This is sufficiently important to be stated and proved separately.

Proposition 3.1.20. *Let $A \subset X$ be any closed set in a metric space. Then for each $x \in X$ define $d_A(x) \doteq \inf\{d(x,a) \,:\, a \in A\}$. The function $d_A : X \to \mathbb{R}$ is continuous. Moreover, $d_A^{-1}(0) = A$.*

Proof. I begin by establishing the continuity of d_A. Let $x \in X$ and let $\varepsilon > 0$ be arbitrarily assigned. Suppose $d_A(x) = k$. Then by the definition of d_A, there is a point $x' \in A$ such that $d(x,x') \le k + \varepsilon/2$. Let $y \in X$ be such that $d(x,y) \le \varepsilon/2$. Then $d(x',y) \le d(x,x') + d(x',y) \le k + \varepsilon$. Hence, $d_A(y) \le k + \varepsilon$, or $|d_A(x) - d_A(y)| \le \varepsilon$ if $d(x,y) \le \varepsilon/2$. This shows that d_A is continuous at x; since x was arbitrary, this shows that $d_A : X \to \mathbb{R}$ is a map.

It remains to show that the set of points on which d_A vanishes is precisely A. Suppose $d_A(y) = 0$. Then given any integer $n > 0$ one can find a point $x_n \in A$ such that $d(y, x_n) < \frac{1}{n}$. But then $\{x_n\}_{n\in\mathbb{N}}$ is a sequence of points of A converging to y and since A is closed $y \in A$. □

We now prove a fundamental property of closed, bounded subsets of $(\mathbb{R}^n, \ell_\infty)$: the *Bolzano–Weierstrass* property.

Theorem 3.1.21 (Bolzano–Weierstrass Theorem).
Let A be any closed and bounded set in $(\mathbb{R}^n, \ell_\infty)$ and let $\mathbf{x} = \{x_n\}$ be a sequence of points of A. Then there is a subsequence $\underline{x} = \{x_{n_k}\}$ which converges (in the ℓ_∞ norm) to a point of A.

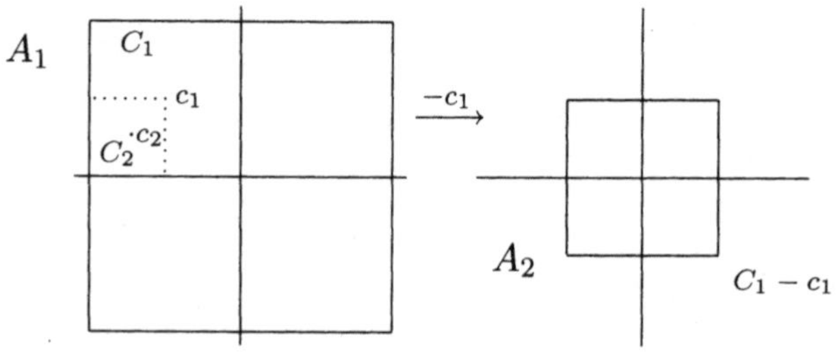

FIGURE 3.1.2. The Bolzano–Weierstrass procedure in $\mathbb{R}^2$

Proof. The proof is an obvious generalization of the proof of the corresponding result for closed, bounded intervals in $\mathbb{R}$: keep halving the set and choosing a portion which contains infinitely many elements of the sequence. I am giving a fairly detailed account since you may have trouble writing out in a formal manner a mathematical argument concerning higher dimensional spaces. You should note, however, that what is being done is not in any way more sophisticated than what you learnt in your earlier course of Calculus. Figure 3.1.2 might be helpful in orienting you to the multidimensional setting.

First observe that each of the sets $H_j = \{(x_1, \ldots, x_n) : x_j = 0\}$ (these are called the *coordinate hyperplanes*) divide the space $\mathbb{R}^n$ into two parts; the "positive" part, P_j where $x_j \geq 0$ and the "negative" part, N_j where $x_j \leq 0$.

Collectively, the hyperplanes H_j will divide $\mathbb{R}^n$ into 2^n parts which I will refer to as the "orthants" of $\mathbb{R}^n$ as explained below. These orthants are conveniently indexed by subsets of $\{1, \ldots, n\}$. If $S \subset \{1, \ldots, n\}$ then the corresponding orthant $\mathcal{O}_S$ is the set:

$$\mathcal{O}_S \doteq \{(x_1, \ldots, x_n) \,:\, x_j \geq 0 \text{ iff } j \in S\}$$

Thus orthants are the higher dimensional analogues of the four quadrants of the plane.

Let $R > 0$ be such that $A \subset \overline{\mathbb{B}}_R(0)$. Notice that since we are using the ℓ_∞-norm, this ball looks like the closed "hyper-cube" centred at the origin, each of whose sides are of length $2R$. I will, for the remainder of this proof, talk about these balls as "cubes". For each subset S of $\{1, \ldots, n\}$ let $A_S = A \cap \mathcal{O}_S \subset \overline{\mathbb{B}}_R(0) \cap \mathcal{O}_S$. Since there are 2^n subsets of $\{1, \ldots, n\}$, at least one of the A_S's will contain infinitely many terms of the sequence $\mathbf{x}$; let A_1 be one such set and choose $x_{n_1} \in A_1$. Let c_1 be the centre of the cube $C_1 = \mathbb{B}_R(0) \cap \mathcal{O}_{S_1}$.

Obviously translation by $-c_1$ will transform the cube C_1 to a cube of radius R centred at the origin. Let A_1' be the set A_{S_1} transformed by the same translation. Now for some $B \subset \{1, \ldots, n\}$ the set $A_1' \cap \mathcal{O}_B$ will contain infinitely many terms of the sequence $x_n - c_1$; let S_2 be such a subset of $\{1, \ldots, n\}$ and choose an $n_2 > n_1$ such that $(x_{n_2} - c_1) \in A_1' \cap \mathcal{O}_{S_2}$. Let C_2 be the cube $B_{R/2}(0) \cap \mathcal{O}_{A_2}$ translated by c_1.

Continuing in this fashion, we obtain a sequence of nested cubes

$$C_1 \supset C_2 \supset \cdots \supset C_n \supset \cdots$$

and points $x_{n_k} \in C_k$ such that:

1. $n_k < n_{k+1}$;
2. C_n has sides of length $R.2^{1-n}$;
3. each C_n contains infinitely many terms of the sequence $\mathbf{x}$.

(Figure 3.1.2 gives a crude depiction of the first two steps in $\mathbb{R}^2$.)

Since x_{n_q} is contained in C_p for $q > p$, the distance between x_{n_r} and x_{n_s} is bounded by $R.2^{-t}$ where $t = \min(r, s) - 1$. Thus the sequence $\underline{\mathbf{x}} = \{x_{n_k}\}$ is a Cauchy sequence in the ℓ_∞ norm and hence is convergent with respect to this norm. Since A is a closed set the limit $\mathbf{x}^* = \lim_{k\to\infty} x_{n_k}$ is a point in A. □

We now explore some of the consequences of this result.

Theorem 3.1.22. *Let A be a closed, bounded set in $(\mathbb{R}^n, \ell_\infty)$ and let $f : A \to \mathbb{R}$ be a map. Then $\{f(A)\}$ is a bounded subset of real numbers and there are points $\underline{x}, \overline{x} \in A$ such that:*

$$\begin{aligned} f(\underline{x}) &= \inf\{f(x) \,:\, x \in A\} \text{ and} \\ f(\overline{x}) &= \sup\{f(x) \,:\, x \in A\}. \end{aligned}$$

Proof. Suppose that $S = \{f(x) : x \in A\}$ is an unbounded set. Then for each positive integer n we can choose points $x_n \in A$ such that $|f(x_n)| > n$. Now by the Bolzano–Weierstrass theorem, there is a subsequence $\{x_{n_k}\}$ of these points which converges to $x^* \in A$. Since f is a map (i. e., it is continuous,) $\lim_k f(x_{n_k}) = f(x^*)$; on the other hand $\lim_k f(x_{n_k})$ does not exist! The contradiction establishes the boundedness of S.

If $f : A \to \mathbb{R}$ were a map that did not attain one of its bounds, say $f(z) \neq M = \sup\{f(x) : x \in A\}$ for any $z \in A$, then the function,

$$f'(x) = \frac{1}{f(x)-M}$$

would be continuous on A but unbounded contradicting what we have established in the previous paragraph! Hence not only is f bounded on A, *f attains its bounds.* □

We can now establish the equivalence of all norms on $\mathbb{R}^n$.

In view of Remark 3.1.9 we only have to establish the following result.

Theorem 3.1.23 (Equivalence of norms).
Any norm ν on $\mathbb{R}^n$ is equivalent to the ℓ_∞-norm.

Proof. One half of our present task was already accomplished in Proposition 3.1.12. It remains to show that if $\nu : \mathbb{R}^n \to \mathbb{R}$ is any norm then we can find a constant $k_\nu > 0$ such that for every $\boldsymbol{x} \in \mathbb{R}^n$ we have the inequality:

$$k_\nu ||x||_\infty \leq \nu(x). \qquad (*)$$

Observe that any $\boldsymbol{x} \in \mathbb{R}^n$ can be written as $\boldsymbol{x} = l_x \cdot \boldsymbol{x}'$ where $l_x \in \mathbb{R}$ and $||\boldsymbol{x}'||_\infty = 1$. Hence if the inequality $(*)$ holds for all vectors belonging to $\mathbf{S}$, the unit sphere of $(\mathbb{R}^n, ||\cdot||_\infty)$, then $(*)$ will continue to hold for all $\boldsymbol{x} \in \mathbb{R}^n$.

Now $\mathbf{S}$ is obviously bounded with respect to the ℓ_∞-norm and it is also closed since it is precisely the set on which the continuous function $\boldsymbol{x} \mapsto ||\boldsymbol{x}||_\infty$ is equal to 1. Hence there is a point $\boldsymbol{z} \in \mathbf{S}$ such that $\nu(\boldsymbol{z}) = k_\nu = \inf\{\nu(x) : x \in \mathbf{S}\}$. Since this vector $\boldsymbol{z}$ is obviously nonzero, k_ν is > 0. Clearly for all $x \in \mathbf{S}$ we have $k_\nu ||\boldsymbol{x}||_\infty \leq \nu(\boldsymbol{x})$ and the result follows. □

The following corollary should be proved by the reader as an easy exercise.

Corollary 3.1.23 (i). *All norms on $\Bbbk^n$ are equivalent to one another.* □

We derive some consequences of this fundamental result.

We have already seen that any linear transformation $T : (\mathbb{R}^n, \ell_\infty) \to F$ taking values in any normed linear space F is continuous. The equivalence of all norms on a finite-dimensional normed linear space implies the next result.

Theorem 3.1.24. *Let E and F be normed linear spaces, E finite-dimensional. Then any linear transformation $T : E \to F$ is a map.*

The following result is an immediate corollary.

Proposition 3.1.25. *Let E and F be finite-dimensional normed linear spaces and $\mathcal{E} = \{e_1, \dots, e_n\}$ and $\mathcal{F} = \{f_1, \dots, f_m\}$ be bases of E and F respectively. Then the isomorphism*

$$\begin{array}{rcl} \phi : L(E,F) & \xrightarrow{\approx} & \mathrm{M}_{m,n}(\Bbbk), \\ \text{defined by } T & \mapsto & \mathrm{M}(T; \mathcal{E}, \mathcal{F}) \end{array}$$

which takes each linear transformation $T : E \to F$ to its matrix with respect to the bases $\mathcal{E}, \mathcal{F}$ is continuous, for any norm on the spaces $L(E,F)$ and $\mathrm{M}_{m,n}(\Bbbk)$.

In particular, the isomorphism $L(E) \cong \mathrm{M}_n(\Bbbk)$ given by representing linear transformations by matrices with respect to any basis of E is a continuous isomorphism between normed linear spaces.

Remarks 3.1.26 (Continuity of matrix functions)

Remark 3.1.26a. Observe that we do not need to look at any specific norms on $L(E,F)$ or $L(E)$, since they are finite-dimensional and hence all norms on them are equivalent. Theorem 3.1.24 implies that ϕ is continuous.

Remark 3.1.26b. Now it is easy to show (the proof is the same as in the corresponding results for functions defined on subsets of $\Bbbk$) that if E is a normed linear space and $f, g : E \to F$ are continuous maps into a normed linear space then $(f+g)$ is also continuous; if $h : E \to \mathbb{R}$ is continuous and f is as above then $h \cdot f$ defined by pointwise scalar multiplication is continuous. Moreover, if $h(x) \neq 0$ then $\frac{1}{h}$ is continuous.

These remarks together with the last two results show that:

Proposition 3.1.27. *If E is a finite-dimensional normed linear space then with respect to any norm on $L(E)$ the trace and determinant are* *continuous functions on $L(E)$. Moreover, if we restrict ourselves to the subset $GL(E) \subset L(E)$, then the "inverting function":*

$$\begin{array}{ccc} GL(E) & \to & GL(E) \\ T & \to & T^{-1} \end{array}$$

is continuous.

Proof. Simply observe that taking a matrix to its $(i, j)^{\text{th}}$ entry is continuous since it is a linear transformation on $\mathrm{M}_n(\mathbb{k})$. Our remarks concerning sums and products of continuous functions, implies the continuity of trace and determinant on $\mathrm{M}_n(\mathbb{k})$ since $\mathrm{A} \mapsto a_{ij}$ are linear transformations and because trace and determinant are obtained by multiplying and adding entries of a matrix. The result follows since $\phi : L(E) \to \mathrm{M}_n(\mathbb{k})$ is continuous.

Finally we observe that if $\mathrm{GL}_n(\mathbb{k})$ denotes the set of nonsingular matrices, regarded as a subset of $\mathbb{k}^{n^2}$, then the function $\boldsymbol{i} : \mathrm{GL}_n(\mathbb{k}) \to \mathrm{M}_n(\mathbb{k})$ which takes a matrix M to its inverse M^{-1} is continuous because it only involves replacing the original elements, m_{ij} by polynomials in the m_{ij}'s and dividing by a non-zero polynomial, $\det \mathrm{M}$, in these same variables. Given this, the reader, as an exercise, should complete the proof. □

Remark. The hypothesis of finite-dimensionality in Theorem 3.1.21 is *essential.*

The crux of the argument establishing equivalence of all norms on finite-dimensional linear spaces was the Bolzano–Weierstrass Theorem. The next example describes an infinite-dimensional Banach space having closed, bounded sets that violate the Bolzano-Weierstrass property: that is this Banach space contains closed bounded sets A such that there are sequences $\mathbf{x}$ in A which do not contain any convergent subsequence.

Example 3.1.28 (The space $\ell_2(\mathbb{R})$).
Let $\ell_2(\mathbb{R})$ be the set of square-summable sequences of real numbers; that is

$$\ell_2(\mathbb{R}) = \left\{ \mathbf{x} = \{x_n\}_{n \geq 1} \ : \ x_n \in \mathbb{R} \text{ and } \sum_{n=1}^{\infty} x_n^2 < \infty \right\}.$$

This is an $\mathbb{R}$-linear space if we define addition and scalar multiplication component-wise; that is, if we set:

$$\mathbf{x} + \mathbf{y} = \{x_n + y_n\} \text{ and } c \cdot \mathbf{x} = \{c \cdot x_n\}.$$

Define an inner product on $\ell_2(\mathbb{R})$ by setting:

$$(*)\ldots\ldots\ldots\ldots\ldots\ldots\ldots\ldots\langle \mathbf{x}, \mathbf{y}\rangle = \sum_n x_n y_n.$$

will see that $\ell_2(\mathbb{R})$ is *complete* wit respect to the induced norm.

Proof. We first show that the formula $(*)$ does define an inner product; the main point is to check that the series on the right hand side of $(*)$ converges. In fact we will show that the series is absolutely convergent. To this end, first observe that if $\mathbf{x} = \{x_n\} \in \ell_2(\mathbb{R})$, then the sequence $\text{abs}(\mathbf{x}) = \{|x_n|\}$ is also in $\ell_2(\mathbb{R})$. We introduce the following *temporary* notation:

> If $\mathbf{x} \in \ell_2(\mathbb{R})$ then for $N > 1$, $\mathbf{x}|N$ is the N-tuple, $(x_1, \ldots, x_N)$. This will be thought of as an element of $\mathbb{R}^N$, whose inner product will be denoted $\langle \cdot, \cdot \rangle_N$.

If $\mathbf{x}, \mathbf{y} \in \ell_2(\mathbb{R})$, then from the Cauchy-Schwarz inequality in $\mathbb{R}^N$, we get:

$$\textstyle\sum_1^N |x_n y_n| = \langle \mathbf{x}(N), \mathbf{y}(N)\rangle_N \leq ||\mathbf{x}(N)|| \cdot ||\mathbf{y}(N)|| \leq ||\mathbf{x}|| \cdot ||\mathbf{y}||.$$

The right-hand side is independent of N and on the left we have a partial sum of a sequence, $\{|x_n y_n|\}$, of non-negative real numbers. Clearly the left-hand side, as a function of N is monotonic increasing and being bounded by the right-hand side, tends to a limit as $N \to \infty$. This proves the (absolute) convergence of the series on the right hand side of $(*)$.

The properties of the inner product can now be easily verified.

The norm on $\ell_2(\mathbb{R})$ is given by $||\mathbf{x}|| = \left(\sum_n x_n^2\right)^{\frac{1}{2}}$

Finally we show that this norm on $\ell_2(\mathbb{R})$ is complete.

Let $\mathbf{x}^{(n)} = \{x_k^{(n)}\}$ be a Cauchy sequence in $\ell_2(\mathbb{R})$. Given any $0 < \varepsilon < 1$, there is an $M = M(\varepsilon)$ such that for $m, n > M$, $||\mathbf{x}^{(m)} - \mathbf{x}^{(n)}|| < \varepsilon$. Then for $m, n > M$, we have:

$$|x_k^{(m)} - x_k^{(m)}|^2 \leq ||\mathbf{x}^{(m)} - \mathbf{x}^{(n)}||^2 < \varepsilon^2 < \varepsilon,$$

showing that for each $k \geq 1$, $\{x_k^{(n)}\}$ is a Cauchy sequence of real numbers. Define, $\xi_k = \lim_n x_k^{(n)}$. It is an easy exercise to check that $\xi = \{\xi_n\}$ is an element of $\ell_2(\mathbb{R})$ and that $\lim_n \mathbf{x}^{(n)} = \xi$.

This shows that the norm on $\ell_2(\mathbb{R})$ is complete. □

We now show that closed, bounded subsets of $\ell_2(\mathbb{R})$ do not necessarily satisfy the Bolzano–Weierstrass property.

For $n \geq 1$ consider the sequence $\mathbf{e}^{(n)}$ whose m^{th} entry is δ_{mn}.

Each of these elements of $\ell_2(\mathbb{R})$ have norm 1 and so we have defined a sequence contained in the unit sphere, $\mathbb{S}_1(0)$, which is a closed and bounded subset of $\ell_2(\mathbb{R})$. But the sequence $\{\mathbf{e}^{(n)}\}$ cannot contain any convergent subsequence since $||\mathbf{e}^{(i)} - \mathbf{e}^{(j)}|| = \sqrt{2}$ for all $i \neq j$.

Exercises

3.1.1 The Parallelogram Identity

Show that in inner product and Hermitian spaces, the norm satisfies the "Parallelogram Identity":

For every $\boldsymbol{x}, \boldsymbol{y} \in E$ we have, the relation,

$$||\boldsymbol{x} + \boldsymbol{y}||^2 + ||\boldsymbol{x} - \boldsymbol{y}||^2 = 2(||\boldsymbol{x}||^2 + ||\boldsymbol{y}||^2)$$

Now prove the converse; that is show that if a norm ν on a $\mathbb{k}$-linear space, E, satisfies the parallelogram inequality, then on E there is an inner product (or Hermitian product, as appropriate) such that ν is the norm induced by this inner product.

(Hint: First suppose we are dealing with a linear space over $\mathbb{R}$ equipped with an inner product. Consider the function

$$f(\boldsymbol{x}, \boldsymbol{y}) = \frac{1}{4} \cdot \{||\boldsymbol{x} + \boldsymbol{y}||^2 - ||\boldsymbol{x} - \boldsymbol{y}||^2\}.$$

Show that $\langle \boldsymbol{x}, \boldsymbol{y} \rangle = f(\boldsymbol{x}, \boldsymbol{y})$. So if $\nu : E \to \mathbb{R}$ satisfies the parallelogram identity, use the above formula and show that f defines an inner product on E.

The proof that all the properties of an inner product are consequences of the parallelogram identity is straightforward except for the bilinearity of f. For this you should first prove that $f(\alpha \cdot \boldsymbol{x}, \boldsymbol{y}) = \alpha \cdot f(\boldsymbol{x}, \boldsymbol{y})$ for $\alpha \in \mathbb{Q}$ and then argue using the continuity of f.

Finally, try to manufacture a formula for the Hermitian product in a normed linear space over $\mathbb{C}$ under the hypothesis that the norm satisfies the "parallelogram identity".)

3.1.2 Limit Points, Closures and Frontiers

If A is a subset of a metric space X, a point $p \in X$ is said to be a *limit point* of A if any open ball $\mathbb{B}_r(p)$ centred at p contains some point of A other than p.

a. Show that A is closed iff A contains all its limit points. For any set A define its *closure*, denoted $\bar{A}$, to be the union of A with all its limit points. Show that the closure of any set is a closed set; indeed, it is the smallest closed set containing A.

b. Show that if X and Y are metric spaces and $f : X \to Y$ is a function between them then f is continuous iff for every subset $A \subset X$, we have the inclusion relation $f(\bar{A}) \subset \overline{f(A)}$. Give an example where the inclusion is strict. Also give an example of a continuous map $f : X \to Y$ such that there is a closed set $A \subset X$ such that $f(A)$ is not closed.

c. If A is any subset of a metric space, define its *frontier*, denoted $\mathrm{Fr}\,(A)$, to be the intersection of the closures of A and its complement. That is, $\mathrm{Fr}\,(A) = (\bar{A} \cap \overline{A^c})$, where the superscript, "c" indicates the complement of a set. What you can say about the set $\mathrm{Fr}\,(\,\mathrm{Fr}\,(A))$?

d. Show that the frontier of a set A is empty iff both A and A^c are closed.

3.1.3 Completions of normed linear spaces

Definition (Isometries between metric spaces).
If X, Y are metric spaces, a function $f : X \to Y$ is said to be an *isometry* if for all $x, x' \in X$, $d_X(x, x') = d_Y\big(f(x), f(x')\big)$.

Let (E, ν) be a normed linear space. A *completion* of E is a complete normed linear space $(\widehat{E}, \widehat{\nu})$ together with an isometry $\lambda_E : E \to \widehat{E}$ such that if $f : E \to F$ is any isometry of E into a complete normed linear space then there is a unique isometry $\widehat{f} : \widehat{E} \to F$ such that $\widehat{f} \circ \lambda_E = f$.

a. Show that if $\lambda_E^1 : E \to \widehat{E}_1$ and $\lambda_E^2 : E \to \widehat{E}_2$ are two completions of E then there is a *bijective* isometry between $\widehat{E}_1$ and $\widehat{E}_2$.

b. Suppose that E is a normed linear space which is not complete. Let $\mathfrak{C}(E)$ be the set of all Cauchy sequences $\mathbf{x} = \{x_n\}$ in E. Define an equivalence relation on $\mathfrak{C}(E)$ by the condition:

   $\mathbf{x} \sim \mathbf{y}$ if for any $\varepsilon > 0$ there is an $N = N(\varepsilon)$ such that $d(x_n, y_n) < \varepsilon$ if $n > N$.

Let $\widehat{E}$ be the set of equivalence classes of $\mathfrak{C}(E)$ and let $\lambda_E : E \to \widehat{E}$ be the function which takes a point $x \in E$ to the equivalence class of the constant sequence $\underline{x}$ all whose terms are equal to x.

c. Show that there is a linear space structure on $\widehat{E}$ arising from the operations:

$$\begin{aligned} \mathbf{x} + \mathbf{y} &\doteq \{x_n + y_n\} \\ \text{and } c{\cdot}\mathbf{x} &\doteq \{c{\cdot}x_n\}. \end{aligned}$$

on the set $\mathfrak{C}(E)$ and that if $[\mathbf{s}]$ represents the equivalence class of a sequence $\mathbf{s}$, $\nu([\mathbf{x}]) = \lim_{n\to\infty} ||x_n||_E$ defines a norm on $\widehat{E}$.

d. Show that $\widehat{E}$ is complete with respect to this norm and that $\lambda_E : E \to \widehat{E}$ is a completion of E.

Remark (Completion of metric spaces).
Notice that modulo obvious changes the same definition and construction yields a complete metric space from any metric space. Indeed by replacing the arbitrary "real number" $\varepsilon > 0$ in the definition of the equivalence relation by arbitrary *rational* numbers one can exhibit $\mathbb{R}$ as the completion of $\mathbb{Q}$: this was Cantor's method of constructing the real numbers as opposed to Dedekind's method of "Dedekind cuts".

3.1.4 Strange Metrics

a. **Bounded Metrics** You may have noticed that I had not defined the notion of *bounded sets* in the context of subsets of metric spaces. This example indicates the reason for this omission.

Let (X, d) be any metric space. Show that the formula:

$$d'(x,y) = \begin{cases} d(x,y) & \text{if } d(x,y) \le 1, \\ 1 & \text{otherwise} \end{cases}$$

defines a metric on X. Let us denote the set X with the metric d', (that is, the metric space (X, d')) by X'. Show that X and X' have the same closed sets but, obviously, every set in X' is contained in a ball of radius 2 and hence bounded!

This procedure for replacing a given metric by a bounded metric (without changing the nature of the closed sets, Cauchy sequences, continuous functions etc.) is a very useful device.

b. Discrete Metrics Let X be any set. Define $d : X \times X \to \mathbb{R}$ by setting,

$$d(x, y) = \begin{cases} 0 & \text{if } x = y, \\ 1 & \text{if } x \neq y. \end{cases}$$

Show that (X, d) is a metric space.

Describe the set of continuous functions from X to $\mathbb{R}$ and from $\mathbb{R}$ to X.

c. p-adic metrics on $\mathbb{Q}$ Let $p > 0$ be any prime. Suppose $q = \pm\frac{a}{b}$ where $a, b \in \mathbb{N}$ have no common factors. Then the p-adic absolute value, $|q|_p$, of q is defined to be p^{-m}, where m is the unique integer such that $q = p^m \cdot \frac{a'}{b'}$ where p does not divide either a' or b'.

Show that $d_p(x, y) = |x - y|_p$ defines a metric on $\mathbb{Q}$ which satisfies the *ultrametric inequality*:

For any $x, y, z \in \mathbb{Q}$, $d_p(x, z) \leq \max\big(d_b p(x, y), d_p(y, z)\big)$.

Describe the set of p-adic *integers*: $\{x \in \mathbb{Q} : ||x||_p \leq 1\}$ and show that these form a `subring` of $\mathbb{Q}$.

d. Balls in Ultrametric Spaces An *ultrametric space* is a metric space (X, d) such that d satisfies the above ultrametric inequality. Show that in any ultrametric space, X, if $x \in X$ and $y \in \mathbb{B}_r(x)$ then $\mathbb{B}_r(y) \subset \mathbb{B}_r(x)$.

3.2 Open sets and Compact sets

I now introduce the class of subsets of normed linear spaces which will serve as the domains of definition for `differentiable functions` in the next chapter. Once again the definitions are given in the context of metric spaces and by specialization to the metric induced by the norm yield the corresponding entities in a normed linear space.

A set $U \subset X$ of a metric space is said to be *open* if given any $p \in U$ there is a $\delta(p) = \delta > 0$, depending on p, such that $\mathbb{B}_\delta(p) \subset U$. Equivalently, for each $p \in U$ and $0 < r < \delta(p)$ any $x \in X$ such that $d(x, p) < r$ is again a point of U.

By convention, the empty set is considered to be an open set.

The notions of open and closed sets in a metric space are obviously modelled on the open and closed intervals of $\mathbb{R}$.

Before giving examples of open sets we prove some fundamental properties of open and.closed subsets of a metric space.

Theorem 3.2.1 (Open sets: their unions, intersections).
Let X be a metric space.

1. *If $\{U_\imath\}_{\imath\in I}$ is an arbitrary collection of open subsets of X then their union is also an open set.*

2. *If $U_1, \ldots, U_p$ are a finite collection of open subsets of X then their intersection, $U = \cap_{k=1}^p U_k$ is also open.*

3. *A set $A \subset X$ is open iff its complement, A^c, is closed.*

Proof. 1. If x is a point in the union of the $U_\imath$'s, then we can choose $\imath_0$ so that $x \in U_{\imath_0}$ and then select $\delta > 0$ so that $\mathbb{B}_\delta(x) \subset U_{\imath_0}$ which will, of course, be contained in the union of the $u_\imath$'s.

2. If $x \in U$, the intersection of $U_1, \ldots, U_p$, then we can choose $\delta_\imath > 0$ such that $\mathbb{B}_{\delta_\imath}(x)$ is contained in $U_\imath$ for each $i = 1, \ldots, p$. Let $\delta = \min\{\delta_1, \ldots, \delta_p\}$. This is > 0 and clearly $\mathbb{B}_\delta(x) \subset U$.

3. Suppose A is a closed subset of X. If $X - A$ is not open then there is a point $x \notin A$ such that for all integers $n > 0$, $\mathbb{B}_{1/n}(x)$ contains points of A; that is there are points $x_n \in A$ such that $d(x, x_n) < 1/n$. Then $\{x_n\}$ is a sequence which converges to x and since A is closed this implies that $x \in A$ contradicting the fact that $x \in X - A$. Hence $X - A$ is open.

Conversely, suppose that A is an open subset of X. Suppose that $\{x_n\}$ is a convergent sequence of points of $X - A$ which converges to x. If $x \in A$ then we can find $\varepsilon > 0$ such that if $d(y, x) < \varepsilon$, then $y \in A$. But since x_n converges to x, there is an $N = N(\varepsilon)$ such that for $n > N$, $d(x, x_n) < \varepsilon$. So if $n > N$, then $x_n \in A$ which is a contradiction. So $x \notin A$, that is $X - A$ is closed. □

By taking complements, the above yields a "dual" result for closed sets in metric spaces.

Theorem 3.2.2 (Closed sets: their unions, intersections).
Let X be a metric space.

1. *If $\{C_i\}$ is an arbitrary collection of closed subsets of X then their intersection $C = \cap_{i\in I} C_i$ is also a closed set.*

2. *If $C_1, \ldots, C_n$ are closed subsets of X then their union is also closed.*

3. C is closed iff C^c is open. □

We now look at some examples of open sets; examples of closed sets have already been provided on page 83.

EXAMPLES OF OPEN SETS AND CLOSED SETS

Example 3.2.3a (Open balls of metric spaces).
The open balls, $\mathbb{B}_r(x)$, in any metric space X are open subsets of X. For, if $y \in \mathbb{B}_r(x)$ then $d(x, y) = t < r$. Choose $0 < \delta < r - t$. If $z \in \mathbb{B}_\delta(y)$ then, by the triangle inequality, $d(x, z) \leq d(x, y) + d(y, z) < r$ and hence $\mathbb{B}_\delta(y) \subset \mathbb{B}_r(x)$. So $\mathbb{B}_r(x)$ is an open subset of X.

This example implies the following characterization of open sets.

Proposition. *If $A \subset X$ is a subset of a metric space then A is open iff A is a union of open balls of X.*

Proof. The "if" part follows from (1) of Theorem 3.2.1 and the example (a) above.

To prove the converse, suppose A is open. Then we can choose for each $x \in A$ a $\delta_x > 0$ such that $\mathbb{B}_{\delta_x}(x) \subset A$. Then

$$A \subset \bigcup_{x \in A} \mathbb{B}_{\delta_x}(x) \subset A.$$

□

In view of Proposition 3.1.20 it follows that a set $U \subset E$ is open iff there is a continuous function $f : E \to \mathbb{R}$ such that $U = \{x \in E : f(x) \neq 0\}$. This leads to an important class of examples.

Example 3.2.3b (General linear groups).
If E is a finite-dimensional normed linear space then

$$GL(E) = \{\alpha \in L(E) : \alpha \text{ is invertible}\}$$

is an open set of $L(E)$.

Proof. This follows since α is invertible iff $\det(\alpha) \neq 0$ and the determinant is a continuous function. □

Notice that $GL(E)$ is a group with respect to the operation of composition; $GL(E)$ is known as the *general linear group* of E and was earlier denoted $\mathrm{GL}_n(\mathbb{k})$ when E was $\mathbb{k}^n$. Let me reiterate, that since in a finite dimensional normed linear space such as $L(E)$ the notions of closed and hence open sets is independent of the choice of norms, the general linear group of E is an open set, irrespective of what norm we choose to impose upon $L(E)$. □

The next result is easy to prove but is of paramount importance since it gives an "ε, δ-free" way of dealing with maps.

Theorem 3.2.4 (A characterization of maps).
Let $f : X \to Y$ be a function between two metric spaces. Then the following conditions are equivalent:

(1). *f is a map;*

(2). *for each open subset, $U \subset Y$, $f^{-1}(U) \subset X$ is an open subset of X;*

(3). *for each closed subset, $C \subset Y$, $f^{-1}(C) \subset X$ is a closed subset of X.*

Proof. We will prove that (1) implies (2) implies (3) implies (1).

(1) implies (2): Let $V \subset Y$ be an open set and suppose $U = f^{-1}(V)$. If U is empty there is nothing to prove; so suppose that $x_0 \in U$. If we denote $f(x_0)$ by y_0, then the openness of V implies that there is a $\varepsilon > 0$ such that $\mathbb{B}_\varepsilon(y_0) \subset V$.

The definition of continuity of f at x_0 implies that there is a $\delta > 0$ such that if $x \in X$ and $d_X(x, x_0) < \delta$, then $d_Y\big(f(x), f(x_0)\big) < \varepsilon$. Hence $\mathbb{B}_\delta(x_0) \subset U$, proving that U is an open subset of X.

(2) implies (3): This is obtained by taking complements.

(3) implies (1): As before let $x_0 \in X$ be any point of X and let $y_0 \doteq f(x_0)$. We will show that (3) implies that f is continuous at x_0. Suppose $\varepsilon > 0$ is given. Note that $D = \{y \in Y : d_Y(y, y_0) \geq \varepsilon\}$ is a closed subset of Y since it is the complement of $\mathbb{B}_\varepsilon(y_0)$. By hypothesis, $C \doteq f^{-1}(D)$ is a closed subset of X which *obviously* does not contain x_0. Since x_0 belongs to the open set $(X - C)$, there is a $\delta > 0$ such that $\mathbb{B}_\delta(x_0) \subset (X - C)$. But then $f\big(\mathbb{B}_\delta(x_0)\big) \subset \mathbb{B}_\varepsilon(y_0)$, proving that f is continuous at x_0.

□

The next result is entirely unsurprising, but note how efficiently it is being proved using the metric-independent formulation of continuity.

Theorem 3.2.5 (Composition of maps).
Let $f : X \to Y$ and $g : Y \to Z$ be maps between spaces. Then $g \circ f : X \to Z$ is also a map.

Proof. If $W \subset Z$ is open then $g^{-1}(W)$ is an open subset of Y since g is a map and hence $f^{-1}(g^{-1}(W))$ is an open subset of X since f is a map. But this last set is precisely $(g \circ f)^{-1}(W)$. Hence the result. □

I will now introduce the fundamental notion of *compact* sets.

Recall that, when in the last section we were trying to establish that all norms on a finite-dimensional Banach space are equivalent, the key steps involved two special properties of closed and bounded subsets of $\mathbb{R}^n$:

1. any sequence in such a set has a convergent subsequence;
2. any real-valued map is bounded on such a set and attains its bounds.

Now these two properties are not always satisfied by closed, bounded subsets of metric spaces, even if the metric spaces themselves are subsets of finite-dimensional normed linear spaces. For example, in the metric space, $X = (a, b] \subset \mathbb{R}$, (for any $0 < \delta < (b-a)$) the set, $C = (a, a+\delta]$, is a closed and bounded subset of X. But consider the functions, $f, g : C \to \mathbb{R}$ defined by:

$$f(x) = \frac{1}{(x-a)}, \quad \text{and } g(x) = x.$$

The functions f and g are both continuous on C but f is unbounded and while g is bounded, it does not attain its inf at any point of C!

In view of the importance of the two properties mentioned above, it would be desirable to identify a class of subsets of *metric spaces* for which these results continue to hold. This is the motivation behind the next definition.

Definition (Compact sets).
Let X be a metric space and A a subset of X. A family of open subsets of X, $\mathfrak{U} = \{U_i \subset X\}_{i \in I}$, is said to be an *open cover* of A if $\bigcup_{i \in I} U_i \supset A$.

A subset K of X is said to be a *compact* set if given any open cover $\mathfrak{U}$ of K there exists a finite number of sets $U_{i_1}, \ldots, U_{i_p} \in \mathfrak{U}$ such that these also form an open cover of K. This is briefly expressed by saying that *any open cover of a compact set contains a finite subcover.* In particular, X is said to be a *compact space* if X is a compact subset of X.

Remark 3.2.6. Notice that the condition defining compactness is motivated by the *Heine-Borel Theorem* which asserts that *any open cover of* $[a, b] \subset \mathbb{R}$ *has a finite subcover.*

Before giving examples of compact sets, it will be convenient to prove a few easy results about compact sets. These proofs will illustrate how the "finite subcover condition" is exploited and the results will enable us to identify a large family of examples of compact sets.

Proposition 3.2.7. *Let* $f : X \to Y$ *be a map of metric spaces and suppose that* $K \subset X$ *is a compact set. Then* $f(K)$ *is a compact subset of* Y.

Proof. Let $\mathcal{V} = \{V_\imath\}_{\imath \in I}$ be any open cover of $L \doteq f(K)$. Since f is a map each of the sets $U_\imath \doteq f^{-1}(V_\imath)$ is open and hence $\mathcal{U} = \{U_\imath\}_{\imath \in I}$ is an open cover of K. By hypothesis there exist $i_1, \ldots, i_p \in I$ such that $\{U_{\imath_1}, \ldots, U_{\imath_p}\}$ forms an open cover of K. Clearly the corresponding sets $V_{\imath_1}, \ldots, V_{\imath_p}$ form an open cover of $f(K) = L$. This proves the compactness of L. □

Remark 3.2.8. Notice that when dealing with open sets and closed sets, I had always specified that, for instance, A is an open (resp. closed) subset of a specific metric space, X; but I have not done so for compact sets.

The reason is simple, but important to note. If (X, d) is a metric space and $S \subset X$ then restricting d to $S \times S$ yields a metric on S and so (S, d) is also a metric space. A set $A \subset S$ may be open (resp. closed) in S but not so in X. Easy examples of such circumstances are furnished by taking $X = \mathbb{R}$, $S = (a, b]$. Then, for all $a < c \leq b$, $A = (a, c]$ is closed in S but not in X and $(c, b]$ is open in S but not in X.

However, compactness is an "absolute" property of a set independent of the space of which it is a subset. This follows from the previous proposition: if $K \subset X$ is compact and X is a subset of a metric space Y, then the inclusion $i : X \to Y$ is a map and hence $i(K) = K \subset Y$ is compact.

Proposition 3.2.9. *Let* X *be a metric space and* $K \subset X$ *a compact set. Then* K *is a closed subset of* X. *Conversely, if* C *is any closed subset of a compact metric space* X *then* C *is a compact set.*

Proof. We will establish that K is closed by showing that $(X - K)$ is open. To this end let y be any point not in K. Then for each $k \in K$ let

$$U_k = \{x \in X : d(x,k) < \tfrac{1}{3}d(y,k)\}$$
$$V_k = \{x \in X : d(x,y) < \tfrac{1}{3}d(y,k)\}.$$

Obviously these are open sets since they are open balls in the space X. For each $k \in K$ the sets U_k and V_k are disjoint and $y \in V_k$ for each $k \in K$. The collection $\mathcal{U} = \{U_k : k \in K\}$ is obviously an open cover of K. Since K is compact there are a finite number of sets $U_{k_1}, \dots, U_{k_p}$ such that $K \subset \cup_{i=1}^{p} U_{k_i}$. Now $V = \cap_{i=1}^{p} V_{k_i}$ is an open set containing y and does not meet K. So if $\delta = \frac{1}{2}\cdot\min(r_1, \dots, r_p)$, where r_m is the radius of the ball, V_{k_m}, then if $d(z,y) < \delta$, then $z \in V$ and hence $z \notin K$. This shows that the complement of K is open.

Now suppose that C is a closed set in a compact space X. If $\mathcal{V}$ is an open cover of C then $\mathcal{V}$ together with $(X - C)$ is an open cover of X. Since X is compact, there is a finite collection, $V_1, \dots, V_p$, of sets from $\mathcal{V}$ which together with $(X - C)$, form an open cover of X. Obviously $V_1, \dots, V_p$ form an open cover of C. So C is a compact set. □

The next result is an immediate corollary of Proposition 3.2.9.

Proposition 3.2.10. *Let K be a compact subset of a metric space X. If X is a subset of a Banach space E, then K is a closed, bounded subset of E.*

Proof. Proposition 3.2.9 shows that K is a closed subset of X and hence Proposition 3.2.7 implies that K is closed in E.

So we only have to show that K is bounded. Consider the family of open balls $\mathcal{B} = \{\mathbb{B}_n(0) : n \in \mathbb{N}\}$ of E. Their union is E and hence, they form an open cover of K. Since K is compact a finite number of these balls cover K and hence $K \subset \mathbb{B}_N(0)$ for some integer $N > 0$, that is K is bounded. □

We now give nontrivial examples of compact sets.

Example 3.2.11 (The compact subsets of $\mathbb{R}^n$).
First observe that since the compactness condition only involves open sets and since open sets are unions of balls of *small* radius, the compact sets of $(\mathbb{R}^n, \ell_\infty)$ are also compact sets of $(\mathbb{R}^n, \nu)$ for any norm, ν on $\mathbb{R}^n$. So I will always work with the ℓ_∞-norm on $\mathbb{R}^n$.

1. **Hypercubes in $\mathbb{R}^n$** Let $I \subset \{1, \dots, n\}$. Then an I-cube in $\mathbb{R}^n$ is a set of the form:

$$\{(x_1, \dots, x_n) : a_i \le x_i \le b_i \text{ for } 1 \le i \le n\}$$

where $a_j < b_j$ if $j \in I$ and $a_j = b_j$ otherwise. An I-cube is also called a k-cube if $|I| = k$. (A k-cube in $\mathbb{R}^n$ is generally called a *hypercube.*) A 0-cube is simply a point and clearly compact; a closed ball in the ℓ_∞-norm of $\mathbb{R}^n$ is an n-cube. A 1-cube in $\mathbb{R}$ is compact by the Heine-Borel Theorem and hence (using Proposition 3.2.7) 1-cubes in any $\mathbb{R}^n$ are compact. We will show by induction that in $\mathbb{R}^n$ all hypercubes are compact.

Proof. Suppose we have shown that all k-cubes in $\mathbb{R}^n$ are compact, where $k \leq p < n$. Let C be a $(p+1)$-cube. We assume, without loss of generality, that

$$C = \{(x_1, \ldots, x_n) \,:\, a_i \leq x_i \leq b_i \text{ and } a_i < b_i \text{ if } 1 \leq i \leq (p+1)\}.$$

I will denote by C_x the subset of points (of C) whose $(p+1)^{\text{th}}$ coordinate, x_{p+1}, satisfies the condition $a_{p+1} \leq x_{p+1} \leq x$. Let $\mathfrak{U} = \{U_\alpha\}_{\alpha \in A}$ be an open cover of C. Since every open set is a union of open balls we will assume that each $U_\alpha = \mathbb{B}(p_\alpha, r_\alpha)$ for points $p_\alpha \in C$ and $r_\alpha > 0$. Let M be the supremum. Let A be the set of $x \in [a_{p+1}, b_{p+1}]$ such that C_x can be covered by a finite number of the U_α's. A is nonempty since $C_{a_{p+1}}$, being a k-cube, is compact by the induction hypothesis. Let $M = \sup A$. We will show that $M = b_{p+1}$.

Suppose, if possible, that $M < b_{p+1}$. Now by the induction hypothesis, the p-cube

$$C'_M = \{(x_1, \ldots, x_n) \in C \,:\, x_{p+1} = M\}$$

is compact.

Choose balls $\mathbb{B}(q_{\alpha_1}, r'_{\alpha_1}), \ldots, \mathbb{B}(q_{\alpha_m}, r'_{\alpha_m})$ each of which is a subset of the set U_α of the family $\mathfrak{U}$ to form a cover C'_M. Without loss of generality we may assume that the centres of these balls are points of C'_M. Let $\delta = \inf(r'_{\alpha_1}, \ldots, r'_{\alpha_m})$.

By the definition of M, the $(p+1)$-cube $C_{M-\delta/2}$ is compact and is obviously covered by the family $\mathfrak{U}$. Suppose $U_{\beta_1}, \ldots, U_{\beta_s}$ cover $C'_{M-\delta/2}$. Then, the collection

$$\{U_{\alpha_k}, U_{\beta_l} \,:\, 1 \leq k \leq m, 1 \leq l \leq s\}$$

cover $C^p_{M+\delta/2}$ contradicting the definition of M. Hence $M = b_{p+1}$ and the induction is complete. □

2. Closed bounded subsets of $\mathbb{R}^n$

Since any bounded set of $\mathbb{R}^n$ is contained in an n-cube, if it is also a closed subset of $\mathbb{R}^n$, then it is closed in the hypercube and hence, by Proposition 3.2.9 is a compact set.

It is clear that we have characterized the compact subsets of $\mathbb{R}^n$ as precisely those sets which are bounded and closed.

Remark 3.2.12 (About infinite-dimensional Banach spaces).
Recall the space $\ell_2(\mathbb{R})$ of Example 3.1.28. This is an infinite-dimensional Banach space in which *not every closed bounded subset is compact*; in fact we will now show that the unit sphere, $\mathbb{S}_1(0)$, centred at the origin of $\ell_2(\mathbb{R})$ is not a compact set. Consider the family

$$\mathcal{B} \doteq \{\mathbb{B}_{\frac{1}{\sqrt{2}}}(x) \;:\; x \in \mathbb{S}_1(0)\}.$$

This is an open cover of the unit sphere (at the origin) of $\ell_2(\mathbb{R})$, but does not contain a finite subcover of the unit sphere. To see this recall the sequence of elements $\mathbf{e}_n \in \ell_2(\mathbb{R})$ defined earlier in Example 3.1.28 on page 88. The distance between any two of the $\boldsymbol{e}_i$'s is $\sqrt{2}$ and hence the triangle inequality implies that no set of the open cover $\mathcal{B}$ can contain more than one of these points. Hence no finite subcollection of $\mathcal{B}$ can cover the unit sphere.

Of course this means, *a fortiori* that the unit ball in $\ell_2(\mathbb{R})$ is not compact. Indeed, it can be shown that the closed unit ball in a Banach space, E, is compact iff E is finite-dimensional.

We now prove that compact sets satisfy the Bolzano–Weierstrass property.

Theorem 3.2.13 (Generalized Bolzano–Weierstrass Theorem).
Let $K \subset X$ be a compact subset of a metric space. If $\mathbf{x} = \{x_n\}$ is any sequence in K, then there is a subsequence $\underline{\mathbf{x}} = \{x_{n_k}\}$ which converges to a point $\underline{\mathbf{x}}^* \in K$.

Proof. The basic idea of the proof is the same as in the proof of the Bolzano–Weierstrass Theorem for closed, bounded subsets of $\mathbb{R}^n$: we try to find progressively smaller compact subsets of K which contain infinitely many elements of the sequence $\mathbf{x}$.

We may assume without loss of generality that no term of the sequence is repeated infinitely many times for in this case we can choose the subsequence to always take this value and this will clearly converge.

Let $\mathcal{U}_1 = \{\mathbb{B}_1(k) : k \in K\}$ be the open covering of K consisting of open balls in X of radius 1 centred at points of K. Since K is compact there is a finite subcover $\mathbb{B}_1(k_1), \ldots, \mathbb{B}_1(k_{m_1})$ and clearly one of these sets, say $\mathbb{B}_1(k_1)$, will contain infinitely many elements of $\mathbf{x}$. Let x_{n_1} be such an element. Now let K_1 be the intersection of K with the closure of this ball. That is, $K_1 = \bar{B}_1(k_1) \cap K$. This is a closed subset of K and hence a compact. Let $\mathcal{U}_2$ be the family of open balls of X with centre at points of K_1 and having radius $\frac{1}{2}$. These constitute an open covering of K_1 and so a finite subcover may be selected from these sets. As before, this implies that there is $k_2 \in K_1$ such that $K_2 = \bar{B}_{\frac{1}{2}}(k_2) \cap K_1$ contains infinitely many elements of $\mathbf{x}$. Choose $n_2 > n_1$ so that $x_{n_2} \in K_2$.

Since K_2 is compact we can cover K_2 by balls of radius $\frac{1}{3}$, select a finite subcover and find a point $k_2 \in K_2$ so that $K_3 = \bar{\mathbb{B}}_{\frac{1}{3}} \cap K_2$ contains infinitely many elements of $\mathbf{x}$.

Continuing the procedure, we can obtain a sequence of compact subsets:

$$K \supset K_1 \supset K_2 \supset \cdots \supset K_m = \left(\bar{\mathbb{B}}_{\frac{1}{m}}(k_m) \cap K_{(m-1)}\right) \supset K_{(m+1)} \supset \cdots$$

and points x_{n_m} of the sequence lying in K_m with $n_m > n_{m-1}$.

Let $\underline{\mathbf{x}} = \{x_{n_m}\}$. Now if $p, q > N$ then $d_x(x_{n_p}, x_{n_q}) < \frac{2}{N}$ and hence $\underline{\mathbf{x}}$ is a Cauchy sequence. Since by the previous result K is closed, this sequence must converge to a point of K. This completes the proof. □

The next result is proved in the same manner as Theorem 3.1.22.

Theorem 3.2.14. *Let X be a metric space and $f : X \to \mathbb{R}$ a map. If K is a compact subspace of X then f is bounded on K and there exist points $\underline{x}$ and $\overline{x}$ of K such that*

$$\begin{aligned} f(\underline{x}) &= \inf_{x \in K} f(x), \\ \textit{and} \quad f(\overline{x}) &= \sup_{x \in K} f(x). \end{aligned}$$ □

Before ending this discussion of compact sets, I will prove two more celebrated results: you have seen the analogue of the second result for the special case of closed intervals $[a, b]$ of $\mathbb{R}$ in your earlier courses.

Theorem 3.2.15 (Lebesgue's Lemma).
Let X be a compact space and let $\mathcal{U} = \{U_\imath\}_{\imath \in I}$ be any open cover of X.

Then there exists an $\varepsilon > 0$ such that any ball of radius $< \varepsilon$ is contained in some $U_{\imath_0}$ of the cover $\mathfrak{U}$. Such a number ε is called a *Lebesgue number* for the cover $\mathfrak{U}$.

Proof. If such an $\varepsilon > 0$ does not exist, then for each integer $n > 0$ we can find a point $x_n \in X$ such that $\mathbb{B}_{\frac{1}{n}}(x_n) \not\subset U_\imath$ for any $i \in I$. Since X is compact there is a subset, $\{x_{n_k}\}$, of these x_n's such that $\lim_{k\to\infty} x_{n_k} = x^*$.

Suppose that $x^* \in U_0$. Then there is a $\delta > 0$ such that $\mathbb{B}_\delta(x^*) \subset U_0$. Choose k so large that $\frac{1}{n_k} < \frac{\delta}{2}$ and $d_X(x^*, x_{n_k}) < \frac{\delta}{2}$. Then $\mathbb{B}_{1/n_k}(x_{n_k}) \subset U_0$, contradicting the choices made. □

A map $f : X \to Y$ between metric spaces is said to be *uniformly continuous* on $A \subset X$, if given $\varepsilon > 0$, we can find a $\delta > 0$ such that if $x', x'' \in A$ and $d_X(x', x'') < \delta$ then $d_Y\big(f(x'), f(x'')\big) < \varepsilon$. If $A = X$, then f is said to be uniformly continuous.

The next result is the most important consequence of Lebesgue's lemma.

Theorem 3.2.16. *Let $f : X \to Y$ be a map of metric spaces. If $K \subset X$ is compact then f is uniformly continuous on K.*

Proof. Let $\varepsilon > 0$ be given. Let $\mathfrak{V}$ be the cover of Y consisting of the open balls of radius $\frac{\varepsilon}{2}$ centred at each point of Y. Let $\mathfrak{U}$ be the open cover of X (and hence, K) consisting of the open sets $\{f^{-1}(V) : V \in \mathfrak{V}\}$. Let $(2 \cdot \delta) > 0$ be a Lebesgue number for the open cover $\mathfrak{U}$ of K. Then if $x', x'' \in K$ and $d_X(x', x'') < \delta$ then there is a $U_0 \in \mathfrak{U}$ such that $x', x'' \in U_0$. Since $f(x')$ and $f(x'')$ then belong to some $V \in \mathfrak{V}$, it follows that $d_Y\big(f(x'), f(x'')\big) < \varepsilon$; this establishes the uniform continuity of f on K. □

Exercises for § 3.2

3.2.1 Elementary Properties

(a) Suppose A, B are open sets of a Banach space E and $U \subset \mathbb{R}$ an open subset of $\mathbb{R}$. Show that the sets,

$$A + B \doteq \{a + b : a \in A,\ b \in B\}$$
$$\text{and } U \cdot A \doteq \{r \cdot a : r \in A, a \in A\}$$

are also open subsets of E.

Will these statements remain true if the word "open" is replaced by "closed" everywhere in the previous two sentences? What if A is open and B closed?

(b) Suppose that A and B are compact subsets of a normed linear space. Show that the set $A+B$ is also compact.

(c) A metric space X is said to be *discrete* if given any point $x \in X$ there is an $r = r_x > 0$ such that there are no points of X at a distance $< r_x$ from x other than x. Show that if X is a discrete metric space then every subset of X is open and closed. Show that if X is discrete and compact then it contains only finitely many points.

3.2.2 Interiors

Let X be a metric space. If $A \subset X$, its *interior*, denoted A°, is the largest open set contained in A. If $x \in A$ then x is called an *interior point* of A if there is a $r > 0$ such that $\mathbb{B}_r(x) \subset A$. Show that A° is the collection of all the interior points of A.

Further show that if A is a subset of X then the closure of A is the complement of the interior of the complement of A; that is,

$$\bar{A} = (\,(A^c)^\circ\,)^c\,.$$

Recall that the frontier of a set A is $\bar{A} \cap \overline{A^c}$. Show that the frontier of an open set U does not meet U.

3.2.3 Convexity and Norms on a Linear Space.

A subset A of a $\mathbb{k}$-linear space is said to be *convex* if whenever $x, y \in A$ the line segment joining x and y, that is, the set $\{t\boldsymbol{x} + (1-t)\boldsymbol{y} \,:\, 0 \leq t \leq 1\}$ is contained in A.

a. Show that if A and B are convex subsets of a linear space then their "sum" $A + B = \{a + b \,:\, a \in A,\, b \in B\}$ is also a convex set. Show that if A is a convex set in a normed linear space then its closure is also convex and so is its interior. Is the union of two convex sets necessarily a convex set?

b. Let E be a normed linear space and U a bounded, convex and open set containing 0_E. Suppose further that U is *balanced*; that is if x belongs to U, then for any scalar λ having absolute value 1, $\lambda \cdot \boldsymbol{x}$ also belongs to U.

Consider the function, by setting $\mu_U(x) = \sup\{\lambda \geq 0 : \lambda x \in U\}$..
(This function is generally called the *Minkowski functional* of U.)
Show that $\nu_U(x) = \frac{1}{\mu_U(x)}$ defines a norm on U.
Identify the unit ball (centred at the origin) of E under this norm.

3.2.4 Norms on Direct Sums

a. Use part b of the previous exercise to show that on $\mathbb{R}^n$, the formula

$$(\ell_p\text{-norm}) \qquad \nu_p(x_1, \ldots, x_n) = (x_1^p + \cdots + x_n^p)^{\frac{1}{p}}$$

defines a norm on $\mathbb{R}^n$ iff $1 \leq p < \infty$.

b. Let $\mathbf{E} = (E_1, \ldots, E_n)$ be a set of normed linear spaces. Show that the following define norms on the direct sum

$$\bigoplus E \doteq E_1 \oplus \cdots \oplus E_n.$$

- (The ℓ_∞-norm.) $||(x_1, \ldots, x_n||_\infty = \max(||x_1||, \cdots, ||x_n||)$.
- (The ℓ_p-norm.) For each $p > 1$, $||(x_1, \ldots, x_n)|| = \{||x_1||^p + \cdots + ||x_n||^p\}^{1/p}$.

c. Show that these norms are equivalent norms on $\bigoplus \mathbf{E}$.

d. Show, more generally that the analogous functions ℓ_r define norms on $\bigoplus \mathbf{E}$ iff $r \geq 1$.

e. Show that if each of the summands is a Banach space, then so is $\bigoplus \mathbf{E}$, for each of these norms.

3.2.5 Continuity and Uniform Convergence

In the last two sections, particularly the present one, you have been exposed to infinite-dimensional linear spaces and you may be still feeling somewhat uncomfortable about them. To convince yourself about these notions really do not require you to acquire new skills but only to learn some new "jargon", review the section on *uniform convergence* from your first analysis course and then prove the following result. We will be assuming the truth of the result in future!

Theorem. *Let X be a metric space and F be a Banach space. Suppose $f_n : X \to F$ is a sequence of maps. Suppose that $\{f_n\}$ converge* uniformly *to $f : X \to F$. (Recall, this means that* given any $\varepsilon > 0$, there is a $N = N(\varepsilon)$ such that if $p > N$, then

$$||f_p(x) - f(x)||_F < \varepsilon \textit{ for all } x \in X.)$$

Then f is a map from X to F.

3.3 Norms on Spaces of Linear Maps

In the title of this section, "Linear Maps" is an abbreviation for *continuous* linear transformations.

In the preamble to Chapter 2, I had mentioned that in order to develop Differential Calculus for functions between open subsets of normed linear spaces, we need to introduce the spaces $\mathfrak{L}_p(E; F)$. I have established in Theorem 3.1.16 a necessary and sufficient condition to determine if a linear transformation between normed linear spaces is continuous. An iteration of this procedure leads to a criterion for deciding if a linear p-function $\mathfrak{L}(E^{[p]}; F)$ is continuous where E, F are normed linear spaces. (Recall that $E^{[p]}$ is our notation for the p-tuple, $(E_1, \ldots, E_p)$ where $E_k = E$ for $k = 1, \ldots, p$.

In this section we will define norms on the sets of continuous linear transformations and continuous linear p-functions between normed linear spaces. This is essential if we are to define higher derivatives of differentiable functions between open subsets of normed linear spaces.

I begin by specifying certain conventions that I will follow throughout the remainder of the book.

- Since we will very frequently have to take limits, it would be very inconvenient to deal with spaces which are not complete. So henceforth we will deal only with Banach spaces. If a normed linear space arises out of some mathematical construct (as, for example, in Exercise 3.1.4) then, of course, a proof of completeness will be necessary and will be provided.

  Sometimes this may force me to state results with unnecessary hypotheses. This is preferable to making complicated statements embracing maximum generality but irrelevant for Calculus: after all, this is not a textbook of Functional Analysis.

- If we are looking at a Banach space E, whose norm is induced from an inner product then such a Banach space will, following standard custom, be called a *Hilbert space.*

Let E, F be Banach spaces, $S : E \to F$ a *continuous* linear transformation. Then we appeal to Theorem 3.1.16 and define the *operator norm*, $||S||_{\text{op}}$, of S by setting:

(Operator norm) .. $\quad ||S||_{\text{op}} = \sup\{||S(\boldsymbol{e})||_F \ : \boldsymbol{e} \in E, ||\boldsymbol{e}||_E = 1\}.$

If E, F are Banach spaces the set of continuous linear transformations from E to F will be denoted $\mathbb{L}(E, F)$. This is obviously a linear space with addition and scalar multiplication being defined pointwise.

We begin by establishing the basic properties of the operator norm that we have just defined.

Theorem 3.3.1 (Properties of the operator norm).
Let E, F be Banach spaces over $\mathbb{k}$. Then for arbitrary $S, T \in \mathbb{L}(E, F)$, and for any $\boldsymbol{x} \in E$ and $\alpha \in \mathbb{k}$ the following relations hold:

(1) $$||S(\boldsymbol{x})|| \leq ||S||_{\text{op}} \cdot ||\boldsymbol{x}||;$$

(2) $$||\alpha T||_{\text{op}} = |\alpha| \cdot ||T||_{\text{op}};$$

(3) $$||S + T||_{\text{op}} \leq ||S||_{op} + ||T||_{\text{op}};$$

(4) $$||T||_{\text{op}} \geq 0, \text{ and } ||T||_{\text{op}} = 0 \text{ iff } T = 0.$$

So we see that $|| \ ||_{op} : \mathbb{L}(E, F) \to \mathbb{R}$ is a norm.

(5) *In fact,* $|| \ ||_{op} : \mathbb{L}(E, F) \to \mathbb{R}$ *is a complete norm on* $\mathbb{L}(E, F)$.

If G is a Banach space and $U \in \mathbb{L}(F, G)$, then

(6) $$||U \circ S||_{\text{op}} \leq ||S||_{\text{op}} \cdot ||U||_{\text{op}}.$$

Note that in these inequalities, the symbols $|| \cdot ||$ denote norms in *different* spaces.

Proof. Keep in mind the ambiguity of notation mentioned above. It would have been more accurate, but pedantic (and ugly) to use $|| \cdot ||_E, || \cdot ||_F$, and three varieties of $|| \ ||_{\text{op}}$. It is better for you to pay attention and figure out for yourself which norm is meant in each case, because such abuses of notation are standard practice.

By definition, $||S||_{\text{op}} = \sup\limits_{\substack{\boldsymbol{x} \in E \\ ||\boldsymbol{x}||=1}} ||S(\boldsymbol{x})||.$

Hence, for each $\boldsymbol{x} \neq 0 \in E$, we have:

$$\begin{aligned} \|S(\boldsymbol{x})\| &= \left\| \|\boldsymbol{x}\| \cdot S\left(\tfrac{\boldsymbol{x}}{\|\boldsymbol{x}\|}\right) \right\| \\ &\leq \|S\|_{\text{op}} \cdot \|\boldsymbol{x}\|, \text{ proving (1).} \end{aligned}$$

To prove the homogeneity property for the operator norm, simply observe that $\|(\alpha \cdot T)(\boldsymbol{x})\| = \|\alpha \cdot T(\boldsymbol{x})\|$
$= |\alpha| \cdot \|T(\boldsymbol{x})\|$ using the homogeneity of the norm on F. Taking suprema over the vectors of norm 1 in E we get (2).

For any $\boldsymbol{x} \in E$, by the subadditivity of the norm on F, we get:

$$\|(S+T)(\boldsymbol{x})\| \leq \|S(\boldsymbol{x})\| + \|T(\boldsymbol{x})\|.$$

(3) follows upon taking the supremum of both sides of the inequality over all vectors of unit length belonging to E.

If the operator norm of $T \in \mathbb{L}(E, F)$ is 0, then for all $\boldsymbol{x}$ of unit length in E, $T(\boldsymbol{x}) = 0$. Since any vector is a scalar multiple of some unit vector (see the argument used to prove (1)) this means that $T = 0$ and proves (4).

So we have established that what we declared to be the "operator norm" does indeed, define a norm on, $\mathbb{L}(E, F)$. We now establish the completeness of the operator norm.

Suppose $\mathbf{T} = \{T_n\}_{n \in \mathbb{N}}$ is a Cauchy sequence in $\mathbb{L}(E, F)$ with respect to the operator norm. Then, for every $\boldsymbol{x} \in E$,

$$\lim_{m,n \to \infty} \|T_m(\boldsymbol{x}) - T_n(\boldsymbol{x})\| \leq \|\boldsymbol{x}\| \cdot \lim_{m,n \to \infty} \|T_m - T_n\| = 0.$$

and hence, for all $\boldsymbol{x} \in E$, $\{T_n(\boldsymbol{x})\}$ is a Cauchy sequence in F. Define $T : E \to F$ by setting $T(\boldsymbol{x}) = \lim_{n \to \infty} T_n(\boldsymbol{x})$. It is easy to check that $T : E \to F$ is a linear transformation. Since $\{T_n\}$ is a Cauchy sequence, given any $\varepsilon > 0$, there is an $N = N(\varepsilon)$ such that if $n > N$ then for all $\boldsymbol{x} \in E$, having norm equal to 1, we have $\|T_m(\boldsymbol{x}) - T_n(\boldsymbol{x})\| < \varepsilon$. Obviously this inequality will continue to hold *a fortiori* for all $\boldsymbol{x} \in \mathbb{B}_1(0_E)$. This means that the continuous functions $T_n : E \to F$ converge uniformly on $\mathbb{B}_1(0_E)$. Hence, the limit function T is continuous on $\mathbb{B}_1(0_E)$ and, hence, at 0_E. Since T is a linear transformation this means that T is continuous and hence an element of $\mathbb{L}(E, F)$.

Finally, we prove (6). From (1) it follows that for all $\boldsymbol{x} \in E$,

$$\|(U \circ S)(\boldsymbol{x})\| \leq \|U\| \cdot \|S(\boldsymbol{x})\| \leq \|U\| \cdot \|S\| \cdot \|\boldsymbol{x}\|$$

(6) follows upon taking the supremum over all vectors of norm 1 in E in the inequality above. □

Henceforth, $\mathbb{L}(E, F)$ will always denote the Banach space of continuous linear transformations from E to F with the operator norm.

Remark 3.3.2 (Concerning Nomenclature).
Note that though continuous linear transformations are called "bounded operators" one does *not* say that a linear transformation between normed linear spaces is *unbounded* if it fails to be continuous. This is because there is a somewhat different notion: that of an *unbounded operator*, between Banach spaces. Unbounded operators will not appear in this book. An example of a discontinuous linear transformation has already been provided in Example 3.1.17.

If E is a Banach space, then the spaces $E' = \mathbb{L}(E, \Bbbk)$ and $E'' = \mathbb{L}(E', \Bbbk)$, are also Banach spaces. These are, respectively, the *continuous dual* and the *continuous double dual* of E. (These are the appropriate "analytic" analogues of the dual and double dual for linear spaces over an arbitrary field.) Elements of E' are called *linear functionals*.

As will be explained in the preamble to Chapter 4, complex Banach spaces fall outside the realm of Differential Calculus. So for the rest of this section, in particular, in our discussion of the Hahn-Banach Theorem, which will follow shortly, I restrict myself to Banach spaces over $\mathbb{R}$.

Example 3.3.3 (An important linear functional).
Let E be a Hilbert space. Then for any $\boldsymbol{x} \in E$, the function $T_{\boldsymbol{x}} : \boldsymbol{y} \mapsto \langle \boldsymbol{x}, \boldsymbol{y} \rangle$ is a *continuous* linear functional.

This is easy to see. For $\boldsymbol{x}, \boldsymbol{y} \in E$, we have $|T_{\boldsymbol{x}}| = |\langle \boldsymbol{x}, \boldsymbol{y} \rangle| \leq ||\boldsymbol{x}|| \cdot ||\boldsymbol{y}||$, by the Cauchy–Schwarz inequality. This shows that $T_{\boldsymbol{x}}(\mathbb{S}_1(0_E)) \subset [-||\boldsymbol{x}||, +||\boldsymbol{x}||]$. The continuity of $T_{\boldsymbol{x}}$ follows from Theorem 3.1.16. In fact, taking suprema of both sides of the above inequality as $\boldsymbol{y}$ ranges over the unit sphere of E, we see that $||T_{\boldsymbol{x}}|| \leq ||\boldsymbol{x}||$ and evaluating on the unit vector $\frac{\boldsymbol{x}}{||\boldsymbol{x}||}$ shows that the norm of $T_{\boldsymbol{x}}$ is equal to the norm of $\boldsymbol{x}$. □

Let E be a Banach space. If $\alpha \in E'$ is a linear functional and $\boldsymbol{x} \in E$ any vector, then from the definition of κ_E, see page 44, we get:

$$\begin{aligned} |\langle \kappa_E(\boldsymbol{x}), \alpha \rangle| &= |\langle \alpha, \boldsymbol{x} \rangle| \\ &\leq ||\alpha|| \cdot ||\boldsymbol{x}||. \end{aligned}$$

This shows that, $\kappa_E(\boldsymbol{x}) : E' \to \mathbb{R}$ defined by $\langle \kappa_E(\boldsymbol{x}), \alpha \rangle = \langle \alpha, \boldsymbol{x} \rangle$ is a continuous linear functional on E' and hence, $\kappa_E(\boldsymbol{x})$ is an element of E''. I will next prove that the linear transformation $\kappa_E : E \to E''$ is an *isometry* ; that is, $||\kappa_E(\boldsymbol{x})|| = ||\boldsymbol{x}||$.

Clearly it is sufficient and, of course, necessary, to show that given any $\boldsymbol{x} \in E$ there is an $\alpha \in E'$ such that $||\alpha|| = 1$ and such that $\langle \alpha, \boldsymbol{x} \rangle = ||\boldsymbol{x}||$

since, in this situation we have:

$$\begin{aligned} ||\kappa_E(\boldsymbol{x})|| &= \sup_{||\alpha||=1} |\langle \alpha, \boldsymbol{x}\rangle| \\ &= ||\boldsymbol{x}||. \end{aligned}$$

For Hilbert spaces (finite-dimensional or not) this is easily done:

Given $\boldsymbol{x} \neq 0$ the linear functional defined by

$$\alpha : \boldsymbol{y} \mapsto \frac{\langle \boldsymbol{x}, \boldsymbol{y}\rangle}{||\boldsymbol{x}||}$$

fits the bill.

For Banach spaces, the result is nontrivial: even if the space involved is finite-dimensional. The proof uses an elegant and ingenious trick which is explained below and, for infinite-dimensional spaces, an application of the Axiom of Choice.

Theorem 3.3.4 (Hahn–Banach).
Let E be any Banach space, E_0 a subspace of E and $\alpha : E_0 \to \mathbb{R}$ a linear functional having norm 1. Then there is a linear functional $\widetilde{\alpha} : E \to \mathbb{R}$ such that $\widetilde{\alpha} \mid E_0 = \alpha$ and such that $||\widetilde{\alpha}|| = 1$.

Proof. Choose any nonzero vector $\boldsymbol{t}_1 \notin E_0$ and let $E_1 = E_0 \oplus \mathbb{R}\cdot\boldsymbol{t}_1$, where $\mathbb{R}\cdot\boldsymbol{t}_1$ is the 1-dimensional normed linear subspace of E spanned by $\boldsymbol{t}_1$. The crucial step is to obtain a continuous extension, α_1, of α to E_1 without increasing the norm. Note that for any $\boldsymbol{x}, \boldsymbol{y} \in E_0$ we have:

$$|\alpha(\boldsymbol{x}) + \alpha(\boldsymbol{y})| = |\alpha(\boldsymbol{x}+\boldsymbol{y})| \leq ||\boldsymbol{x}+\boldsymbol{y}|| \leq ||\boldsymbol{x}|| + ||\boldsymbol{y}||.$$

Since the right side of the inequality is positive it will continue to hold without the absolute value on the left side. Using the triangle inequality on the right side of this, we get

$$\alpha(\boldsymbol{x}) + \alpha(\boldsymbol{y}) \leq ||\boldsymbol{x} - \boldsymbol{t}_1|| + ||\boldsymbol{y} + \boldsymbol{t}_1||$$

from which we get:

$$\alpha(\boldsymbol{x}) - ||\boldsymbol{x} - \boldsymbol{t}_1|| \leq ||\boldsymbol{y} + \boldsymbol{t}_1|| - \alpha(\boldsymbol{y}).$$

Since the left side is independent of $\boldsymbol{y}$ it is bounded. Let M be the supremum of the left side as $\boldsymbol{x}$ varies over E_0. Now define $\alpha_1 : E_1 \to \mathbb{R}$ by setting

$$\alpha_1(\boldsymbol{x} + c\boldsymbol{t}_1) = \alpha(\boldsymbol{x}) + cM \quad \text{for all } \boldsymbol{x} \in E_0 \text{ and } c \in \mathbb{R}.$$

α_1 is clearly a linear transformation; we only have to show that $||\alpha_1|| = 1$; this will prove continuity as well as the fact that the extension procedure has not increased the norm.

The definition of M implies two inequalities:

$$\alpha(\boldsymbol{x}) - M \leq ||\boldsymbol{x} - \boldsymbol{t}_1||, \text{ and}$$
$$\alpha(\boldsymbol{y}) + M \leq ||\boldsymbol{y} + \boldsymbol{t}_1||.$$

These hold for all $\boldsymbol{x}, \boldsymbol{y} \in E_0$. Now we substitute $s^{-1}\boldsymbol{z}$ for $\boldsymbol{x}$ and $\boldsymbol{y}$ (where $s > 0$ and $\boldsymbol{z} \in E_0$) in the above and multiply by s and obtain:

$$\alpha_1(\boldsymbol{z} - s\boldsymbol{t}_1) = \alpha(\boldsymbol{z}) - sM \leq ||\boldsymbol{z} - s\boldsymbol{t}_1||, \text{ and}$$
$$\alpha_1(\boldsymbol{z} + s\boldsymbol{t}_1) = \alpha(\boldsymbol{z}) + sM \leq ||\boldsymbol{z} + s\boldsymbol{t}_1||.$$

Now substituting $-\boldsymbol{z}$ for $\boldsymbol{z}$ in each of the above inequalities we get,

$$-\alpha_1(\boldsymbol{z} + s\boldsymbol{t}_1) = -\alpha(\boldsymbol{z}) - sM \leq ||-\boldsymbol{z} - s\boldsymbol{t}_1|| = ||\boldsymbol{z} + s\boldsymbol{t}_1||,$$
$$-\alpha_1(\boldsymbol{z} - s\boldsymbol{t}_1) = -\alpha(\boldsymbol{z}) + sM \leq ||-\boldsymbol{z} + s\boldsymbol{t}_1|| = ||\boldsymbol{z} - s\boldsymbol{t}_1||.$$

The last four inequalities, together imply:

$$|\alpha_1(\boldsymbol{z} + s\boldsymbol{t}_1| \leq ||\boldsymbol{z} + s\boldsymbol{t}_1|| \text{ for all } s \in \mathbb{R}.$$

We have thus extended α to E_1 without increasing the norm; if $E_1 \neq E$ then choose $\boldsymbol{t}_2 \notin E_1$ and extend α_1 to $E_2 = E_1 \oplus \mathbb{R}\cdot\boldsymbol{t}_2$ without increasing the norm. Continuing in this way we can prove the result for finite-dimensional E by induction. For infinite dimensional spaces we have to take recourse to the Axiom of Choice. □

The following immediate corollary should be thought of as a geometric reformulation of the Hahn-Banach theorem.

Theorem 3.3.5. 1. *Let $\boldsymbol{x} \neq 0$ be any element of a Banach space. Then there exists a $\alpha \in E'$ such that $||\alpha|| = 1$ and $\alpha(\boldsymbol{x}) = ||\boldsymbol{x}||$.*

2. *This implies that $\kappa_E : E \to E''$ is a* norm-preserving injection *of Banach spaces.*

Proof. Let E_0 be the 1-dimensional space spanned by $\boldsymbol{x}$.

The function $\phi : E_0 \to \mathbb{R}$ defined by $\phi(t \cdot \boldsymbol{x}) = t \cdot ||\boldsymbol{x}||$ is a linear transformation of norm 1. Hence by the Hahn-Banach theorem there is a linear functional $\widetilde{\phi} : E \to \mathbb{R}$ such that $\widetilde{\phi}|E_0 = \phi$ and $||\widetilde{\phi}|| = 1$. Choosing $\alpha = \widetilde{\phi}$ proves the first assertion.

For the second part, note that by definition,

$$||\kappa_E(\boldsymbol{x})|| = \sup\{|\langle\theta, \boldsymbol{x}\rangle| \, : \, \theta \in E' \text{ and } ||\theta|| = 1\}.$$

Choosing $\theta = \alpha$ as in (1), shows that $||\kappa_E(\boldsymbol{x})|| = ||\boldsymbol{x}||$. □

We now extend the idea behind the definition of the operator norm of continuous linear transformations to define a suitable notion of continuous *multilinear functions* between Banach spaces.

Let $E_1, \ldots, E_p$ and F be Banach spaces. Then we have seen (Exercise 3.2.4) that the direct sum $E = (E_1 \oplus \cdots \oplus E_p)$ is a Banach space with respect to the "ℓ_∞-norm", defined by the formula:

$$||(\boldsymbol{x}_1, \ldots, \boldsymbol{x}_p)||_\infty = \max_{1 \le k \le p} ||\boldsymbol{x}_k||_{E_k}.$$

Multilinear functions are functions defined on $E_1 \times \cdots \times E_p$ which is the underlying set of the linear space $E_1 \oplus \cdots \oplus E_p$. For the purpose of studying continuity, we will regard multilinear functions as being defined on the Banach space $E_1 \oplus \cdots \oplus E_p$, with the ℓ_∞-norm. If F is a Banach space, it now makes sense to ask if a multilinear function $f : E_1 \times \cdots \times E_p \to F$ is continuous or not.

If $\mathbf{E} = (E_1, \ldots, E_p)$ is an ordered p-tuple of Banach spaces, and F another Banach space then $\mathbb{M}(\mathbf{E}; F)$ will denote the set of *continuous* multilinear functions from $\times\mathbf{E} \to F$.

The first task is to establish a criterion for the continuity of a multilinear function $\phi : E_1 \times \cdots \times E_p \to F$ between Banach spaces.

The continuous p-linear functions will be called p-linear *maps*.

Proposition 3.3.6. *Let $\phi : E = E_1 \times \cdots \times E_p \to F$ be a multilinear function. Then ϕ is continuous iff there is a real number $a > 0$ such that we have:*

$$(\star) \qquad ||\phi(\boldsymbol{x}_1, \ldots, \boldsymbol{x}_p))||_F \le a||\boldsymbol{x}_1||_{E_1} \cdots ||\boldsymbol{x}_p||_{E_p}.$$

for all $\boldsymbol{x_i} \in E_i$, $(i = 1, \ldots, p)$.

Proof. I will write out the proof for $p = 2$. The general case can be proved with somewhat more hard work along the same lines; no new ideas are involved, only the notation becomes more complicated.

We first prove the sufficiency of the condition. To show continuity of ϕ at $(\boldsymbol{x}, \boldsymbol{y}) \in E$ we estimate the norm of $\phi(\boldsymbol{x} + \boldsymbol{h}, \boldsymbol{y} + \boldsymbol{k}) - \phi(\boldsymbol{x}, \boldsymbol{y})$ for

$(\boldsymbol{h}, \boldsymbol{k})$ having small norm as follows.

$$\begin{aligned}\phi(\boldsymbol{x}+\boldsymbol{h}, \boldsymbol{y}+\boldsymbol{k}) - \phi(\boldsymbol{x}, \boldsymbol{y}) &= (\, \phi(\boldsymbol{x}+\boldsymbol{h}, \boldsymbol{y}+\boldsymbol{k}) - \phi(\boldsymbol{x}+\boldsymbol{h}, \boldsymbol{y}) \,) \\ &\quad + (\, \phi(\boldsymbol{x}+\boldsymbol{h}, \boldsymbol{y}) - \phi(\boldsymbol{x}, \boldsymbol{y}) \,) \\ &= \phi(\boldsymbol{x}+\boldsymbol{h}, \boldsymbol{k}) + \phi(\boldsymbol{h}, \boldsymbol{y}).\end{aligned}$$

So we have the inequality

$$||\phi(\boldsymbol{x}+\boldsymbol{h}, \boldsymbol{y}+\boldsymbol{k}) - \phi(\boldsymbol{x}, \boldsymbol{y})|| \leq ||\phi(\boldsymbol{x}+\boldsymbol{h}, \boldsymbol{k})|| + ||\phi(\boldsymbol{h}, \boldsymbol{y})||.$$

So from the hypothesis, we get

$$||\phi(\boldsymbol{x}+\boldsymbol{h}, \boldsymbol{y}+\boldsymbol{k}) - \phi(\boldsymbol{x}, \boldsymbol{y})|| \leq a(||\boldsymbol{x}+\boldsymbol{h}|| \cdot ||\boldsymbol{k}|| + ||\boldsymbol{h}|| \cdot ||\boldsymbol{y}||).$$

Now suppose that $||\boldsymbol{h}||, ||\boldsymbol{k}|| \leq \delta$ where $0 < \delta < 1$. Then $||\boldsymbol{x}+\boldsymbol{h}|| < ||\boldsymbol{x}|| + 1$ and we get

$$||\phi(\boldsymbol{x}+\boldsymbol{h}, \boldsymbol{y}+\boldsymbol{k}) - \phi(\boldsymbol{x}, \boldsymbol{y})|| \leq a(||\boldsymbol{x}|| + 1 + ||\boldsymbol{y}||)\delta.$$

Hence the left side can be made arbitrarily small by choosing $||(\boldsymbol{h}, \boldsymbol{k})|| = \max(||\boldsymbol{h}||, ||\boldsymbol{k}||) = \delta$ suitably small.

Now suppose that ϕ is a continuous multilinear transformation. Since ϕ is continuous at $(0,0) \in E_1 \times E_2$ and since $\phi(0.0) = 0$ there is an $r > 0$ such that for $(\boldsymbol{x}, \boldsymbol{y}) \in \overline{\mathbb{B}}_r(0,0)$ we have $||\phi(\boldsymbol{x}, \boldsymbol{y})|| \leq 1$. Now consider any point $(\boldsymbol{z}_1, \boldsymbol{z}_2) \in E$ with both the entries nonzero. Then if we put $\boldsymbol{x}_1 = r\boldsymbol{z}_1/||\boldsymbol{z}_1||$ and $\boldsymbol{x}_2 = r\boldsymbol{z}_2/||\boldsymbol{z}_2||$ then $(\boldsymbol{x}_1, \boldsymbol{x}_2) \in \overline{\mathbb{B}}_r(0,0)$. Hence, $||\phi(\boldsymbol{x}_1, \boldsymbol{x}_2)|| \leq 1$. But

$$\phi(\boldsymbol{x}_1, \boldsymbol{x}_2) = \phi(\boldsymbol{z}_1, \boldsymbol{z}_2) \cdot \frac{r^2}{||\boldsymbol{z}_1||.||\boldsymbol{z}_2||}$$

and hence $||\phi(\boldsymbol{z}_1, \boldsymbol{z}_2)|| \leq a||\boldsymbol{z}_1||.||\boldsymbol{z}_2||$ with $a = 1/r^2$. Finally if either $\boldsymbol{z}_1$ or $\boldsymbol{z}_2$ is zero, then $\phi(\boldsymbol{z}_1, \boldsymbol{z}_2) = 0$ and hence the desired inequality continues to hold. This completes the proof. □

This was hard work! It will turn out when we switch to studying continuity of linear p-functions, life is much easier.

REMARKS 3.3.7 (ON THE CONTINUITY OF MULTILINEAR FUNCTIONS)

Remark 3.3.7a. In view of the multilinear property, in order to assert that $\phi : \mathbf{E} = E_1 \times \cdots \times E_p \to F$ is continuous, we only need that the inequality $(\star)$ of page 112 is valid for the *unit sphere* (with respect to the ℓ_∞-norm) of $\boldsymbol{E} = E_1 \oplus \ldots E_p$. This is the set,

$$\mathbb{S}_1(0; \boldsymbol{E}) \doteq \{(\boldsymbol{x}_1, \ldots, \boldsymbol{x}_p) \in E_1 \times \cdots \times E_p \; : \; ||\boldsymbol{x}_i||_{E_i} = 1 \text{ for } i = 1, \ldots, p\}.$$

To see this notice that if $(\boldsymbol{x}_1, \ldots, \boldsymbol{x}_p) \in E_1 \times \cdots \times E_p$ and $0 \neq r \in \mathbb{R}$, then $||\phi(r\boldsymbol{x}_1, \ldots, r\boldsymbol{x}_p)||_F = r^p ||\phi(\boldsymbol{x}_1, \ldots, \boldsymbol{x}_p)||_F$ and the right side of $(\star)$ is also multiplied by a factor of r^p.

Remark 3.3.7b. Notice that when $p = 1$ the previous remark gives the usual criterion for continuity of linear transformations: that they should be bounded on the unit sphere of the domain.

Definition 3.3.8 (The norm of a continuous p-linear function).
Let $E_1, \ldots, E_p$ be an ordered p-tuple of Banach spaces and F a Banach space. Then we define the *operator norm* of a continuous multilinear function, $\phi : E_1 \times \cdots \times E_p \to F$ by the formula
(†)................ $||\phi|| = \sup_{||\boldsymbol{x}_i||_{E_i}=1} ||\phi(\boldsymbol{x}_1, \ldots, \boldsymbol{x}_p)||_F.$

Lemma 3.3.9. *The formula (3.3.8) defines a norm on the linear space of continuous multilinear transformations,* $\mathbb{M}_p(E_1 \times \ldots E_p; F)$.

Proof. We begin by showing that the supremum on the right-hand side of (†) exists. Since ϕ is continuous at $0_{\boldsymbol{E}}$ there is an $r > 0$ such that for all $||(\boldsymbol{x}_1, \ldots, \boldsymbol{x}_p)||_{\ell_\infty} \leq r$, we have $||\phi(\boldsymbol{x}_1, \ldots, \boldsymbol{x}_p)||_F \leq 1$. Then from the multilinearity of ϕ it follows that the right hand side of (†) is bounded by r^{-p} and hence the supremum is being taken over a bounded set of real numbers. Hence, $||\phi||$ is well-defined.

If $\phi, \Psi \in \mathbb{M}_p(\times\mathbf{E}; F)$, then

$$\begin{aligned}||\phi + \Psi|| &= \sup_{i=1,\ldots,p} \{||\phi(\boldsymbol{x}_1, \ldots, \boldsymbol{x}_p) + \Psi(\boldsymbol{x}_1, \ldots, \boldsymbol{x}_p)|| \; : ||\boldsymbol{x}_i||_{E_i} = 1\} \\ &\leq \sup_{i=1,\ldots,p} \{||\phi(\boldsymbol{x}_1, \ldots, \boldsymbol{x}_p)|| \; : ||\boldsymbol{x}_i||_{E_i} = 1\} \\ &\quad + \sup_{i=1,\ldots,p} \{||\Psi(\boldsymbol{x}_1, \ldots, \boldsymbol{x}_p)|| \; : ||\boldsymbol{x}_i||_{E_i} = 1\} \\ &\leq ||\phi|| + ||\Psi||.\end{aligned}$$

Positivity and homogeneity are easily checked. □

Example 3.3.10 (Composition of linear maps).
Let E, F, G be Banach spaces. Consider the "composition function"

$$\Gamma : \mathbb{L}(E, F) \times \mathbb{L}(F, G) \to \mathbb{L}(E, G)$$

defined by $\Gamma(\alpha, \beta) = \beta \circ \alpha$. By (6) of Theorem 3.3.1 $||\Gamma(\alpha, \beta)|| \leq ||\alpha|| ||\beta||$. Hence, Γ is a continuous bilinear function whose norm is ≤ 1.

We now turn to the study of continuity properties of linear p-functions. As before, let $E_1, \dots, E_p$ and F be Banach spaces. We inductively define:

$$\mathbb{L}_p(\mathbf{E}, F) \doteq \begin{cases} \mathbb{L}(E_1, F) & \text{if } p = 1, \\ \mathbb{L}(E_p, \mathbb{L}_{p-1}(E_1, \dots, E_{p-1}; F)) & \text{if } p \geq 2. \end{cases}$$

Notice that the $\mathbb{L}$'s are Banach spaces *by definition* and their norms are completely determined by the norms of $E_1, \dots, E_p$'s and F.

The next result sums up the properties and relationship between the normed linear space structures of $\mathbb{M}_p(\times\mathbf{E}; F)$ and $\mathbb{L}_p(\mathbf{E}; F)$. (We are using $\mathbf{E}$ to denote the p-tuple $(E_1, \dots, E_p)$.)

Theorem 3.3.11 (The Law of Exponents III).
The restrictions of the "exponential law bijections"

$$\Xi_p : \mathbb{L}_p(\mathbf{E}; F) \to \mathbb{M}_p(\times\mathbf{E}, F)$$
$$\text{and } \xi_p : \mathbb{M}_p(\times\mathbf{E}, F) \to \mathbb{L}_p(\mathbf{E}; F)$$

are norm-preserving isomorphisms. In particular, $\mathbb{M}_p(\times\mathbf{E}; F)$ is also a Banach space.

Proof. To facilitate the exposition I introduce some *temporary* notation.

- For $k \leq p$ the norm in the spaces $\mathbb{L}_k(E_1, \dots, E_k; F)$, will be denoted $||\cdot||_{[k]}$;
- the norm in $\mathbb{M}_k(E_1 \times \cdots \times E_k, F)$ by $||\cdot||_{(k)}$.

Then using this notation, for any $A \in \mathbb{L}_p(E_1, \dots, E_p; F)$

$$\begin{aligned} ||A||_{[p]} &= \sup_{||e_p||=1} ||A(e_p)||_{[p-1]} \\ &= \sup_{\substack{||e_p||=1 \\ ||e_{p-1}||=1}} ||A(e_p : e_{p-1})||_{[p-2]} \\ &= \dots \\ &\vdots \quad \vdots \quad \text{after } p \text{ steps,} \\ &= \dots \end{aligned}$$

$$(1) \quad ||A||_{[p]} = \sup\{||A(e_p : e_{p-1} : \cdots : e_1)||_F \, : \, ||e_j|| = 1, 1 \leq j \leq p\}.$$

On the other hand, from the definition of Ξ, we have

$$\Xi(A)(e_1, \dots, e_p) = A(e_p : \cdots : e_1),$$

and hence taking the supremum over the unit sphere of $\times\mathbf{E}$ we get

$$(2) \qquad ||\Xi(A)||_{(p)} = \sup\{||A(e_p : \cdots : e_1) : ||e_i||_{E_i} = 1\}$$

From (1) and (2), we see that Ξ is an isometry. A similar argument shows that ξ is an isometry. □

Exercises for § 3.3

3.3.1 Equivalent Norms

Let $E_1, \ldots, E_p, F$ and $\widehat{E}_1, \ldots, \widehat{E}_p, \widehat{F}$ be the same set of linear spaces but with equivalent norms. Show that the sets $\mathbb{L}_p(\mathbf{E}; F)$ and $\mathbb{L}_p(\widehat{\mathbf{E}}; \widehat{F})$ have the same underlying set and the norms on them are also equivalent.

3.3.2 Norms on $L(\mathbb{R}^n)$

Let E be a finite-dimensional space so that $\mathbb{L}$ and L are identical.

a. Show that if ν is any norm on $L(E)$ then there is a constant $K = K(\nu) \geq 1$ such that for all $S, T \in L(E)$ we have the inequality:

$$\nu(S \circ T) \leq K\nu(S)\nu(T).$$

b. Let $E = \mathbb{R}^2$ with the standard Euclidean norm, $||\cdot||_2$. Determine the operator norms of the linear transformations whose matrices with respect to the standard basis are $\left[\begin{smallmatrix} 1 & 0 \\ 2 & 1 \end{smallmatrix}\right]$ and $\left[\begin{smallmatrix} 0 & \alpha \\ 0 & 0 \end{smallmatrix}\right]$.

c. Let E and F be the linear space $\mathbb{R}^n$ with the ℓ_1-norm and ℓ_∞-norms respectively. Let $\mathrm{A} \in \mathrm{M}_n(\mathbb{R})$ and let T_A be the linear transformation whose matrix with respect to the standard basis is A. Express the operator norm of $T_\mathrm{A} \in L(E, F)$ in terms of the entries of A. Also calculate the operator norm of T_A regarded as an element of $L(F, E)$.

d. Let $\mathbb{R}^m$ and $\mathbb{R}^n$ be considered as Hilbert spaces with the standard inner-product, let A be a $(n \times m)$ matrix and represent a linear transformation $T(\mathrm{A}) : \mathbb{R}^m \to \mathbb{R}^n$ via the equation $T(\mathrm{A})(v) = \mathrm{A}v$. Show that if $||\ \ ||$ denotes the operator norm then $||T(\mathrm{A})||^2$ is equal to the largest eigenvalue of the symmetric matrix A^tA.

Formulate (and prove) the corresponding result for general inner-product spaces and for unitary spaces.

3.3.3 Examples of Continuous Duals

a. Let $\mathbf{c}_0$ be the set of sequences $\boldsymbol{x} = \{x_n\}_{n\in\mathbb{N}}$ of real numbers such that $\lim_{n\to\infty} x_n = 0$. (Since 'sequences' are the vectors in this instance, they get the "bold italics" hitherto reserved for vectors!)

Show that the formula:

$$||\boldsymbol{x}|| = \sup_{n\in\mathbb{N}} |x_n|$$

defines a complete norm on $\mathbf{c}_0$.

Let $\boldsymbol{e}_m = \{\delta_{mn}\}$ be the sequence with zero everywhere except at the mth spot where we have 1. Show that if $\boldsymbol{x} = \{x_n\} \in \mathbf{c}_0$ then the sequence $s_n = \sum_{k=1}^{n} x_k \boldsymbol{e}_k$ converges in $\mathbf{c}_0$ to $\boldsymbol{x}$.

b. Let $\alpha \in \mathbf{c}_0' \doteq \mathbb{L}(\mathbf{c}_0, \mathbb{R})$; that is, α is a linear functional on $\mathbf{c}_0$. Suppose that $\langle \alpha, \boldsymbol{e}_n \rangle = \xi_n$. Show that $\sum_{n=1}^{\infty} \xi_n$ is an absolutely convergent series and that the formula:

$$||\alpha|| = \sum_{n=1}^{\infty} |\xi_n|$$

defines the operator norm on $\mathbf{c}_0'$.

Let ℓ_1 be the linear space of sequences $\mathbf{y} = \{y_n\}$ of real numbers such that $\sum_n y_n$is absolutely convergent. Show that

$$||\mathbf{y}|| = \sum_{n=1}^{\infty} |y_n|$$

defines a complete norm on ℓ_1. Finally show that $(\mathbf{c}_0)'$ is isomorphic as a Banach space to ℓ_1.

c. Let ℓ_∞ be the linear space of *bounded* sequences of real numbers with the "sup-norm": $||\mathbf{x}|| = \sup_n |x_n|$. Show that ℓ_∞ is a Banach space and that $(\ell_1)' = \ell_\infty$.

d. Finally show that $\kappa_{\mathbf{c}_0} : \mathbf{c}_0 \to (\mathbf{c}_0)'' = \ell_\infty$ is the 'inclusion' of $\mathbf{c}_0$ in ℓ_∞. So $\kappa_{\mathbf{c}_0}$ is not surjective.

This example shows that reflexivity does not hold for the process of "continuous double dualization" of infinite-dimensional Banach spaces.

e. Let $\ell_2(\mathbb{R})$ be the Banach, (in fact,) Hilbert space of sequences of "square summable" real numbers. Show that the continuous dual, $\ell_2(\mathbb{R})' = \ell_2(\mathbb{R})$ and that the function κ, in this instance, is the identity map. Thus $\ell_2(\mathbb{R})$ is a "reflexive Banach space".

3.4 Infinite Series

Some of you may have been surprised by the omission of any mention of "convergent series" in the first section of this chapter; after all, traditionally convergence of sequences and series are discussed at the same time. The reason I did not introduce this earlier is that the most interesting series of real or complex numbers are "power series" which, of course, do not make sense in an arbitrary Banach space. But we now know that if V is a Banach space then so is $E \doteq \mathbb{L}(V, V) = \mathbb{L}(V)$ and we can define powers of elements of E by defining the product of two operators to be their composition. We will soon find that very interesting power series arise in such spaces.

I will begin with some definitions which should not cause any surprise. Let $\mathbf{x} = \{x_n : n \geq k\}$ (k some integer) be a sequence of elements of a Banach space E. For each $n \geq k$ we define the sequence of *partial sums*, $s_n(\mathbf{x})$, by setting $s_n(\mathbf{x}) = \sum_{k \leq \imath \leq n} x_\imath$.
We say that the infinite series associated to $\mathbf{x}$ *converges* if the sequence of partial sums is a convergent sequence in E. In this case, we define the limit of the sequence of partial sums to be the *sum* of the infinite series defined by the sequence $\mathbf{x}$ and denote this limit by $\sum_k^\infty \boldsymbol{x}_n$. Thus

$$\sum_{n \geq k}^{\infty} x_n = \lim_{N \to \infty} s_N(\mathbf{x}) \quad \text{provided the limit exists.}$$

The series associated to $\mathbf{x}$ is said to be *absolutely convergent* if $\{||x_n||\}$ is a sequence of real numbers that form a convergent series. You can prove as in the real or complex case that if $\sum_n x_n$ converges absolutely then so does any rearrangement and the sum of the series is independent of the way we rearrange the terms. However there is no analogue of "Riemann's Theorem":

if a series of real numbers converges, but does not converge absolutely, then given an arbitrary $l \in \mathbb{R}$, one can always find a rearrangement, such that the rearranged series converges to l.

The following results are proved simply by replacing the elements $\boldsymbol{x}_n$ by their norms and appealing to the corresponding results for series of real numbers.

Proposition 3.4.1 (The Ratio Test). *Let $\{\boldsymbol{x}_n\}_{n\geq k}$ be a sequence in a Banach space E and suppose that $\overline{\lim_{n\to\infty}}\left(\frac{||\boldsymbol{x}_n||}{||\boldsymbol{x}_{n-1}||}\right) < 1$. Then the series $\sum_{n\geq k}\boldsymbol{x}_n$ converges absolutely.* □

Proposition 3.4.2 (The Root Test). *Let $\{\boldsymbol{x}_n\}_{n\geq k}$ be a sequence in a Banach space E and suppose that $\overline{\lim_{n\to\infty}}||\boldsymbol{x}_n||^{\frac{1}{n}} < 1$. Then the series $\sum_{n\geq k}\boldsymbol{x}_n$ converges absolutely.* □

Proposition 3.4.3 (The M-test of Weierstrass). *Let $\mathbf{x} = \{\boldsymbol{x}_n\}_{n\geq 1}$ be a sequence in a Banach space and suppose there is a constant $M > 0$ and a sequence of positive real numbers $\{c_n\}_{n\geq N}$ such that for each $n \geq N$, $||\boldsymbol{x}_n|| \leq M.c_n$ and the series $\sum_{n\geq 1} c_n$ is convergent. Then $\sum_{n\geq 1}\boldsymbol{x}_n$ converges absolutely.* □

We now look at some examples of convergent series.

Example 3.4.4a (The Neumann Series).
I begin by noting two simple facts:

1. If $t > 0$ then $(1+t)^n \geq 1 + nt > 1$,
   and

2. if $0 < x < 1$ say, $x = \frac{1}{1+t}$ for some $t > 0$, then $x^n < \frac{1}{1+nt} \to 0$ as $n \to \infty$.

Now suppose that $E = \mathbb{L}(V)$, where V is a Banach space and suppose that $A \in E$ has norm < 1.

From (6) of Theorem 3.3.1 we know that for any $S, T \in E$, $||S \circ T|| \leq ||S|| \cdot ||T||$ and hence

$$\lim_{n\to\infty} ||A^n|| \leq \lim_{n\to\infty} ||A||^n = 0$$

showing that $A^n \to 0$ as $n \to \infty$.

Now elementary algebra shows that

$$(1_V - A)(1_V + A + A^2 + \cdots + A^n) = (1_V - A^{n+1}) \to 1_V \text{ as } n \to \infty.$$

Hence, if $A \in E = \mathbb{L}(V)$ has operator norm < 1, then the "Neumann Series" $\sum_0^\infty A^n$ converges absolutely to an inverse of $(1_V - A)$.

Notice that the basic underlying theme is the "geometric series" which is high school mathematics. □

The simple-minded idea behind the geometric series can be useful in a very wide context. Here is an amusing instance.

Example 3.4.4b (A Fake Neumann Series).
Let $A : E \to F$ and $B : F \to E$ be continuous linear transformations between Banach spaces. Suppose that $(1_F - AB)$ has a continuous inverse $X \in \mathbb{L}(F)$. Then $Y = (BXA + 1_E) \in \mathbb{L}(E)$ is a continuous inverse of $(1_E - BA) : E \to E$.

Proof. Before performing the simple algebraic verification we give below the 'clue' explaining how this formula for $(1_E - BA)^{-1}$ arises.

Using formal algebra, *ignoring convergence related issues* we get:

$$\begin{aligned} (1_E - BA)^{-1} &= 1_E + BA + BABA + BABABA + \cdots \\ \text{and hence, } (1_E - BA)^{-1} - 1_E &= B(1_F + AB + ABAB + \cdots)A \\ &= B(1_F - AB)^{-1}A \\ \text{So finally we get } (1_E - BA) &= BXA + 1_E. \end{aligned}$$

We are hoping that this 'formal' argument, which would be valid if the power series converged will yield the inverse that we are looking for.

The terms on the right hand side of the last equation are continuous linear transformations.

For arbitrary $e \in E$, apply BXA to $\big(e - BA(e)\big)$. This yields:

$$\begin{aligned} BXA\big(e - BA(e)\big) &= B(1_F - AB)^{-1}\big(A(e) - ABA(e)\big) \\ &= BA(e). \end{aligned}$$

Hence $Y\big(e - BA(e)\big) = e$. This shows that Y is a left inverse of $(1_E - BA)$. A similar calculation shows that Y is a right inverse also. □

Example 3.4.4c (The Exponential Series).
Let $A \in E = \mathbb{L}(V)$ be any continuous linear transformation of a Banach space V into itself and consider the sequence $\{\frac{A^n}{n!}\}_{n \geq 0}$. An application of the ratio test immediately shows that $\sum_{n \geq 0} \frac{A^n}{n!}$ is absolutely convergent. The sum is called the *exponential* of the operator A and I will denote this sum by $\mathbf{e}^A$ or $\mathbf{e}(A)$. The operation $A \mapsto \mathbf{e}^A$ has many interesting properties which are of fundamental importance. Some of these will be explored in the exercises at the end of this and subsequent sections.

Example 3.4.4d. There is also an obvious analogue of the familiar *logarithmic series* for $\log(1_E + A)$ when $A \in \mathbb{L}(E)$ has norm < 1. (See the exercises at the end of this section.) However, this series is not particularly useful device for explicitly studying the 'inverse' of the exponential function.

The following explicit (though *ad hoc*) construction of the logarithm of an orthogonal matrix is quite useful and worth noting since it can be generalized to a far wider context.

Proposition 3.4.5. *Let* A *be a* (2×2) *orthogonal matrix of determinant* 1. *Then* A *is of the form:*

$$\mathrm{A} = \begin{bmatrix} \cos\theta & -\sin\theta \\ \sin\theta & \cos\theta \end{bmatrix}.$$

for some $0 \le \theta < 2\pi$.

The skew symmetric matrix $\Lambda(\mathrm{A}) = \begin{bmatrix} 0 & -\theta \\ \theta & 0 \end{bmatrix}$ *satisfies the relation* $\mathbf{e}^{\Lambda(\mathrm{A})} = \mathrm{A}$. □

Remark. To see why such a procedure works, first observe that the effect of the linear transformation $T(\mathrm{A}) : \mathbb{R}^2 \to \mathbb{R}^2$ is to rotate each vector in a counterclockwise direction through an angle θ. If we identify $\mathbb{R}^2$ with $\mathbb{C}$ by identifying the unit vector $\boldsymbol{e}_2$ with the complex number $\sqrt{-1}$ then the action of $T(\mathrm{A})$ is the same as multiplication by $e^{\sqrt{-1}\cdot\theta}$. The "logarithm" of this is, of course, $\sqrt{-1}\cdot\theta$.

Now the matrix $\left[\begin{smallmatrix} 0 & -1 \\ 1 & 0 \end{smallmatrix}\right]$ satisfies the relation $\mathrm{J}^2 = -1_{\mathbb{R}^2}$; that is J acts like $\sqrt{-1}$. So it is not surprising that $(\theta\cdot\mathrm{J})$ satisfies the relation, $\mathbf{e}^{\theta\mathrm{J}} = \mathrm{A}$.

So far I have not proved any theorems concerning infinite series: all proofs have been assigned to you as exercises. This was reasonable since the results stated so far are familiar to you in some form and their proofs are easily adapted from the proofs which apply when the Banach space happens to be $\mathbb{R}$ or $\mathbb{C}$. I will now establish a result which is almost certainly new to you. It is useful in its own right and you will see its real (and more powerful form) when you learn the theory of the Lebesgue integral.

Theorem 3.4.6 (The Dominated Convergence Theorem).
Let $\{\boldsymbol{x}_{\alpha,n}\}$ *be a family of elements of a Banach space,* E, *indexed by ordered pairs* $(\alpha, n) \in \mathbb{N}\times\mathbb{N}$. *Suppose that* $\{r_n\}$ *is a sequence of real numbers such that the following conditions obtain:*

DCT 1: $||\boldsymbol{x}_{\alpha n}|| \le r_n$ *for all* $\alpha \in \mathbb{N}$.

DCT 2: *For fixed* n, $\lim_{\alpha\to\infty} \boldsymbol{x}_{\alpha,n} = \xi_n$.

(*as* $\alpha \to \infty$) *to* $\xi_n \in E$.

DCT 3: *The series* $\sum_{n\ge 0} r_n$ *converges.*

Then the two series $\sum_{n\geq 0} \boldsymbol{x}_{\alpha,n}$ and $\sum_{n\geq 0} \xi_n$ converge and

$$\lim_{\alpha\to\infty} \left(\sum_{n\geq 0} \boldsymbol{x}_{\alpha,n} \right) = \sum_{\alpha\geq 0} \xi_n.$$

Proof. Let $\varepsilon > 0$ be an arbitrary positive number. Then by DCT 3, there is an $n_0 = n_0(\varepsilon)$ such that

$$\sum_{n=n_0}^{N} r_n < \varepsilon \quad \text{for all } N > n_0.$$

Then from DCT 1,

$$\left\| \sum_{n=n_0}^{N} \boldsymbol{x}_{\alpha,n} \right\| \leq \sum_{n=n_0}^{N} ||\boldsymbol{x}_{\alpha,n}|| \leq \sum_{n=n_0}^{N} r_n < \varepsilon$$

for all $N > n_0$. This shows that $\sum_{n\geq 0} \boldsymbol{x}_{\alpha,n}$ converges for all $\alpha \in \mathbb{N}$. Now observe that DCT 2 also implies that

$$||\xi_n|| = \lim_{\alpha\to\infty} ||\boldsymbol{x}_{\alpha,n}|| \leq r_n.$$

So we have, in fact, established the convergence of $\sum_{n=1}^{\infty} \xi_n$.

It remains to show that summing over n commutes with taking limits with respect to α. Note that the above inequalities imply that

$$\left\| \sum_{n=1}^{\infty} \boldsymbol{x}_{\alpha,n} - \sum_{n=1}^{\infty} \xi_n \right\| \leq \left\| \sum_{n=1}^{n_0-1} \boldsymbol{x}_{\alpha,n} - \sum_{n=1}^{n_0-1} \xi_n \right\| + 2\varepsilon.$$

On the right side we have a difference of two *finite* sums inside the norm and hence we can interchange the sum and taking limits over α. By taking superior limits of both sides we get:

$$\overline{\lim_{\alpha\to\infty}} \left\| \sum_{n=1}^{\infty} \boldsymbol{x}_{\alpha,n} - \sum_{n=1}^{\infty} \xi_n \right\| \leq 2\varepsilon.$$

Note that it is necessary to take superior limit (rather than limit) since we do not know that the left side, in fact, converges. This proves the result since the left side is nonnegative and ε arbitrary. □

Here is an important application of the Dominated Convergence Theorem.

Proposition 3.4.7. *Let E be a Banach space and $\{A_n \in \mathbb{L}(E)\}_{n\in\mathbb{N}}$ a sequence of continuous linear transformations such that $\lim_{n\to\infty} A_n = A$.*

If 1_E denotes the identity map of E, then

$$\lim_{n\to\infty}\left(1_E + \frac{A_n}{n}\right)^n = \mathbf{e}^A.$$

Proof. We rewrite the binomial expansion above as follows:

$$\left(1_E + \frac{A_n}{n}\right)^n = \sum_{\alpha=0}^{n} \frac{n!}{\alpha!(n-\alpha)!}\left(\frac{A_n}{n}\right)^\alpha$$

$$= \sum_{\alpha=0}^{n}\left(1-\frac{1}{n}\right)\left(1-\frac{2}{n}\right)\cdots\left(1-\frac{\alpha-1}{n}\right)\frac{A_n^\alpha}{\alpha!}.$$

Now define

$$A_{\alpha,n} = \begin{cases}\left(1-\frac{1}{n}\right)\left(1-\frac{2}{n}\right)\cdots\left(1-\frac{\alpha-1}{n}\right)\frac{A_n^\alpha}{\alpha!} & \text{if } \alpha \le n\\ 0 & \text{if } \alpha > n.\end{cases}$$

Since the A_n's form a convergent sequence they are bounded; choose M so that $||A_n|| \le M$ for all n. Then, the following:

1. $||A_{\alpha,n}|| \le \dfrac{M^\alpha}{\alpha!}$ for all $(\alpha, n) \in \mathbb{N}\times\mathbb{N}$;

2. $\lim_{n\to\infty} A_{\alpha,n} = \dfrac{A^\alpha}{\alpha!}$ for all $\alpha \in \mathbb{N}$.

3. $\sum_{\alpha\ge 0} \dfrac{M^\alpha}{\alpha!} = e^M$ is a convergent series.

The result now follows from the definition of the exponential series and the DCT. □

Further applications of the DCT will be found in the exercises that follow.

Perhaps it is worth mentioning that these exercises require a little more mathematical virtuosity than the ones that have appeared so far in this book. But they are extremely important results. In particular, part (d) is the crucial step in constructing `Lie Groups` corresponding to `Lie subalgebras`. Any further comments would take us too far afield.

Exercises for § 3.4

3.4.1 On the Exponentials of Linear Transformations

a. Establish that for any Banach space V and any two continuous linear transformations, $A, B \in \mathbb{L}(V) \doteq E$ which commute (i. e., $AB = BA$) we have:

$$\mathbf{e}^{(A+B)} = \mathbf{e}^A \mathbf{e}^B$$

Deduce that for every $A \in E$, $\mathbf{e}^A$ has an inverse in E, that is, show that there is a $A' \in E$ such that $A' \cdot \mathbf{e}^A = \mathbf{e}^A \cdot A' = 1_V$.

b. Show that for any $A, B \in \mathbb{L}(E)$ the following inequalities hold:

$$\begin{aligned} ||\mathbf{e}^{(A+B)} - \mathbf{e}^A|| &\leq ||A - B|| \cdot \sum_{n=0}^{\infty} \left(\frac{1}{n!} \sum_{r=0}^{n-1} ||A||^{n-(r+1)} ||B||^r \right) \\ &\leq |(||A|| - ||B||)| \cdot e^{(||A||+||B||)}. \end{aligned}$$

Hint : Write $(A^n - B^n)$ in the form

$$\sum_{r=0}^{n-1} (A^{n-r} B^r - A^{n-(r+1)} B^{r+1}) = \sum_{r=0}^{n-1} A^{n-(r+1)} (A - B) B^r.$$

c. Show that for any $A \in E$,

$$\lim_{n \to \infty} n(\mathbf{e}^{A/n} - 1)^n = A.$$

d. Use this and the dominated convergence theorem to show that for any two $A, B \in \mathbb{L}(V)$ we have the identities:

$$\begin{aligned} \lim_{n \to \infty} (\mathbf{e}^{A/n} \mathbf{e}^{B/n})^n &= \mathbf{e}^{(A+B)} \text{ and} \\ \lim_{n \to \infty} \left(\mathbf{e}^{A/\sqrt{n}} \mathbf{e}^{B/\sqrt{n}} \mathbf{e}^{-A/\sqrt{n}} \mathbf{e}^{-B/\sqrt{n}} \right)^n &= \mathbf{e}^{[A,B]}, \end{aligned}$$

where $[A, B]$ denotes the *commutator* of A and B, that is, $[A, B] = AB - BA$.

3.5 Connectedness and Density

In this section I will discuss two notions which have little in common with one another except that arguments using these two ideas will be used to conclude that, under certain circumstances, two maps defined on a metric space are, in fact, equal to one another.

Let X be a metric space. X is said to be *connected* if, whenever A and B are disjoint open sets of X whose union is X one of them is X itself and the other is the empty set. There is a closely related notion which is more geometrically motivated. A metric space is said to be *path-connected* if given any two point x_0 and x_1 in X there is a map $\lambda : \mathbf{I} = [0,1] \to X$ such that $\lambda(0) = x_0$ and $\lambda(1) = x_1$. I will refer to such a map λ as a *path* in X from x_0 to x_1.

EXAMPLES OF CONNECTED SETS

Example 3.5.1a (Intervals $\subset \mathbb{R}$ are path-connected).
Any interval $I = \langle \alpha, \beta \rangle \subset \mathbb{R}$ is path-connected. (Here 'chevron' are used to denote any of '[', '(', ')' or ']'. and α (resp. β) can be $-\infty$ (resp. ∞). For if $x < y$ are points of I then $\lambda(t) = \big(x + (y-x)t\big)$ is a path from x to y.

Example 3.5.1b (Intervals are connected).
Any interval $I \subset \mathbb{R}$ is connected. The word 'interval' has the same meaning as in the previous example.

Proof. For suppose $I = A \cup B$ where A and B are disjoint, nonempty open sets whose union is $\mathbf{I} = \langle \alpha, \beta \rangle$ say. Let $a \in A$ and $b \in B$ with $a < b$, say. Then the set $\{x \in A : x < b\}$ is obviously bounded above. Let ξ be its supremum. Then $\xi \notin A$ since A being open there would be a $\delta > 0$ such that $(\xi - \delta, \xi + \delta) \subset A$ and $\xi + \delta < b$. This contradicts the definition of ξ. Now suppose that $\xi \in B$. Then for all $\delta > 0$ there will be a point $a' \in A$ such that $\xi - \delta < a'$ which would imply that B is not open. This contradiction establishes that $\langle \alpha, \beta \rangle$ is connected. $\square$

If X is a metric space and $A \subset X$ a subset of X, then I will say that A is connected (resp. path-connected) if A regarded as a metric space is connected (resp. path-connected.) Note that *A is connected iff it is not contained in the disjoint union of two open sets of X unless one of them contains A.*

Here are some easily proved observations on the notions of connectedness and path-connectedness.

Remarks 3.5.2 (On connected and path-connected spaces)

Remark 3.5.2a (An alternative definition of connectedness).
In the definition of connectedness we can replace the phrase "open set" by "closed set".

The following is an alternative definition of connectedness:

A metric space X is connected if the only sets that are both open and closed *are X and $\emptyset$.* □

Remark 3.5.2b (Connected sets have connected closures).
Observe that if A is a connected subset of a space X then its closure, $\bar{A}$, is also connected. Recall that $\bar{A}$ is the smallest closed set containing A and so if $\bar{A} = K_1 \cup K_2$ where K_i are disjoint closed subsets of X the connectedness of A will imply that one of the K's contains A and hence $\bar{A}$. □

Remark 3.5.2c. Note that the converse of the above is not true: that is, $\bar{A}$ can be connected even if, A is not connected. The simplest example is furnished by the subset $\mathbb{Q} \subset \mathbb{R}$. which regarded as a subspace of $\mathbb{R}$ is not connected: for example

$$\mathbb{Q} = \left(\mathbb{Q} \cap (-\infty, \sqrt{2}\right) \cup \left(\mathbb{Q} \cap (\sqrt{2}, \infty)\right)$$

expresses $\mathbb{Q}$ as a disjoint union of open subsets. On the other hand, the closure of $\mathbb{Q}$ is all of $\mathbb{R}$ which we saw is a connected space.

The properties of connected and path-connected sets that we will use later in the book are described in the next result.

Theorem 3.5.3.
1. *Let X be a connected metric space and $f : X \to Y$ any map. Then* Image $f = f(X)$ *is a connected subset of* Y.
2. *A path-connected space is also a connected space.*

Proof. Suppose Image f is not connected. Then there are disjoint open subsets A and B of Y such that Image $f \subset A \cup B$. Since f is continuous $U = f^{-1}(A)$ and $V = f^{-1}(B)$ are open and disjoint and $X = U \cup V$. Hence one of them must be all of X and hence either A or B contains Image f. The result follows from the remark made immediately after the definition of connected sets.

Let X be a path-connected space and suppose $X = A \cup B$ where A and B are nonempty and disjoint open subsets of X. Let $a \in A$ and $b \in B$ and λ a path in X from a to b. Then $\lambda^{-1}(A)$ and $\lambda^{-1}(B)$ are nonempty, disjoint open subset of **I** whose union is **I**, contradicting the fact that **I** is connected. □

Remark 3.5.4. The first part and Example 3.5.1b implies the Intermediate Value Theorem of elementary analysis.

The following simple result will be used repeatedly.

Proposition 3.5.5. *Let $f, g : X \to Y$ be maps between metric spaces and suppose that X is connected. If $C_{f,g} = \{x \in X : f(x) = g(x)\}$, the set on which f and g coincide, is open and nonempty then $f(x) = g(x)$ for all $x \in X$.*

Proof. If $p \notin C_{f,g}$ let $d_Y\big(f(p), g(p)\big) = 3\varepsilon$. Since f, g are maps, we can find $\delta_f, \delta_g > 0$ such that

$$d_Y\big(f(p), f(x)\big) < \varepsilon \text{ if } d_X(p, x) < \delta_f$$
$$\text{and } d_Y\big(g(p), g(x)\big) < \varepsilon \text{ if } d_X(p, x) < \delta_g.$$

If $\delta = \min(\delta_f, \delta_g)$ and $x \in \mathbb{B}_\delta(p)$ then,

$$\begin{aligned} 3\varepsilon &= d_Y(f(p), g(p)) \le d_Y(f(p), f(x)) + d_Y(f(x), g(x)) + d_Y(g(x), g(p)) \\ &\le \varepsilon + d_Y(f(x), g(x)) + \varepsilon. \end{aligned}$$

So $d_Y(f(x), g(x)) > \varepsilon$. Hence $\mathbb{B}_\delta(p)$ is contained in the complement of $C_{f,g}$. So $C_{f,g}$ is closed.

Since $C_{f,g}$ is a nonempty open set (by hypothesis) which is also closed the connectedness of X implies $C_{f,g} = X$. □

I now introduce the notion of dense sets. If A is a subset of a space X, then we say that A is *dense in* X or that A is a *dense subset* of X if, given any point $x \in X$ and any $\varepsilon > 0$ there is a point $a \in A$ such that $d_X(x, a) < \varepsilon$.

The most obvious example is of course the set of rational numbers $\mathbb{Q} \subset \mathbb{R}$ which is a dense subset of $\mathbb{R}$. The following properties of dense sets are being listed without proof. The reader should verify them as exercises.

Proposition 3.5.6 (Properties of dense sets).

1. *A is a dense subset of X* iff *the closure of A is X.*

2. *If A is dense in X and $A \subset A'$ then A' is also dense in X.*

3. *If $f, g : X \to Y$ are maps between spaces, A a dense subset of X and $f \mid A = g \mid A$ then $f = g$.* □

Remarks 3.5.7 (On density)

Remark 3.5.7a. Notice that by virtue of (1) in Proposition 3.5.6, if X is a subset of a *finite-dimensional* Banach space and $A \subset X$, with the induced metric from X, then the question of A being dense in X is independent of the norm on the linear space.

Remark 3.5.7b. It is (3) of Proposition 3.5.6 that will be particularly useful in the rest of this section.

We now come to the last major result to be proved in this section. This is a result which, in the older literature, is sometimes referred to by the rather grand but confusing title of "The Principle of Irrelevance of Algebraic Inequalities": I will simply use PIAI to refer to this result.

Although the PIAI is a straightforward generalization of the elementary fact that a polynomial, $p(x) = a_0x^n + a_1x^{n-1} \cdots + a_n$, in one variable, can have only a finite number of roots, (unless, of course, all the a_i's are zero,) it is an extremely powerful tool and perhaps the grandiose title by which it was earlier known is justified.

At this point I urge you to recall from Chapter 1 the definition of polynomial functions. (See page 31).

Theorem 3.5.8. *Let E be a finite-dimensional normed linear space over $\mathbb{R}$ and suppose $p : E \to \mathbb{R}$ is a polynomial. Then the set $N_p = \{t \in E : p(t) \neq 0\}$ is dense in E.*

Proof. We can (and will) identify E with $\mathbb{R}^n$ and work with the ℓ_∞-norm on $\mathbb{R}^n$.

We proceed by induction on n. If $n = 1$ then elementary algebra tells us that a polynomial can have only finitely many roots and hence N_p contains all of $\mathbb{R}$ except perhaps a finite number of points. It is clear that N_p is dense.

Now suppose we have proved the theorem for all normed linear spaces of dimension $\leq n$. Let $p : \mathbb{R}^{n+1} \to \mathbb{R}$ be a polynomial map. I will denote the coordinates in $\mathbb{R}^{n+1}$ by $(t_0, \ldots, t_n)$; the reason for this change will become clear in a moment. To show that N_p is dense in $\mathbb{R}^{n+1}$, it will suffice to show that for any $x = (x_0, x_1, \ldots, x_n) \in \mathbb{R}^{n+1}$ and $\varepsilon > 0$ we can find a point $y \in N_p$ such that $||x - y|| < \varepsilon$.

If $p(x_0, x_1, \ldots, x_n) \neq 0$ then there is nothing more to prove. So we suppose that $p(x_0, x_1, \ldots, x_n) = 0$. Now, consider the function $p' : \mathbb{R}^n \to \mathbb{R}$ defined by $p'(t_1, \ldots, t_n) = p(x_0, t_1, \ldots, t_n)$. It is easy to check that it is a polynomial function. Since it has a zero at the point $x' = (x_1, \ldots, x_n)$, by the induction hypothesis, there is an $y' = (y_1, \ldots, y_n)$ such that $||x' - y'|| < \varepsilon$ and $p'(y') \neq 0$. If $y = (x_0, y_1, \ldots, y_n)$ then $||x - y|| < \varepsilon$ and $p(y) = p'(y') \neq 0$, as required. □

Remark 3.5.9. We have seen (Proposition 3.1.20) that any closed set in $\mathbb{R}^n$ is the set of zeros of a continuous function $\mathbb{R}^n \to \mathbb{R}$, thus given any open set U, there is a continuous function $f_U : \mathbb{R}^n \to \mathbb{R}$ such that $U = \{x \in \mathbb{R}^n : f(x) \neq 0\}$. Polynomials are very special kinds of maps: the set on which a polynomial is nonzero must be a dense subset of $\mathbb{R}^n$.

Here are some interesting examples of the use of the previous theorem.

1. Consider the linear space $\mathrm{M}_n(\mathbb{R})$, of $(n \times n)$ matrices with entries from $\mathbb{R}$. The determinant function $\det : \mathrm{M}_n(\mathbb{R}) \to \mathbb{R}$ is clearly a polynomial function. Hence $\mathrm{GL}_n(\mathbb{R}) = \{\mathrm{A} : \det(\mathrm{A}) \neq 0\}$ is a dense subset of $\mathrm{M}_n(\mathbb{R})$.

2. If you now review the discussion about symmetric polynomials in the last section of Chapter 1 you will see that the previous result in fact implies that the semisimple matrices are dense in $\mathrm{M}_n(\mathbb{C})$.

Now the "Principle" itself. It is, in fact, a trivial corollary of the previous theorem.

Theorem 3.5.10 (PIAI).
Let E be a finite-dimensional normed linear space over $\mathbb{R}$. Let $\phi, \psi : E \to X$ be two maps of spaces and $g : E \to \mathbb{R}$ a polynomial function. If $\phi(x) = \psi(x)$ for all $x \in E$ such that $g(x) \neq 0$ then $\phi = \psi$.

Proof. The set of points at which ϕ and ψ agree is a closed subset of E. On the other hand, by hypothesis, it contains the set $N_g = \{x \in E : g(x) \neq 0\}$, which by the previous result is dense in E and hence its closure is E. It follows that ϕ and ψ agree at all points of E. □

Actually I have cheated a little bit: the PIAI is a purely algebraic result (no metric or norm need be invoked.)

REMARKS 3.5.10 (THE ALGEBRAIC APPROACH TO PIAI)

Remark 3.5.10a (The algebraic formulation of the PIAI).
Let E be a finite-dimensional linear space over an arbitrary infinite *field* $\mathbf{k}$, *$\phi : E \to \mathbf{k}$ a polynomial function and $f : E \to F$ a polynomial function from E taking values in a linear space over* $\mathbf{k}$. *If $f(p) = 0$ for all $p \in E$ such that $\phi(p) \neq 0$, then f is identically zero.*

Remark 3.5.10b. I haven't defined polynomial functions over arbitrary fields, but I trust you will understand what is being meant: these polynomials are finite **k**-linear combinations of functions of the form $(t_1, \ldots, t_m) \mapsto t_1^{a_1} \cdots t_m^{a_m}$ where a_i are nonnegative integers.

We now move on to the applications of the PIAI. Let $E = \mathrm{M}_n(\mathbb{R})$ be the normed linear space of $(n \times n)$ matrices with entries from $\mathbb{R}$ or $\mathbb{C}$. (Recall, that the choice of norm is immaterial in deciding questions concerning density in a finite-dimensional normed linear space.) We had indicated at the end of Chapter 1 that there is a polynomial $\Delta : E \to \mathbb{R}$ such that if $\Delta(\mathrm{A}) \neq 0$ then the matrix A possesses n distinct eigenvalues. This implies that

Proposition 3.5.11. *For all $n \geq 0$, the set of semisimple matrices is dense in $M_n(\mathbb{R})$.*

Recall, that *a semisimple matrix is similar to a diagonal matrix.*

Theorem 3.5.12. *For any matrix* $\mathrm{A} \in \mathrm{M}_n(\mathbb{R})$, $\det(\mathbf{e}^{\mathrm{A}}) = e^{(\mathrm{Trace\,A})}$.

Proof. We first observe, that both sides are continuous functions $E \to \mathbb{R}$. Next, note that if $\mathrm{B} = \mathrm{GAG}^{-1}$ where G is a nonsingular matrix, then $\mathrm{B}^m = \mathrm{GA}^m\mathrm{G}^{-1}$ for all $m \geq 0$ and hence $\mathbf{e}^{\mathrm{B}} = \mathrm{G}\mathbf{e}^{\mathrm{A}}\mathrm{G}^{-1}$. Hence the left hand side of the above equation does not change its value when A is replaced by a similar matrix. Now the same is true of the right hand side since the trace of a matrix is invariant under similarity.

If A is a diagonal matrix the equality above is obviously valid. Hence the two continuous functions agree on the set of semisimple matrices. The result follows from Proposition 3.5.11. □

As a second application of the PIAI we compute the determinant of the Kronecker product of two square matrices or equivalently the determinant of the tensor product of two endomorphisms of finite-dimensional linear spaces.

Proposition 3.5.13 (Determinants of Kronecker products).
Let $\mathrm{A} \in \mathrm{M}_p(\mathbb{R})$ *and* $\mathrm{B} \in \mathrm{M}_q(\mathbb{R})$. *Then*

$$\det(\mathrm{A} \otimes \mathrm{B}) = (\det \mathrm{A})^q \cdot (\det \mathrm{B})^p.$$

Proof. Let $A, A' : \mathbb{R}^p \to \mathbb{R}^p$ and $B, B' : \mathbb{R}^q \to \mathbb{R}^q$ be the linear maps corresponding to matrices A, A′ and B, B′ respectively. Recall that with appropriate choice of bases of the underlying linear spaces $\mathbb{R}^p, \mathbb{R}^q$ and $\mathbb{R}^p \otimes \mathbb{R}^q \approxeq \mathbb{R}^{pq}$ the matrix of $A \otimes B$ is the Kronecker product

$$\begin{bmatrix} a_{11}\mathrm{B} & a_{12}\mathrm{B} & \cdots & . & a_{1p}\mathrm{B} \\ \vdots & & a_{ij}\mathrm{B} & & \vdots \\ a_{p1}\mathrm{B} & a_{p2}\mathrm{B} & \cdots & & a_{pp}\mathrm{B} \end{bmatrix}$$

from which, using the rules of multiplying partitioned matrices and the definition of the linear transformation

$$A \otimes B : \mathbb{R}^p \otimes \mathbb{R}^q \longrightarrow \mathbb{R}^p \otimes \mathbb{R}^q$$

we get the following relations

$$\begin{aligned} \mathrm{M}(A' \circ A) \otimes \mathrm{B} &= (\mathrm{A}' \otimes 1_q) \odot (\mathrm{A} \otimes \mathrm{B}), \\ \mathrm{M}(A \circ A') \otimes \mathrm{B} &= (\mathrm{A} \otimes \mathrm{B}) \odot (\mathrm{A}' \otimes \mathrm{I}_q), \\ \mathrm{M}\Big(A' \otimes (B \circ B')\Big) &= (\mathrm{A} \otimes \mathrm{B}) \odot (\mathrm{I}_p \otimes B'), \\ \mathrm{M}\Big(A \otimes (B' \circ B)\Big) &= (\mathrm{I}_p \otimes \mathrm{B}') \odot (\mathrm{A} \otimes \mathrm{B}). \end{aligned}$$

where on the right side we have used $\odot$ to represent matrix multiplication.

An immediate corollary is that if G is a nonsingular matrix then $(\mathrm{G} \otimes \mathrm{I}_q)^{-1} = (\mathrm{G}^{-1} \otimes I_q)$. It now follows, that if $\mathrm{A}' = \mathrm{GAG}^{-1}$ then for any B we have $\mathrm{A}' \otimes \mathrm{B} = (\mathrm{G} \otimes \mathrm{I}_q)(\mathrm{A} \otimes \mathrm{B})(\mathrm{G}^{-1} \otimes \mathrm{I}_q)$. So both sides of the equality to be established are unaltered if we replace A by a similar matrix. Now for each fixed B, $\mathrm{A} \mapsto \det(\mathrm{A} \otimes \mathrm{B})$ is a continuous function from $\mathrm{M}_p(\mathbb{R})$ to $\mathbb{R}$. From the representation of $A \otimes B$ as the Kronecker product of two matrices it is obvious that the equality we seek to prove is indeed true when A is a diagonal matrix and hence when A is semisimple. Such matrices are dense and hence the result follows in the general situation by appealing to the PIAI. □

Exercises for §3.5

3.5.1 Connectedness and path-connectedness

a. Let

$$\left.\begin{aligned} X_1 &= \{(0, x_2) \in \mathbb{R}^2 \,:\, -1 \leq x_2 \leq 1\} \\ X_2 &= \{(x_1, x_2) \in \mathbb{R}^2 \,:\, 0 < x_1 < 1, x_2 = \sin \tfrac{1}{x_1} \end{aligned}\right\}.$$

Prove that $S = (X_1 \cup X_2) \subset \mathbb{R}^2$, sometimes called the "topologists' sine curve", is a connected metric space, which is not path-connected.

(HINT: Show that the closure of X_2 contains X_1.)

b. A metric space X is said to be *locally path-connected* if for each $x \in X$ there is an $\epsilon = \epsilon(x) > 0$ such that for each $\delta \leq \epsilon$, the ball $\mathbb{B}_\delta(x)$ is path-connected.

Show that X is path-connected iff it is locally path-connected and connected.

3.5.2 A CURIOUS RESULT

Let M be a $(2n \times 2n)$ matrix with a block decomposition of the form:

$$\mathrm{M} = \begin{bmatrix} \mathrm{A} & \mathrm{B} \\ \mathrm{C} & \mathrm{D} \end{bmatrix},$$

where $\mathrm{A}, \mathrm{B}, \mathrm{C}, \mathrm{D}$ are $(n \times n)$ matrices.

Suppose that (i) A and C commute, i. e. $\mathrm{AC} - \mathrm{CA} = 0$, and (ii) A is nonsingular.

(a) Show that $\det(\mathrm{M}) = \det(\mathrm{AD} - \mathrm{CB})$. . . . . (∗)

(HINT: First suppose that A is nonsingular. Left multiply by $\left[\begin{smallmatrix} \mathrm{A}^{-1} & 0 \\ -\mathrm{C} & \mathrm{A} \end{smallmatrix}\right]$).

(b) Use the PIAI to show that the result continues to hold if condition 2 (the *inequality*, $\det \mathrm{A} \neq 0$,) is omitted.

(c) Show by a counterexample that condition 1 may not be omitted without invalidating (∗).

3.5.3 THE SPECTRA OF AB AND BA

Show that for any two matrices $\mathrm{A}, \mathrm{B} \in \mathrm{M}_n(\mathbb{C})$, $\Sigma(AB) = \Sigma(BA)$.

Since the spectrum is completely determined by the coefficients of the characteristic polynomial, it suffices to show that for all $\mathrm{A}, \mathrm{B} \in \mathrm{M}_n(\mathbb{C})$, $c_i(\mathrm{AB}) = c_i(\mathrm{BA})$, where $c_k(\mathrm{M})$ is the coefficient of T^{n-k} in the characteristic polynomial of an $(n \times n)$ matrix, M.

If A is nonsingular then $\mathrm{A}^{-1}(\mathrm{AB})\mathrm{A} = \mathrm{BA}$ and therefore AB and BA are similar.

Using this and the PIAI conclude that the spectra of AB and BA are always the same.

3.5.4 The Cayley–Hamilton Theorem

Given $A \in M_n(\mathbb{k})$ the characteristic polynomial of A is the polynomial

$$\chi_A(T) \doteq \det(TI_n - A) = T^n + \sum_{k=1}^{n} c_k(A)T^{n-k}.$$

Using the PIAI and the fact that semisimple matrices are dense, show that every matrix satisfies its own characteristic polynomial, that is

$$A^n + \sum_{k=1}^{n} c_k(A)A^{n-k} = 0.$$

Remark. In fact, the Cayley–Hamilton Theorem holds for matrices over an *arbitrary* field $\mathbf{k}$. If $\mathbf{k}$ is infinite we use the algebraic PIAI and if $\mathbf{k}$ is finite we can embed it in an infinite field, like the field of rational functions, $\mathbf{k}(t)$. Then PIAI implies the result for matrices over $\mathbf{k}(t)$ and hence the Theorem holds, in particular, for matrices over $\mathbf{k}$.

The next exercise introduces a new set of "functors" of matrices. It is known variously as the Grassmann powers, higher compounds or exterior powers of matrices. I don't really discuss this procedure but simply *define* the operation. The end result is a celebrated though (no doubt, for you, surprising) method of computing determinants via traces!

3.5.5 Exterior powers and the determinant of $I - A$

Let A be a square matrix of order n. For each $1 \leq k \leq n-1$, choose an ordering for the set K of k-tuples $\{1 \leq i_1 < i_2 \cdots < i_k \leq n\}$. The kth exterior power of A is the square matrix of order ${}^nC_k$ whose $(I, J)^{\text{th}}$ element for $I, J \in K$ is the determinant of the $k \times k$ "minor" obtained by choosing the $i_1, \ldots, i_k$ rows and $j_1, \ldots, j_k$ columns of A. This matrix is denoted by $A^{[k]}$ or $\lambda^k(A)$.

a. Show that if A has rank r then $\lambda^r(A)$ has rank 1.

b. Use the PIAI to prove that

$$\det(I - A) = 1 + \sum_{r=1}^{n}(-1)^r \operatorname{Trace} \lambda^r(A).$$

4
Calculus

I have now done all the spadework necessary for building up a form of the Differential Calculus which will be applicable to functions between open subsets of Banach spaces.

In Section 1, I will introduce the notion of differentiable functions and their derivatives. The derivative of some interesting and important functions are computed: these will convey to you the power and usefulness of this new *avatar* of the derivative.

Although a few "theorems" are proved, these are really the analogues of the "Rules of Differentiation" like the formula which describes the derivative of a product of two differentiable functions, say u and v, in terms of the derivatives u' and v', (that is, the formula: $(u \cdot v)' = u' \cdot v + u \cdot v'$) and the chain rule for differentiating the composition of differentiable functions. I also introduce the so-called `partial derivatives` of a function defined on open subsets of $\mathbb{R}^n$ and relate them to the derivative of a differentiable function in our sense.

At the end of Section 1, I present a rather crude version of the integration of functions defined on closed, bounded intervals of $\mathbb{R}$ and taking values in a Banach space. The treatment is somewhat sketchy; but since what is being done is really a *simplified* version of the Riemann integral, I do not think you will find this troublesome.

Section 2 is devoted to the Mean-Value Theorem. Contrary to standard usage the *truly fundamental* result of Calculus is this Mean-Value

Theorem. It also happens to be the only result which, in the present formulation of the subject, is somewhat different from the result you learnt in your first course. (The difference, though, is superficial as will be pointed out.) Several applications of the Mean-Value Theorem are presented; the main one yields a necessary and sufficient condition, in terms of the behaviour of the partial derivatives of the function, for a function to be `continuously differentiable` in our sense.

In Section 3, I discuss higher derivatives and prove several versions of Taylor's formula. Although, most of this material will not be used in the later chapters it is worth the trouble to master this section, for only then will you realize that the approach I have adopted does indeed yield a version of the Differential Calculus in which *all* the classical theorems have adequate generalizations. The traditional approach to "functions of several variables" might lead one to believe that only maps defined on open subsets of $\mathbb{R}$ possess higher derivatives or that for $\mathbb{R}$-*valued* mappings defined on subsets of $\mathbb{R}^n$ one can only define the first and second derivatives but not go any further!

Section 4 and the short subsection within, may be omitted on a first reading since the material is a little more abstract. It is used only in the §5.2, which is devoted to a discussion of Ordinary Differential Equations.

Although this book is primarily an exposition of the Differential Calculus of "Functions of several variables", the methods used apply to functions defined on open sets of infinite-dimensional Banach spaces without any modification. It should be noted however that the proofs of theorems that have been given here, would not be simplified by even a single line by assuming that the Banach spaces involved are finite-dimensional. So while you are welcome to make the assumption of finite-dimensionality, if you so wish, it will not make things easier. In fact, there is always the temptation, when dealing with linear transformations of finite-dimensional spaces, to use matrices. Yielding to such temptations in the context of Calculus makes life extremely difficult and perhaps it is better to remove this possibility by ignoring altogether the dimension of the spaces involved.

One final point needs to be emphasized. *All the linear spaces in this chapter will be linear spaces over* $\mathbb{R}$. Let me elaborate on this.

I had mentioned in the preamble to Chapter 2 that the derivative of a differentiable function is a linear transformation between normed linear spaces. My disregard of complex scalars amounts to requiring that the derivative (even between $\mathbb{C}$-linear spaces) to be only an $\mathbb{R}$-linear transformation.

There is a theory of differentiable functions for functions defined on $\mathbb{C}$-linear spaces where one explicitly requires the derivative to be a $\mathbb{C}$-linear transformation. But this subject, variously known as the theory of `analytic` or `holomorphic` functions, is much deeper and has little resemblance with Differential Calculus.

4.1 Differentiation and Derivatives

At the introductory level, the notion of the derivative of a real-valued function of a real variable is motivated by discussing the geometric problem of trying to draw a tangent to the graph of a function. We discuss this anew to motivate our definition of the derivative of maps between open sets of Banach spaces.

So, suppose that $f : (a, b) \to \mathbb{R}$ is a function. Then given a point $\xi \in (a, b)$ we wish to find a straight line in the plane which passes through the point $\big(\xi, f(\xi)\big)$ and which "just touches" the curve given by the graph of f. As you know an answer to this problem leads one to defining the derivative as a limit of the ratio $\dfrac{f(\xi+\delta)-f(\xi)}{\delta}$ as δ tends to 0.

The well-known geometric interpretation of this is that the derivative of the function f at ξ is the tangent of the angle made by this tangent line with the x-axis. More precisely, the equation of the tangent at $(\xi, f(\xi))$ is

$$y(x) = f(\xi) + (x - \xi)f'(\xi).$$

The tangent line of the graph of f at ξ, if it exists, is the best straight-line approximation to the graph of f at the point $(\xi, f(\xi))$. In other words:

> the function $x \mapsto f(\xi) + (x - \xi)f'(\xi)$ is the *best linear approximation* to f near the point ξ.

Now a real number, such as the derivative of f at the point ξ, may also be thought of as a linear transformation of $\mathbb{R}$ into itself, since

$$L(\mathbb{R}, \mathbb{R}) \approxeq \mathbb{R} \text{ via } \alpha \mapsto \alpha(1).$$

This trivial observation, that a real number is also a linear transformation of $\mathbb{R}$ into itself, is the guiding principle in obtaining an appropriately general notion of the derivative of a function.

I will begin by introducing some jargon and notation. Let E, F be Banach spaces and $U \subset E$ an open set containing the origin. A function $f : U \to F$ is said to be *of order o* if given any $\varepsilon > 0$ there is a $r = r(\varepsilon)$ such that if $h \in U$ and $||h|| \leq r$ then $||f(h)|| \leq \varepsilon||h||$. I shall write f is $o(\boldsymbol{h})$ or $f = o(h)$ to describe this situation. One says "f is of order "little $o(h)$" or simply f is "little o".

Now let x^0 be any point in a Banach space E. Suppose f, g are functions defined on subsets of E which contain an open ball centred at x^0. If f, g take values in a Banach space F, f and g are said to be *tangent* to one another at x^0 if the function $h \mapsto f(x^0 + h) - g(x^0 + h)$, which is defined for $||\boldsymbol{h}||$ sufficiently small, is $o(h)$. This means that for all $\varepsilon > 0$ we can find a $r > 0$ such that if $||\boldsymbol{h}|| < r$, then

$$||f(x^0 + h) - g(x^0 + h)|| \leq \varepsilon||\boldsymbol{h}||.$$

The crucial ingredient I require in order to define the notion of the derivative of a function is the following simple lemma.

Lemma 4.1.1. *Let $f : U \to F$ be a map of an open set of a Banach space E into a Banach space F and suppose that $x^0 \in U$. Then there can be at most one continuous linear transformation $\Lambda \in \mathbb{L}(E, F)$ such that f and the function $x \mapsto f(x^0) + \Lambda(\boldsymbol{x} - \boldsymbol{x}^0)$ are tangent to one another at x^0.*

Proof. Suppose that $\Lambda_1, \Lambda_2 \in \mathbb{L}(E, F)$ are two continuous linear transformations such that $f(x^0) + \Lambda_\imath(\boldsymbol{x} - \boldsymbol{x}^0)$ and f are tangent at x^0. Then given any $\varepsilon > 0$ there is an $r > 0$ such that for all $\boldsymbol{h}$ having norm $< r$ we have the two inequalities:

$$||f(x^0 + h) - f(x^0) - \Lambda_1(\boldsymbol{h})||_F \leq \varepsilon||\boldsymbol{h}||_E,$$
$$||f(x^0 + h) - f(x^0) - \Lambda_2(\boldsymbol{h})||_F \leq \varepsilon||\boldsymbol{h}||_E.$$

This shows that $||\Lambda_1(\boldsymbol{h}) - \Lambda_2(\boldsymbol{h})||_F \leq 2\varepsilon||\boldsymbol{h}||_E$, for all $\boldsymbol{h} \in \mathbb{B}_r(0_E)$. The homogeneity of the norm and linearity of $(\Lambda_1 - \Lambda_2)$ imply that this last inequality will hold for arbitrary $\boldsymbol{h} \in E$ and hence $||(\Lambda_1 - \Lambda_2)|| = 0$.. So $\Lambda_1 = \Lambda_2$. □

Remark 4.1.2 (On notation).
Observe, that the arguments of linear transformations or the norm function are being written using "vectorial font" thus: $\boldsymbol{h}, \boldsymbol{x}, \boldsymbol{y} \ldots$. For exam-

ple, in the expression:

$$||f(x^0 + h) - f(x^0) - \Lambda(\boldsymbol{h})|| \leq \varepsilon ||\boldsymbol{h}||$$

the arguments of f in spite of being elements of a linear space are not being treated as vectors: $x^0 + h$ is simply a point in the domain of f but the argument of the Λ's have to be a vector since the Λ's are linear transformations.

A *continuous linear transformation* $\Lambda \in \mathbb{L}(E, F)$ is said to be the *derivative* of f at x^0 if

$$f(x^0 + h) - f(x^0) - \Lambda(\boldsymbol{h}) = o(h)$$

Notice the use of the definite article before "derivative": Lemma 4.1.1 ensures that if f has a derivative at x^0 then it is unique. If $U \subset E$ is an open subset of the Banach space E and $f : U \to F$ a map which is differentiable at $x^0 \in U$ I will denote the derivative of f at x^0 by $\mathbf{D}f(x^0)$. I will say that f is *differentiable* on U or that f is a differentiable map defined on U if f is differentiable at each point of U.

There are a few situations in which I use a slightly different notation and terminology.

1. If $E = \mathbb{R}$, then $\mathbf{D}f(x^0) \in L(\mathbb{R}, F)$ can be identified with an element of F via the correspondence $\mathbf{D}f(x^0) \mapsto \mathbf{D}f(x^0)(1)$. In this case f will be called a *differentiable curve* in F and the vector $f'(x^0) = \mathbf{D}f(x^0)(1)$ will be called the *velocity* of f at the point x^0. If $[a, b]$ is a closed interval in $\mathbb{R}$ and a map $f : [a, b] \to F$ is differentiable on (a, b) and the limits

$$\lim_{t \to 0+} \frac{f(a + t) - f(a)}{t} \text{ and } \lim_{t \to 0+} \frac{f(b) - f(b - t)}{t}$$

exist, then f is said to be a differentiable curve defined on $[a, b]$.

2. If on the other hand, the range F of a differentiable function is $\mathbb{R}$ then the derivative $\mathbf{D}f(x^0)$ is an element of E', the *continuous* dual of E. It is known as the *differential* of f at x^0 and denoted $\mathbf{d}f(x^0)$. Thus,

$$\langle \mathbf{d}f(x^0), \boldsymbol{h} \rangle \doteq \mathbf{D}f(x^0)(\boldsymbol{h}), \text{ for all } \boldsymbol{h} \in E.$$

I will often write $\mathbf{d}f(x^0)(\boldsymbol{v})$ instead of using the Linear Algebraic convention of $\langle \mathbf{d}f(x^0), \boldsymbol{v} \rangle$.

3. If E is a Hilbert space then a further identification is commonly used. Recall, (see page 110,) that given any linear functional λ on a Hilbert space E there is a unique $\boldsymbol{x}_\lambda \in E$ such that for all $\boldsymbol{v} \in E$ $\langle \lambda, \boldsymbol{v} \rangle = \boldsymbol{x}_\lambda \odot \boldsymbol{v}$, here $\odot$ is temporarily used to denote the inner product in E. Using this identification the differential $\mathbf{d}f(x^0)$ corresponds to an element $\nabla f(x^0)$ defined by $\mathbf{d}f(x^0)(\boldsymbol{v}) = \nabla f(x^0) \odot \boldsymbol{v}$ for all $\boldsymbol{v} \in E$. This vector $\nabla f(x^0)$ is known as the *gradient* of f at x^0. The function $U \to E$ which takes x to $\nabla f(x)$ is sometimes called the **gradient vector field** of the **scalar field** $x \mapsto f(x)$. (This is physicists' jargon; vector fields will be defined in § 4.4.)

REMARKS 4.1.3 (ON THE NOTION OF DERIVATIVE)

Remark 4.1.3a. If you wish to think about finite-dimensional Banach spaces only, then you can ignore the continuity condition imposed on the derivative of a map, since in this situation all linear transformations will be continuous.

Remark 4.1.3b. I wish to point out that when $E = F = \mathbb{R}$ our new definition of the derivative reduces to the classical notion of "differential quotient".

First note that in this case the norm is the absolute value. Suppose that

(*) .... $|f(x^0+t) - f(x^0) - \Lambda_l(\boldsymbol{t})|$ is $o(t)$ for some $\Lambda_l \in \mathbb{L}(\mathbb{R},\mathbb{R})$.

Let $\Lambda_l(1) = l \in \mathbb{R}$. From the definition of $\mathbf{D}f(x^0)$ we get:

$$\lim_{t\to 0}\left|\frac{f(x^0+t)-f(x^0)}{t} - l\right| = \lim_{t\to 0}\left|\frac{f(x^0+t)-f(x^0)-lt}{t}\right|.$$

Hence the classical derivative exists iff f is differentiable in the sense I have just defined and in this case, the value of the classical "differential coefficient" $\frac{df}{dx}$ at x^0 is $\mathbf{D}f(x^0)(1)$.

So for real-valued differentiable functions defined on open intervals, the two notions of the derivative are essentially identical.

Remark 4.1.3c (Differentiability implies continuity).
In defining the notion of derivatives and differentiable functions I had assumed that $f : U \to F$ is continuous on U. This is, in fact, unnecessary. If a function $f : U \to F$ is differentiable at $x \in U$ then it is necessarily continuous at x.

To see this, suppose that $f : U \to F$ is any function and for some $x \in U$ there is a continuous linear transformation $\Lambda : E \to F$ such that

$f(x+h) - f(x) - \Lambda(\boldsymbol{h}) = o(\boldsymbol{h})$. Then if $\varepsilon > 0$ is given we can find a $0 < r < 1$ such that if $||\boldsymbol{h}|| < r$ then the point $x + h \in U$ and

$$||f(x+h) - f(x) - \Lambda(\boldsymbol{h})|| \leq ||\boldsymbol{h}||.$$

Consequently, $||f(x+h) - f(x)|| \leq (||\Lambda|| + 1)||\boldsymbol{h}||$.

So if $||\boldsymbol{h}|| < \min\left(r, \frac{\varepsilon}{(||\Lambda||+1)}\right)$, then we have $||f(x+h) - f(x)|| \leq \varepsilon$ and hence f is continuous at x. □

So we see that *differentiability at a point implies continuity at that point.* This is why I have not bothered to talk about differentiable *functions* and only talked of differentiability of maps. The attendant loss of generality is almost entirely illusory!

I will now give some elementary examples of the computation of derivatives. In all these examples the technique is to estimate $||f(x^0 + h) - f(x^0)||$ and attempt to find a continuous linear transformation $\Lambda : E \to F$ such that $||f(x^0 + h) - f(x^0) - \Lambda(\boldsymbol{h})||$ is $o(h)$ as $||h|| \to 0$; then Lemma 4.1.1 assures us that Λ *is* the derivative of f at x^0.

Examples of Derivatives

Example 4.1.4a (The derivative of a constant map).
Suppose $f : U \to F$ is a constant map from an open set $U \subset E$ to F. It is trivially checked that f is differentiable on U and for all $x^0 \in U$, $\mathbf{D}f(x^0) = 0 \in \mathbb{L}(E, F)$.

Example 4.1.4b (The derivative of a Bounded Linear Operator).
Let $\Lambda : E \to F$ be a continuous linear transformation between Banach spaces. Since $\Lambda(\boldsymbol{x} + \boldsymbol{h}) - \Lambda(\boldsymbol{x}) = \Lambda(\boldsymbol{h})$ for all $\boldsymbol{x}, \boldsymbol{h} \in E$ it follows that Λ is differentiable on E and that the derivative of Λ at every $x \in E$ is Λ itself.

This is the analogue of the familiar 1-dimensional formula:

$$\frac{d}{dx}(c \cdot x) = c.$$

Example 4.1.4c (The derivative of a bilinear gmap).
A bilinear map is, of course, a *continuous* bilinear function, $B : E_1 \times E_2 \to F$ where E_1, E_2 and F are Banach spaces.

For brevity, I will write "$[\boldsymbol{x}, \boldsymbol{y}]$" for "$B(\boldsymbol{x}, \boldsymbol{y})$" during the computations which follow. For any $(x, y) \in E_1 \times E_2$, $\boldsymbol{h} \in E_1$ and $\boldsymbol{k} \in E_2$, there are relations:

$$
\begin{aligned}
&[\boldsymbol{x}+\boldsymbol{h}, \boldsymbol{y}+\boldsymbol{k}]-[\boldsymbol{x}, \boldsymbol{y}] \\
&\quad=([\boldsymbol{x}+\boldsymbol{h}, \boldsymbol{y}+\boldsymbol{k}]-[\boldsymbol{x}+\boldsymbol{h}, \boldsymbol{y}]) \\
&\quad\quad+([\boldsymbol{x}+\boldsymbol{h}, \boldsymbol{y}]-[\boldsymbol{x}, \boldsymbol{y}]) \\
&\quad=[\boldsymbol{x}+\boldsymbol{h}, \boldsymbol{k}]+[\boldsymbol{h}, \boldsymbol{y}] \\
&\quad=([\boldsymbol{x}, \boldsymbol{k}]+[\boldsymbol{h}, \boldsymbol{y}])+[\boldsymbol{h}, \boldsymbol{k}].
\end{aligned}
$$

Observe that $\Lambda : (\boldsymbol{h}, \boldsymbol{k}) \mapsto [\boldsymbol{x}, \boldsymbol{k}] + [\boldsymbol{h}, \boldsymbol{y}]$ is a linear transformation from $E_1 \oplus E_2 \to F$, whose continuity follows from the continuity of $B : E_1 \times E_2 \to F$. Now observe that if $(\boldsymbol{h}, \boldsymbol{k}) \in E_1 \times E_2$ is regarded as a vector, $\boldsymbol{v} \in E_1 \oplus E_2$, then, with respect to the ℓ_∞-norm of $E_1 \oplus E_2$, $||\boldsymbol{v}|| = \max(||\boldsymbol{h}||, ||\boldsymbol{k}||)$. Hence, $||B(\boldsymbol{v})|| \leq ||B|| \cdot ||\boldsymbol{v}||^2 = o(v)$. It follows from Lemma 4.1.1 that B is differentiable at all points $(x, y) \in E_1 \times E_2$ and that the derivative of B is given by the formula:

$$\mathbf{D}B(x, y)(\boldsymbol{h}, \boldsymbol{k}) = B(\boldsymbol{x}, \boldsymbol{k}) + B(\boldsymbol{h}, \boldsymbol{y}). \qquad \square$$

As a particular case we get:

Example 4.1.4d. If E is a Hilbert space over $\mathbb{R}$, ν the associated norm and $M(x) = \nu(x)^2 = \langle \boldsymbol{x}, \boldsymbol{x} \rangle$, then $\mathbf{D}M(x)(\boldsymbol{v}) = 2\langle \boldsymbol{x}, \boldsymbol{v} \rangle$.

This follows immediately from the previous example and the fact that the inner product is a *symmetric* bilinear form.

This result should be thought of, as a generalization of the formula

$$\frac{d}{dx}(x^2) = 2x. \qquad \square$$

The next result involves a more substantial computation.

Proposition 4.1.5. *Let E be a Banach space and let $\mathbf{GL}(E)$ be the set of continuous linear transformations which also possess a continuous inverse. Then,*

1. *$\mathbf{GL}(E)$ is an open subset of $\mathbb{L}(E)$, and*

2. *the function $\phi : \mathbf{GL}(E) \to \mathbb{L}(E)$ defined by $\phi(X) = X^{-1}$ is differentiable and its derivative is given by the formula $\mathbf{D}\phi(X)(\boldsymbol{H}) = -X^{-1}\boldsymbol{H}X^{-1}$.*

Proof. 1. I will denote the identity map of E by 1_E. Recall that if $A \in \mathbb{L}(E)$ has norm < 1 then $(1_E - A)$ has a continuous inverse which can be represented as the sum of the infinite series $\sum_{k \geq 0} A^k$ where $A^0 = 1_E$. Moreover, the relation $||AB|| \leq ||A|| \cdot ||B||$ which holds for arbitrary elements $A, B \in \mathbb{L}(E)$ implies that $||(1_E - A)^{-1}|| \leq (||1_E - A||)^{-1}$.

So if $X \in \mathbf{GL}(E)$ and $||X^{-1}|| = r$, then for all $\boldsymbol{h} \in \mathbb{B}_{1/r}(0) \subset \mathbb{L}(E)$ the series $\sum_{k \geq 0}(X^{-1}h)^k$ converges to the inverse of $(1_E - X^{-1}\boldsymbol{h})$. It is easy to check that $(1_E - X^{-1}\boldsymbol{h})^{-1}X^{-1}$ is a two-sided inverse of $(X - \boldsymbol{h})$. This shows that if $X \in \mathbf{GL}(E)$ then the ball of radius $1/||X||$ centred at X is contained in $\mathbf{GL}(E)$ and hence $\mathbf{GL}(E)$ is an open subset of $\mathbb{L}(E)$.

2. The proof of differentiability of ϕ will, as usual, involve actually computing its derivative.

Let $X \in \mathbf{GL}(E)$. Choose $\boldsymbol{h} \in \mathbb{L}(E)$ so that $||\boldsymbol{h}|| < 1/||\mathbf{X}^{-1}||$.

First note that $(X+h)^{-1} = \big(X(1_E+X^{-1}h)\big)^{-1} = \big(1_E+X^{-1}h\big)^{-1}X^{-1}$. Now we can compute as follows.

$$\begin{aligned} ||(X+\boldsymbol{h})^{-1} - X^{-1} - X^{-1}\boldsymbol{h}X^{-1}|| &= ||(1_E + X^{-1}\boldsymbol{h})^{-1}X^{-1} - X^{-1} \\ &\qquad - X^{-1}\boldsymbol{h}X^{-1}|| \\ &\leq ||(1_E + X^{-1}h)^{-1} - 1_E - X^{-1}\boldsymbol{h}||\cdot||X^{-1}|| \\ &\leq \Big|\Big|1_E + \sum_{k=0}^{\infty}(X^{-1}h)^k - 1_E - X^{-1}\boldsymbol{h}\Big|\Big|\cdot||X^{-1}|| \\ &\leq \frac{\sum_{k\geq 2}||(X^{-1}\boldsymbol{h})^k||}{||1_E - X^{-1}\boldsymbol{h}||^{-1}}\cdot||X^{-1}|| = o(h). \end{aligned}$$

Hence, $\phi(X+H) - \phi(X) + X^{-1}HX^{-1} = o(H)$, and therefore

$$\mathbf{D}\phi(X)(\boldsymbol{H}) = -X^{-1}HX^{-1}.$$

□

The formula for $\mathbf{D}\phi$ we have just derived, should be regarded as the noncommutative and higher-dimensional generalization of the formula:

$$\frac{d}{dx}\left(\frac{1}{x}\right) = -\left(\frac{1}{x^2}\right).$$

Remark. Notice that most of the work was done in proving the openness of $\mathbf{GL}(E)$ for an arbitrary Banach space. This is, of course, much easier in the finite-dimensional situation. (See Example 3.2.3b). However, the actual estimates used in computing the derivative did not involve in any way the dimension of E and would not have been simplified by requiring E to be finite-dimensional.

I will now develop a few techniques which are useful in computing the derivatives of complicated functions. The first shows that the composition of differentiable maps is again a differentiable map and gives a formula for the derivative of the composite map.

Theorem 4.1.6 (The Chain Rule).
Let E, F and G be three Banach spaces and suppose that $U \subset E, V \subset F$ are open subsets. Let $f : U \to F$ and $g : V \to F$ be differentiable maps and suppose that $f(U) \subset V$.

Then the composite function $\phi \doteq gf : U \to G$ is differentiable at x^0 and the derivative is given by the formula

$$\text{(Chain Rule:)} \qquad \mathbf{D}\phi(x^0) = \mathbf{D}g\big(f(x^0)\big)\mathbf{D}f(x^0).$$

In other words, "the derivative of the composition is the composition of the derivatives".

Proof. For convenience, I will write y^0 for $f(x^0)$. Choose $r > 0$ so that $\mathbb{B}_r(x^0) \subset U$ and $\mathbb{B}_r(y^0) \subset V$. Let $0 < \varepsilon < 1$ be given. Then the differentiability conditions on f and g imply that if $||\boldsymbol{h}|| < r$, $||\boldsymbol{k}|| < r$ then,

$$f(x^0 + h) - f(x^0) = \mathbf{D}f(x^0)(\boldsymbol{h}) + o_1(h) \tag{1}$$

$$g(y^0 + k) - g(y^0) = \mathbf{D}g(y^0)(\boldsymbol{k}) + o_2(k) \tag{2}$$

where $||o_1(h)|| \leq \varepsilon||\boldsymbol{h}||$ and $||o_2(k)|| \leq \varepsilon||\boldsymbol{k}||$ if r has been chosen suitably small.

Let $||\mathbf{D}f(x^0)|| = A$ and $||\mathbf{D}g(y^0)|| = B$. Then, provided $||\boldsymbol{h}|| < r$, the right side of (1) can be estimated as follows:

$$||\mathbf{D}f(x^0)(\boldsymbol{h}) + o_1(h)|| \leq (A + \varepsilon)||\boldsymbol{h}|| \leq (A+1)||\boldsymbol{h}||.$$

So if $||\boldsymbol{h}|| < r/(A+1)$, then ϕ is defined at $(x^0 + h)$ and we may proceed to make the estimates necessary to compute the derivative of ϕ.

$$\begin{aligned}
\phi(x^0 + h) - \phi(x^0) &= g(f(x^0 + h)) - g(y^0), \\
&= g\big(y^0 + \mathbf{D}f(x^0)(\boldsymbol{h}) + o_1(h)\big) - g(y^0), \\
&= \mathbf{D}g(y^0)\big(\mathbf{D}f(x^0)(\boldsymbol{h}) + o_1(h)\big) + o_2(l), \\
&= \mathbf{D}g(y^0) \circ \mathbf{D}f(x^0)(\boldsymbol{h}) + \mathbf{D}g(y^0)(o_1(h)) + o_2(l),
\end{aligned}$$

where $l = \mathbf{D}f(x^0)(\boldsymbol{h}) + o_1(h)$. This yields the inequality:

$$\begin{aligned}
||\phi(x^0 + h) - \phi(x^0) - \mathbf{D}g(y^0)\mathbf{D}f(x^0)(\boldsymbol{h})|| &\leq B\varepsilon||\boldsymbol{h}|| + \varepsilon(A+1)||\boldsymbol{h}|| \\
&= (A + B + 1)\varepsilon||\boldsymbol{h}||
\end{aligned}$$

which shows that $\mathbf{D}\phi(x^0) = \mathbf{D}g(y^0)\mathbf{D}f(x^0)$. □

The "chain rule" has myriad applications, some of the more important ones are described below.

Proposition 4.1.7 (The derivative of vector-valued functions).
Let E and $F_1, \dots, F_n$ be Banach spaces, $U \subset E$ an open set and for $i = 1, \dots, n$ let $f_i : U \to F_i$ be maps. Let $F = F_1 \oplus \cdots \oplus F_n$ be the Cartesian product of the F_i's with the linear space structure of the direct sum and the ℓ_∞-norm. Then the "product map" $f : U \to F$ defined by $f(x) = \big(f_1(x), \dots, f_n(x)\big)$ is differentiable at $x^0 \in U$ iff each f_i is differentiable at x^0. If this is so, then for $\boldsymbol{h} \in E$

$$\mathbf{D}f(x^0)(\boldsymbol{h}) = \big(\mathbf{D}f_1(x)(\boldsymbol{h}), \dots, \mathbf{D}f_n(x^0)(\boldsymbol{h})\big). \tag{4.1.1}$$

Proof. For each $i = 1, \dots, n$, the differentiability of f_i at x^0 implies that

$$f_i(x^0 + h) - f_i(x^0) - \mathbf{D}f_i(x^0)(\boldsymbol{h}) = o(h).$$

Since F has the ℓ_∞-norm, if $\mathbf{D}f(x^0)$ is defined by the formula 4.1.1 then

$$f(x^0) - f(x^0 + \boldsymbol{h}) - f(x^0) - \mathbf{D}f(x^0)(\boldsymbol{h}) = o(h).$$

To prove the converse first observe that for $i = 1, \dots, n$, $f_i = \pi_i \circ f$ where π_i is the projection of F onto the i^{th}component; this is a continuous linear transformation. The chain rule now implies that the f_i's are differentiable if f is differentiable. □

Proposition 4.1.8 (The derivative of sums of functions).
Let U be an open set in the Banach space E and let $f_1, \dots, f_n : U \to F$ be maps into a Banach space. If each of the f_k's are differentiable at $x^0 \in U$ then so is the sum of these functions and

$$\mathbf{D}(f_1 + \cdots + f_n)(x^0) = \mathbf{D}f_1(x^0) + \cdots + \mathbf{D}f_n(x^0).$$

Proof. Let $F_k = F$ for each $k = 1, \dots, n$. The previous result implies that if the f_k's are differentiable then so is the product map. The proof is completed by observing that the function $F_1 \oplus \cdots \oplus F_n \to F$ which adds all the components is a continuous linear transformation and appealing to the chain rule. □

Further applications of the chain rule follow. It is worth noting that the difficulty of giving "detailed proofs", or even exact statements of some of these examples would persist, when we deal with functions defined on an open interval of $\mathbb{R}$ which take values in a finite-dimensional vector space (even if this was $\mathbb{R}$ itself). Since, in this last case you have learned to live with a certain amount of imprecision, I am requiring you to take a similar tolerant attitude here.

Examples of derivative computations

Example 4.1.9a (The "product rule").

Let $U \subset E$ be an open set in a Banach space and suppose that $f_k : U \to F_k$, $(k = 1, 2)$ are maps taking values in Banach spaces.

Suppose that $B : F_1 \times F_2 \to V$ is a continuous bilinear function. (I will write $B(\boldsymbol{y}_1, \boldsymbol{y}_2)$ as $[\boldsymbol{y}_1, \boldsymbol{y}_2]$ for convenience.) Define $F : U \to V$ by $F(x) = [f_1(x), f_2(x)]$. If f_1 and f_2 are differentiable at $x^0 \in U$ then so is F and for any $\boldsymbol{h} \in E$ we have:

$$\text{(4.1.2)} \qquad \mathbf{D}F(x^0)(\boldsymbol{h}) = [\mathbf{D}f_1(x^0)(\boldsymbol{h}), f_2(x^0)] + [f_1(x^0), \mathbf{D}f_2(x^0)(\boldsymbol{h})].$$

Proof. By Proposition 4.1.7 $x \mapsto \big(f_1(x), f_2(x)\big)$ is differentiable at x^0. Now f is the composition of this map with the continuous bilinear function B. Since continuous bilinear functions have been shown to be differentiable the result follows from the chain rule.

Note that the above result is the analogue of $(u \cdot v)' = u' \cdot v + u \cdot v'$. □

Example 4.1.9b (The derivative of the norm in a Hilbert space). Let E be a Hilbert space and $\nu(x) = \sqrt{\langle x, x\rangle}$ the associated norm. Then $\nu(x) = \phi(M(x))$ where $\phi : \mathbb{R}_+ \to \mathbb{R}$ is the square-root function, whose derivative you have known for a long time. Then, using the computation in Example 4.1.4d and the chain rule we get:

$$\text{(4.1.3)} \qquad \mathbf{D}\nu(x)(\boldsymbol{v}) = \frac{\langle \boldsymbol{x}, \boldsymbol{v}\rangle}{||\boldsymbol{x}||}$$

Remark 4.1.9d. Note that *no norm* can be differentiable at the origin. *The classical analogue of Equation 4.1.3 is the well-known (and trivial) formula:*

$$\frac{d}{dx}(|x|) = \frac{x}{|x|} \text{ for } x \neq 0 \ .$$

Examples of Derivatives (contd.)

Example 4.1.9c (Multiple Products).
Let U be an open subset of a Banach space, E, and for $i = 1, \dots, n$ suppose that $f_i : U \to F_i$ are differentiable maps into Banach spaces, F_i.

If $A : F_1 \times \cdots \times F_n \to F$ is a continuous n-linear function then the function $\phi : U \to F$ defined by $\phi(x) = A\big(f_1(x), \dots, f_n(x)\big)$ is differentiable and

$$\mathbf{D}\phi(x)(\boldsymbol{h}) = \sum_{i=1}^{n} \phi\big(f_1(x), \dots, f_{(i-1)}(x), \mathbf{D}f_i(x)(\boldsymbol{h}), f_{(i+1)}(x), \dots, f_n(x)\big)$$

for each $x \in U$ and $\boldsymbol{h} \in E$.

Proof. This is being left as an exercise. Look at the proof of the "Product Rule" and use as a guide the section in Chapter 3 where norms on the spaces of multilinear functions are defined. □

The examples which follow are illustrations of the power of the product rule combined with the Chain Rule.

Example 4.1.9d (Rational functions of differentiable maps).
If $f_i : U \to \mathbb{k}$ are real- or complex-valued differentiable functions, defined on an open set of a Banach space E, then so is any function of the form

$$\phi(x) = \big(f_1(x)\big)^{m_1} \cdots \big(f_n(x)\big)^{m_n}$$

where the m_i's are non-negative integers. If $f_\alpha(x^0) \neq 0$ at $x^0 \in U$ we can allow the exponent m_α to be negative and the resulting function will be differentiable at x^0. The proof is a routine but tiresome application of the chain and product rules. □

Particular cases of the above yield important results.

Example 4.1.9e (Determinants and inverses).
From the previous example it follows that the determinant, $\det : \mathrm{M}_n(\mathbb{k}) \to \mathbb{k}$, and the inversion map: $I : \mathrm{GL}_n(\mathbb{k}) \to \mathrm{M}_n(\mathbb{k})$ which inverts a matrix are differentiable maps.

Example 4.1.9f (The inversion map revisited).
I had pointed out that the major portion of the effort in establishing Proposition 4.1.5, was expended in proving that $\mathbf{GL}(E)$ was open in $\mathbb{L}(E)$. If we assume, in addition, that the inversion map is differentiable

– this has been just done for the finite dimensional situation – there is a very elegant and easy proof of the formula for the derivative of $\phi : X \mapsto X^{-1}$.

From the definition, we know that $X \cdot \phi(X) = 1_E$. So using the product rule we get:

$$\begin{aligned} \boldsymbol{H} \cdot X^{-1} + X \cdot \mathbf{D}\phi(X)(\boldsymbol{H}) &= 0 \\ \text{and hence } \mathbf{D}\phi(X)(\boldsymbol{H}) &= -X^{-1} \cdot H \cdot X^{-1}. \end{aligned} \qquad \square$$

Example 4.1.9g (The Gram-Schmidt procedure).
Recall that if $\mathrm{M} \in \mathrm{GL}_n(\mathbb{R})$ then one can also regard the columns of M, $\{\gamma_i(\mathrm{M})\}_{i=1,\ldots,n}$ as an ordered n-tuple of vectors of $\mathbb{R}^n$. The fact that M is nonsingular implies that these columns, in fact, constitute a basis of $\mathbb{R}^n$. Now the Gram-Schmidt procedure applied to this ordered basis yields an ordered orthonormal basis

$$\{\epsilon_1(\mathrm{M}), \ldots, \epsilon_n(\mathrm{M})\}$$

which we can regard as an orthogonal matrix. Thus the Gram-Schmidt process yields a function $\Psi : \mathrm{GL}_n(\mathbb{R}) \to \mathrm{O}(n) \subset \mathrm{M}_n(\mathbb{R})$. The fact that the entries of $\Psi(\mathrm{M})$ are obtained from those of M by multiplication, addition and division by the lengths of nonzero vectors implies that $\Psi : \mathrm{GL}_n(\mathbb{R}) \to \mathrm{O}(n) \subset \mathrm{M}_n(\mathbb{R})$ is a differentiable map from $\mathrm{GL}_n(\mathbb{R})$ to $\mathrm{M}_n(\mathbb{R})$. A similar result holds for the Gram-Schmidt process applied to a complex nonsingular matrix.

Computing the derivative of the map Ψ of the last example is, of course, difficult. Special cases, however may be computed quite easily and provide interesting information. (One such computation of this kind is given as an exercise at the end of this section.)

From these examples it should be clear to you that though we have been able to compute the derivatives of some interesting maps, there is no general procedure for computing the derivative of a differentiable map. Each of my computations depended either on a new trick or on the formal rules of differentiation. However, if $f : U \to F$ is a map from an open subset U of a Banach space E which is differentiable at $x^0 \in U$ then for any $\boldsymbol{v} \in E$, it is quite easy to give a recipe for computing the value of $\mathbf{D}f(x^0)$ on the vector $\boldsymbol{v}$. Since U is open we can find $\varepsilon > 0$ such that $\{x^0 + tv \ : \ |t| < \varepsilon\} \subset U$ and so for values of t sufficiently near 0, the curve $\alpha_{\boldsymbol{v}} : t \mapsto x^0 + t\boldsymbol{v}$ takes values in U. Now consider the function $\phi_{\boldsymbol{v}}(t) \doteq f(\alpha_{\boldsymbol{v}}(t))$ which is defined for $t \in (-\delta, \delta)$ if $\delta > 0$ is suitably small. The limit:

$$\phi'_{\boldsymbol{v}}(0) \doteq \lim_{t\to 0} \frac{\phi_{\boldsymbol{v}}(t) - \phi_{\boldsymbol{v}}(0)}{t}$$

is called the *directional derivative* (along $\boldsymbol{v}$) of f at x^0. (This is also sometimes called the *Gateux* derivative.)

From the chain rule, we get $\phi'_{\boldsymbol{v}}(0) = \mathbf{D}f(\alpha_v(0))\left(\frac{d}{dt}(x^0 + t\boldsymbol{v})\Big|_{t=0}\right)$

$$(\star) \qquad \text{or, } \phi'_v(0) = \mathbf{D}f(x^0)(\boldsymbol{v}).$$

Notice that $(\star)$ presupposes the existence of the derivative of f at x^0. The directional derivative of f along some (or even all!) $\boldsymbol{v} \in E$ may exist even if $\mathbf{D}f(x^0)$ does not exist.

Remark 4.1.10. Notice that if $\beta : (-\delta, \delta) \to U$ is any curve whose velocity at $t = 0$ is $\boldsymbol{v}$ then the velocity of $(f \circ \beta)$ at $t = 0$ will also be equal to $\mathbf{D}f(x^0)(\boldsymbol{v})$.

The technique described in the above remark will be used frequently and so I introduce a bit of jargon which will save a lot of words of explanation.

Let U and x^0 be as before. Given a vector $\boldsymbol{v} \in E$ we will say that a curve α in U *represents* $\boldsymbol{v}$ *as a* tangent vector *at* x^0 if

TV1: α is defined for some open interval $(-\delta, \delta) \subset \mathbb{R}$;

TV2: $\alpha(t) \in U$ for $|t| < \delta$;

TV3: $\alpha(0) = x^0$ and $\alpha'(0) = \boldsymbol{v}$.

I formally record what we have shown above in the next proposition.

Proposition 4.1.11. *Let $U \subset E$ be an open subset of the Banach space E. Suppose that $f : U \to F$ is a map which is differentiable at $x^0 \in U$. If $\boldsymbol{v} \in E$ and $\alpha_{\boldsymbol{v}}$ is a curve which represents $\boldsymbol{v}$ as a tangent vector at x^0 then*

$$\mathbf{D}f(x^0)(v) = \frac{d}{dt}(f \circ \alpha_{\boldsymbol{v}})(0). \qquad \square$$

I formally define for future use (in § 4.4) the notions of *tangent vectors* at a point p and the *tangent space* at p as follows.w

Definition 4.1.12 (Tangent vectors and tangent spaces).
Let $p \in E$ be a point in the Banach space E. Two differentiable curves,

α, β defined on an open interval $I \subset \mathbb{R}$ containing 0 will be regarded as *tangentially equivalent at* p, (denoted $\alpha \sim_p \beta$,) if $\alpha(0) = \beta(0) = p$ and, $\alpha'(0) = \beta'(0)$.

It is obvious that $\sim_p$ is an equivalence relation. An equivalence class of curves starting at p will be called a *tangent vector* at p. The set of tangent vectors at p will be denoted T_pE.

Henceforth, the argument of the derivative of a function will be considered to be a tangent vector.

As we shall see in § 4.4, tangent vectors and certain collections of tangent vectors (known as vector fields) play many different roles. This is why I have chosen a distinctive notation (the bold italics) for them.

The following is easily proved.

Proposition 4.1.13. *The function, τ_p, which takes a tangent vector at p to the common value of the derivative (at 0) of the equivalent curves is a bijection between T_pE and E.*

Proof. Given $\boldsymbol{v} \in E$ and $p \in E$, the curve $\alpha_{\boldsymbol{v}} : (-1, 1) \to E$ defined by $\alpha_{\boldsymbol{v}}(t) = p + tv$ satisfies the relation $\tau_p(\alpha_{\boldsymbol{v}}) = \boldsymbol{v}$ and if any other curve passing through p at time $t = 0$ has the same derivative at $t = 0$ then it is equivalent to $\alpha_{\boldsymbol{v}}$. So τ_p is a bijection. □

As an application of these ideas, we perform a very useful computation. Recall that if $\mathrm{A} \in \mathrm{M}_n(\mathbf{k})$ and $\mathrm{A} = (a_{ij})$, then the *cofactor* matrix is the $(n \times n)$ matrix whose $(i, j)^{\text{th}}$ entry is $(-1)^{i+j}$ times $\mathrm{A}_{[ij]}$, where $\mathrm{A}_{[ij]}$ is the determinant of the $\big((n-1) \times (n-1)\big)$ matrix obtained by deleting the ith row and j^{th} column of A.

The transpose of the cofactor matrix, sometimes called the *adjoint of* A, will be denoted $\mathrm{A}^\dagger$.

Proposition 4.1.14 (The derivative of det).
Let $\Delta : \mathrm{M}_n(\Bbbk) \to \Bbbk$ be the determinant function. Then,

$$\mathbf{D}\Delta(\mathrm{X})(\boldsymbol{H}) = \mathrm{Trace}\,(\mathrm{X}^\dagger \mathrm{H})$$

Proof. Note that in the formulae above and in those which follow, matrices occur either simply as a matrix or as a tangent vector and the choice of font reflects this.

I have already pointed out that $\Delta \doteq \det : M_n(\Bbbk) \to \Bbbk$ is a differentiable map. So we use Proposition 4.1.11 to compute the directional derivatives along any tangent vector. In the proof, I will only deal with the case $\Bbbk = \mathbb{R}$. The case of complex matrices is being left as an exercise.

For $1 \le i, j \le n$, let $E_{ij} \in M_n(\mathbb{R})$ be the matrix with 1 at the $(i,j)^{\text{th}}$ place and zeros everywhere else. Then $t \mapsto (X + tE_{ij})$ is a curve which represents E_{ij} as a tangent vector at X. We now calculate, as follows.

$$\begin{aligned}\mathbf{D}\Delta(X)(\boldsymbol{E}_{ij}) &= \frac{d}{dt}(\det(X + tE_{ij}))\Big|_{t=0} \\ &= \frac{d}{dt}\left(\sum_{k=1}^{n}(-1)^{l+k}(x_{lk} + t\delta_{il}\delta_{jk})X_{[lk]}\right)\Big|_{t=0} \\ &= (-1)^{i+j}X_{[ij]}.\end{aligned}$$

Since the E_{ij}'s form a basis of $M_n(\mathbb{R})$ and $\mathbf{D}\Delta(X)$ is a linear map, we get:

$$\mathbf{D}\Delta(X)(\boldsymbol{H}) = \sum_{1 \le i,j \le n} (-1)^{i+j} X_{[ij]} h_{ij},$$

where h_{ij} is the $(i,j)^{\text{th}}$ element of the matrix H. The result now follows from the well-known (and easily checked) fact that, if A, B are $(n \times n)$ matrices, then

$$\text{Trace}\,(A^t B) = \sum_{i,j=1}^{n} a_{ij} b_{ij}.$$

□

I will now discuss the notion of the differential and gradient of $\mathbb{R}$-valued differentiable maps. Let E be a Euclidean space of dimension n, $U \subset E$ an open set and $f : U \to \mathbb{R}$ a differentiable map. If $\mathcal{V} = \{\boldsymbol{v}_1, \dots, \boldsymbol{v}_n\}$ is an orthonormal basis of E, then each point of U is completely determined by n real numbers since this basis induces an isometry of E with $\mathbb{R}^n$ which carries the basis vector, $\boldsymbol{v}_k$, to the kth standard vector of $\mathbb{R}^n$. So any function $f : U \to \mathbb{R}$ may be thought of as a function which depends on the values of n arguments, each of which is a real number. If $x^0 = \sum_{k=1}^{n} x_k^0 \boldsymbol{v}_k$, then I will write $f(x_1^0, \dots, x_n^0)$ for the value of f at x^0. More simply, one says that f is a "function of n variables". Now if f is differentiable at $x^0 \in U$, its derivative or differential at x^0, $\mathbf{d}f(x^0)$, is an element of E^*. It is natural to look for an expression of $\mathbf{d}f(x^0)$ as a linear combination of the basis vectors dual to $\mathcal{E}$.

I now do this but to simplify the notation, I will assume that $E = \mathbb{R}^n$ and that $\mathcal{E}$ is the standard basis. There is no loss of generality in this simplification. Note that for $1 \leq i \leq n$, the curve

$$\gamma_i : t \mapsto (x_1^0, \ldots, x_{(i-1)}^0, x_i^0 + t, x_{(i+1)}^0, \ldots x_n^0)$$

is contained in U if the domain is restricted to $(-t, t)$ for $|t|$ sufficiently small and represents the ith standard basis vector, $\boldsymbol{e}_i$, as a tangent vector at x^0. Hence

$$\begin{aligned}\mathbf{d}f(x^0)(\boldsymbol{e}_i) &= \lim_{t\to 0}\left(\frac{f(x_1^0, \ldots, x_i^0 + t, \ldots, x_n^0) - f(x_1^0, \ldots, x_i^0, \ldots, x_n^0)}{t}\right) \\ &\doteq \frac{\partial f}{\partial x_i}(x^0)\end{aligned}$$

The last quantity is known as the ith *partial derivative of f at the point* x^0. Now a vector $\boldsymbol{x} = \sum_{i=1}^n x_i \boldsymbol{e}_i$ is represented as a column vector $(x_1, \ldots, x_n)$ and so in terms of the dual basis, the differential $\mathbf{d}f(x^0)$ is the row vector

$$\left[\frac{\partial f}{\partial x_1}(x^0) \quad \cdots \quad \frac{\partial f}{\partial x_n}(x^0)\right].$$

Notice that simply using the definition of the gradient we see that:

$$\nabla f(x^0) = \left(\frac{\partial f}{\partial x_1}, \frac{\partial f}{\partial x_2}, \ldots, \frac{\partial f}{\partial x_n}\right)_{x^0}$$

Here we are using the space-saving notation for column vectors and, the subscript x^0 indicates that the partial derivatives are computed at the point $x^0 \in U$.

Remark 4.1.15 (On differentials).
In some books, $\{\mathbf{d}x_1, \ldots, \mathbf{d}x_n\}$ is used to denote the basis dual to the standard basis of $\mathbb{R}^n$. This leads to the formula:

$$(*) \ldots\ldots\ldots\ldots\ldots\ldots \quad \mathbf{d}f(x^0) = \sum_{j=1}^n \frac{\partial f}{\partial x_j}(x^0)\mathbf{d}x_j.$$

In old-fashioned books, this formula is sometimes accompanied by statements like:

> "$\mathbf{d}f(x^0)$ is the infinitesimal increment in the value of the function f when the arguments x_k^0 are given infinitesimal increments dx_k."

This makes some sort of sense to a physicist but you should not take this seriously. You should remember that all we have, in formula $(*)$, is an expression of the differential $\mathbf{d}f(x^0) \in (\mathbb{R}^n)^*$ in terms of the basis dual to the standard basis of $\mathbb{R}^n$.

Now suppose that $U \subset \mathbb{R}^n$ is open and that $f : U \to \mathbb{R}^m$ a differentiable map. Then we can write $f(x) = \sum_{j=1}^{m} f_j(x)\boldsymbol{e}_j(m)$ where $\{\boldsymbol{e}_1(m), \dots, \boldsymbol{e}_m(m)\}$ is the standard basis of $\mathbb{R}^m$. So we can think of f as the "vector-valued function" obtained from the m real-valued functions $f_j : U \to \mathbb{R}$. which, by Proposition 4.1.7, are differentiable at each $x \in U$. Proposition 4.1.7 implies that if $\boldsymbol{v} = (v_1, \dots, v_n)$, then,

$$\mathbf{D}f(x)(\boldsymbol{v}) = \begin{bmatrix} \mathbf{d}f_1(x)(\boldsymbol{v}) \\ \vdots \\ \mathbf{d}f_m(x)(\boldsymbol{v}) \end{bmatrix} = \begin{bmatrix} \frac{\partial f_1}{\partial x_1} & \cdots & \frac{\partial f_1}{\partial x_n} \\ \vdots & \frac{\partial f_i}{\partial x_j} & \vdots \\ \frac{\partial f_m}{\partial x_1} & \cdots & \frac{\partial f_m}{\partial x_n} \end{bmatrix}_x \odot \begin{bmatrix} v_1 \\ \vdots \\ v_n \end{bmatrix}$$

where the '$\odot$' stands for matrix multiplication and the subscript 'x' indicates that all partial derivatives are computed at x. (For the sake of clarity, I have, temporarily, abandoned my space-saving device of writing column vectors as a row of numbers enclosed by parentheses.) From this we conclude that the matrix of the linear transformation $\mathbf{D}f(x) : \mathbb{R}^n \to \mathbb{R}^m$ in terms of the standard bases of $\mathbb{R}^n$ and $\mathbb{R}^m$ is

$$J(f;x) \doteq \begin{bmatrix} \frac{\partial f_1}{\partial x_1} & \cdots & \frac{\partial f_1}{\partial x_n} \\ \vdots & \frac{\partial f_i}{\partial x_j} & \vdots \\ \frac{\partial f_m}{\partial x_1} & \cdots & \frac{\partial f_m}{\partial x_n} \end{bmatrix}_x$$

which is referred to as the *Jacobian* of f at the point x. The Jacobian, at the point $x^0 \in U$, is sometimes denoted $\dfrac{\partial(f_1, \dots, f_m)}{\partial(x_1, \dots, x_n)}(x^0)$.

I summarize the above findings in the following result which is stated in the context of maps defined on an open set of a Euclidean space and taking values in another Euclidean space. I will explain a little later why I state this result so carefully and in such a special situation.

Theorem 4.1.16 (The Jacobian Formula).
Let E and F be Euclidean spaces and suppose that $\mathcal{E} = \{\boldsymbol{e}_1, \dots, \boldsymbol{e}_n\}$ and $\mathcal{F} = \{\boldsymbol{f}_1, \dots, \boldsymbol{f}_m\}$ are orthonormal bases of E and F respectively. In what follows these bases will be used to write points of U and F as n-tuples and m-tuples. Let $U \subset E$ be an open set and $\phi : U \to F$ be a

differentiable map. If

$$\phi(x) = \phi\Big(\sum_{k=1}^{n} x_k \boldsymbol{e}_k\Big) = \sum_{l=1}^{m} \phi_l(x) \boldsymbol{f}_l$$

or, alternatively:

$$\phi(x_1, \ldots, x_n) = \big(\phi_1(x_1, \ldots, x_n), \ldots, \phi_m(x_1, \ldots, x_n)\big)$$

then the $\phi_l(x)$ are differentiable real-valued functions on U and the matrix of $\mathbf{D}\phi(x^0)$, relative to the bases $\mathcal{E}$ and $\mathcal{F}$, is the Jacobian matrix:

$$J(\phi; x^0) = \begin{bmatrix} \dfrac{\partial \phi_1}{\partial x_1} & \cdots & \dfrac{\partial \phi_1}{\partial x_n} \\ \vdots & \dfrac{\partial \phi_i}{\partial x_j} & \vdots \\ \\ \dfrac{\partial \phi_m}{\partial x_1} & \cdots & \dfrac{\partial \phi_m}{\partial x_n} \end{bmatrix}_{x^0}$$

where the central element is the $(i, j)^{\text{th}}$ entry and all the partial derivatives are evaluated at the point x^0. □

Those of you who still prefer to think in terms of matrices rather than linear transformations may now be heaving a sigh of relief. Now that you have learnt the recipe for writing down the matrix of the derivative of a map with respect to the standard bases, you may think that for the purposes of computations, there is no further need to think of the derivative as a linear transformation. Unfortunately this is not the case and the difficulties of using the Jacobian as a surrogate for the derivative are outlined below.

REMARKS 4.1.17 (JACOBIANS VS. DERIVATIVES)

Remark 4.1.17a. The first thing to note is that the Jacobian may exist at a point where the function is *not differentiable!*

Consider, for example, the function $f : \mathbb{R}^2 \to \mathbb{R}$ defined by:

$$f(x, y) = \begin{cases} \dfrac{xy}{x^2 + y^2} & \text{if } (x, y) \neq (0, 0) \\ 0 & \text{if } x = y = 0. \end{cases}$$

It is clear that f vanishes on the x- and y-axes and therefore $\frac{\partial f}{\partial x}(0,0)$ and $\frac{\partial f}{\partial y}(0,0)$ are both equal to 0, so the Jacobian, $J(f;(0,0))$ exists. But the function is *not even continuous*! At each point (δ, δ), no matter how small a $\delta > 0$ we select, f takes the value $\frac{1}{2}$. So if f were differentiable at $(0,0)$ this would contradict the result that if the derivative exists at a point then the function has to be continuous at that point.

Remark 4.1.17b. The Jacobian cannot really be used as a computational tool except in trivial situations. For instance, try to compute the Jacobian of the function which takes a nonsingular matrix to its inverse. (I doubt if you can do this twice, for the (3×3) case, and get the same answer each time!) And even after the computation is over can one really tell from an inspection of this huge $(3^2 \times 3^2)$ matrix what it really does to a (3×3) matrix? Is it possible to see that this huge array is in fact a straightforward (though noncommutative) generalization of multiplication by the number, $-\left(\frac{1}{x}\right)^2$?

Remark 4.1.17c. You should be wary of one other pitfall which awaits you when you read classical treatments of Calculus. In many contexts attention is restricted to functions defined on open subsets of $\mathbb{R}^n$ and taking values in $\mathbb{R}^m$ where these are simply regarded as the spaces of n-tuples and m-tuples of real numbers respectively. Such functions are then written in the form

$$f(x_1, \ldots, x_n) = \big(f_1(x_1, \ldots, x_n), \cdots, f_m(x_1, \ldots, x_n)\big).$$

Such functions are called (vector-valued) functions of n real variables. Instead of defining the derivative as a linear transformation, the Jacobian matrix is either defined to be the derivative or implicitly the Jacobian is treated as the derivative.

This may seem a harmless and innocuous practice but consider the danger posed by the following example. As you know, the nonzero vectors of $\mathbb{R}^2$ may be represented uniquely by polar coordinates (r, θ) where $r > 0$ and $0 \leq \theta < 2\pi$. Now if we consider the map f which takes the point with polar coordinates (r, θ) to $(r\cos\theta)\cdot \boldsymbol{e}_1 + (r\sin\theta)\boldsymbol{e}_2$, where the $\boldsymbol{e}_k$'s are standard basis vectors of $\mathbb{R}^2$, we get, of course, the identity map of $\mathbb{R}^2_\times$.

But the traditional way of writing this map is $f(r, \theta) = (r\cos\theta, r\sin\theta)$. If you are not careful, you may mechanically compute partial derivatives and decide that since the Jacobian of f at (r, θ) is the matrix:

$$\begin{bmatrix} \cos\theta & -r\sin\theta \\ \sin\theta & r\cos\theta \end{bmatrix}.$$

it represents the derivative of f, which is absurd since f is the restriction of the identity map of $\mathbb{R}^2$ to $\mathbb{R}^2_\times$.

I give this example since many books on applied mathematics use the notation

$$\frac{\partial(x, y)}{\partial(r, \theta)} = \begin{bmatrix} \cos\theta & -r\sin\theta \\ \sin\theta & r\cos\theta \end{bmatrix}$$

and declare that this matrix represents the "transformation of vectors from Cartesian to polar coordinates". In § 4.4 I will indicate precisely the very important role played by this matrix in the transformation of Cartesian to polar coordinates.

I hope you now appreciate why I wrote the "Jacobian formula" in the context of Euclidean spaces and specific choices of orthonormal bases. I can now explain a remark I had made on page 135.

I had said that the old-fashioned treatment of "functions of several variables may lead some students to think that for real-valued functions of n variables ($n > 1$) the first and second derivatives can be defined but no further. (These two derivatives appear as $(n \times n)$ matrices known as the Jacobian and the Hessian respectively.)

Remark 4.1.18 (About Hessians). Suppose that we are dealing with real-valued differentiable functions defined on an open subset, $U \subset \mathbb{R}^n$. Such a function is expressed as "a function of n variables" by using coordinates arising from an orthonormal basis of $\mathbb{R}^n$.

Then the first derivative, at a point x^0, is the differential, $\mathbf{d}f(x^0) \in (\mathbb{R}^n)^*$. The duality in the inner product space is used to identify $\mathbf{d}f(x)$ with the gradient $\nabla f(x)$.

In this way, the first derivative of a function $f : U \to \mathbb{R}$ appears as a function

$$\nabla f : (x_1, \ldots, x_n) \mapsto \left(\frac{\partial f}{\partial x_1}, \cdots, \frac{\partial f}{\partial x_n}\right).$$

This, of course, is a function $\mathbb{R}^n \supset U \xrightarrow{\nabla f} \mathbb{R}^n$. By the Jacobian Formula ∇f has a derivative whose "Jacobian matrix" is called the *Hessian* of

f:

$$\operatorname{Hess}(f;x^0) = \begin{bmatrix} \frac{\partial}{\partial x_1}\left(\frac{\partial f}{\partial x_1}\right) & \cdots & \frac{\partial}{\partial x_n}\left(\frac{\partial f}{\partial x_1}\right) \\ \vdots & \frac{\partial}{\partial x_j}\left(\frac{\partial f}{\partial x_i}\right) & \vdots \\ \frac{\partial}{\partial x_1}\left(\frac{\partial f}{\partial x_1}\right) & \cdots & \frac{\partial}{\partial x_n}\left(\frac{\partial f}{\partial x_n}\right) \end{bmatrix}_{x^0}$$

where the subscript at the bottom indicates the partial derivatives are being computed at the point x^0.

The functions:

$$\left\{\frac{\partial^2 f}{\partial x_i \partial x_j} \doteq \frac{\partial}{\partial x_i}\left(\frac{\partial f}{\partial x_j}\right)\right\}_{1\leq i,j\leq n}$$

are known as the *second* partial derivatives of f.

We will acquaint ourselves with the *genuine* second (and higher) derivatives, later in this chapter.

Primitives and Integrals ·

I will end this section with a brief discussion of a variant of the Riemann integral which is applicable to a certain class of functions defined on compact intervals, $[a,b] \subset \mathbb{R}$, and taking values in a Banach space. My aim is to construct *primitives* or *antiderivatives* of such functions. I will be somewhat sketchy in my exposition since you are already familiar with the techniques. All I will be doing is to apply, in a new context, the ideas that are used in defining the Riemann integral of a real-valued function.

Let E be a Banach space and $f : [a,b] \to E$ a function. A function $\phi : [a,b] \to E$ is said to be a *primitive* (or antiderivative) of f if there is a set $S \subset [a,b]$ which is *at most countable* such that for every $t \in \big((a,b) - S\big)$, the function ϕ is differentiable at t and $\phi'(t) = f(t)$. My objective is to construct primitives for a class of functions which includes all maps of $[a,b]$ into E.

As you might surmise, the idea is to define the "indefinite Riemann integral" of f; this ought to serve as a primitive when f is a map; recall the "`Fundamental Theorem of Calculus`".

For any interval $[a, b] \subset \mathbb{R}$ a *partition* $\mathcal{P}$ is a finite number of points $a = t_0 < t_1 < \cdots < t_n = b$. A function $f : [a, b] \to E$ taking values in a Banach space is said to be a *step function* with respect to such a partition if for each $0 \le i < n - 1$, the restriction $f|(t_i, t_{(i+1)})$ is a constant. The value of f on this open interval is denoted $f(i, \mathcal{P})$. Note that no constraints are imposed on the function at the points t_i which constitute the partition and hence f is not necessarily a map. Since such a step function is a map iff it is also continuous at the points of $\mathcal{P}$, a step function is continuous iff it is a constant.

If $\mathcal{P}$ and $\mathcal{P}'$ are two partitions of $[a, b]$ then $\mathcal{P}$ is said to be *finer* than $\mathcal{P}'$ if $\mathcal{P}' \subset \mathcal{P}$. Observe that if f is a step function with respect to a certain partition of $[a, b]$ then it is also a step function with respect to any finer partition. Clearly given any two partitions of $[a, b]$ their union, after suitably rearranging the points in an increasing order, yields a partition finer than either.

If f and $\mathcal{P}$ are as above then we define the $\mathcal{R}$-integral or, more simply, the *integral* of f over the interval $[a, b]$ by the formula:

$$\int_a^b f(x)dx = \sum_{i=1}^{n} (t_i - t_{(i-1)}) f\big((i-1), \mathcal{P}\big).$$

Suppose f is a step function with respect to two distinct partitions of $[a, b]$ then it is clear that the value of the integral is the same when computed using either of the partitions. Moreover, if f and g are step functions with respect to different partitions the union of these partitions (we must rearrange the points so that they are in increasing order) is a partition, for which both f and g are step functions. It is clear that $(f+g)$ is a step function with respect to this new partition. Clearly if f is a step function (with respect to $\mathcal{P}$), then for any $c \in \mathbb{R}$ and $\Lambda : E \to F$ a continuous linear transformation between Banach spaces, the functions $(c \cdot f)$ and $(\Lambda \circ f)$ are also step functions defined on $[a, b]$.

We have the following relations between the integrals of the various step functions we have been discussing.

$$(1) \qquad \int_a^b \big(f(x) + g(x)\big)dx = \int_a^b f(x)dx + \int_a^b g(x)dx,$$

$$(2) \qquad \int_a^b (c \cdot f)(x)dx = c \cdot \int_a^b f(x)dx,$$

$$(3) \qquad \int_a^b (\Lambda \circ f)(x)dx = \Lambda\Big(\int_a^b f(x)dx\Big)$$

for any continuous linear transformation $\Lambda : E \to F$;

$$(4) \qquad \left|\left| \int_a^b f(x)dx \right|\right| \leq \int_a^b ||f(x)||dx$$

The proofs are trivial because the integral of any step function is a finite $\mathbb{R}$-linear combination of vectors in some Banach space.

Remark 4.1.19. Notice that the usual upper (resp. lower) Riemann sums of a real-valued function defined on $[a, b]$ are examples of $\mathcal{R}$-integrals of step functions.

A function $f : [a, b] \to E$ is said to be *regulated* if there is a countable set $S_f \subset [a, b]$ and a sequence of partitions $\mathcal{P}$ such that, for all n, $\mathcal{P}_n \subset S_f$ and there are step functions f_n with respect to $\mathcal{P}_n$ such that the f_n converge *uniformly* to f on $[a, b] - S_f$.

Proposition 4.1.20. *Any continuous function $f : [a, b] \to E$ is a regulated function.*

Proof. Let $\varepsilon > 0$ be given. Since $[a, b]$ is compact, the map f is uniformly continuous. (See Theorem 3.2.16 on page 103.) Choose an integer $M > 0$ so that $||f(t) - f(t')||_E < \varepsilon$ whenever $|t - t'| < (b - a)/M$. Now, for each $n \geq M$, consider the partitions $\mathcal{P}_n$ given by:

$$t_0^n = a \text{ and } t_j^n = \left(a + j \cdot \tfrac{b-a}{n}\right), \text{ where } 1 \leq j \leq n.$$

For each such partition, $\mathcal{P}_m$, and $j = 1, \ldots, m$, choose a point $v_j^m \in E$ from the set $f([t_{(j-1)}^m, t_j^m]) \subset E$. Define a step function, ϕ^m (with respect to the partition $\mathcal{P}_m$) by setting: $\phi^m(j, \mathcal{P}_m) = v_j^m$ and setting the value of ϕ^m arbitrarily on the points of the partition $\mathcal{P}_m$. Then clearly:

$$||f(t) - \phi^m(t)||_E < \varepsilon \text{ for all } t \in \big([a, b] - \mathcal{P}_m\big).$$

Let S_f be the union of the partitions $\{\mathcal{P}_n : n \geq M\}$. Obviously the ϕ^m's converge uniformly to F on $([a, b] - S_f)$. □

Henceforth, if $f : [a, b] \to E$ is a regulated function and $\{f_n\}$ a sequence of step functions which converge uniformly to f on the complement of a countable subset of $[a, b]$ I will simply say that *the step functions f_n converge to f* uniformly. The following lemma allows us to define the integral of any regulated function.

Lemma 4.1.21. *Let $\phi : [a, b] \to E$ be a regulated function and $\{f_n\}$, $\{g_n\}$ two sequences of step functions which converge uniformly to ϕ.*

Then

$$\lim_{n\to\infty}\int_a^b f_n(x)dx = \lim_{n\to\infty}\int_a^b g_n(x)dx$$

Proof. Let $\varepsilon > 0$ be given. Choose $N(\varepsilon) = N$ so that for all $n > N$ we have $||\phi(t) - f_n(t)||, ||\phi(t) - g_n(t)|| < \varepsilon/2$. Then using any partition for which both f_n, g_n are step functions we see from the inequality (4) on page 157 that

$$\left|\left|\int_a^b f_n(x)dx - \int^a g_n(x)dx\right|\right| = \left|\left|\int_a^b (f_n(x) - g_n(x))dx\right|\right|$$
$$\leq \int_a^b ||f_n(x) - g_n(x)||dx \leq (b-a)\varepsilon.$$

The result follows. □

If f is a regulated function and $\{f_n, \mathcal{P}_n\}$ a sequence of step functions and corresponding partitions such that f_n's converge uniformly to f then we define the *integral of* f over the interval $[a, b]$ by the formula

$$\int_a^b f(x)dx = \lim_{n\to\infty}\int_a^b f_n(x)dx.$$

You will notice that we are merely reformulating the definitions of the Riemann integral in a slightly disguised form.

By taking limits over sequences of step functions, the formulae (1)–(4) on page 157 immediately yield the following properties of the integrals of regulated functions.

Theorem 4.1.22 (Properties of the Integral).

1. *If $f, g : [a, b] \to E$ are regulated functions taking values in a Banach space E, then $(f + g)$ is a regulated function and*

$$\int_a^b \big(f(x) + g(x)\big)dx = \int_a^b f(x)dx + \int_a^b g(x)dx.$$

2. *If $f : [a, b] \to E$ is a regulated function and $\Lambda : E \to F$ a continuous linear transformation then $\Lambda \circ f$ is also a regulated function and*

$$\int_a^b (\Lambda \circ f)(x)dx = \Lambda\Big(\int_a^b f(x)dx\Big).$$

Note that this includes the case of multiplying a regulated function by a scalar.

3. *If* $f : [a, b] \to E$ *is a regulated function, then*

$$\int_a^b ||f(x)||_E dx \geq \left|\left| \int_a^b f(x)dx \right|\right|_E . \qquad \square$$

Remark 4.1.23. Property (2) in the previous theorem is particularly useful. Note that it implies, in particular, that for any continuous linear functional $\lambda : E \to \mathbb{R}$ and for every regulated function f on $[a, b]$ we have the relation

$$\int_a^b (\lambda \circ f)(x)dx = \lambda\left(\int_a^b f(x)dx\right).$$

But the left-hand side of this equation is just the Riemann integral of a $\mathbb{R}$-valued function. Since the Hahn-Banach theorem implies that two vectors in E are equal iff every continuous linear functional takes the same value on these vectors, we see that, *in principle, we can determine the integral of a regulated function by evaluating the Riemann integral of certain* $\mathbb{R}$*-valued functions.*

Note that if $f : [a, b] \to E$ is a regulated function and $c \in (a, b)$ then the restrictions of f to $[a, c]$ and $[c, b]$ are also regulated functions. Moreover,

$$\int_a^c f(x)dx + \int_c^b f(x)dx = \int_a^b f(x)dx.$$

If for $s < t \in [a, b]$ we define

$$\int_t^s f(x)dx = -\int_s^t f(x)dx$$

then for all $x, y, z \in [a, b]$ we have

$$\int_x^y f(x)dx + \int_y^z f(x)dx = \int_x^z f(x)dx$$

This leads to the familiar "anti-derivative" property of the integral, which we now explain briefly.

Let $f : [a, b] \to E$ be a regulated function and suppose $x^0 \in [a, b]$ is any point.

Define $\mathbf{P}(f; x^0) : [a, b] \to E$ by the integral: $\mathbf{P}(f; x^0)(t) = \displaystyle\int_{x^0}^t f(x)dx.$

$\mathbf{P}_{(f;x^0)}$ is called the *primitive* of f with respect to the point x^0.

Theorem 4.1.24 (The fundamental theorem of Calculus).
If $f : [a, b] \to E$ is a map of a closed interval into a Banach space E then for any $x^0 \in [a, b]$ the primitive $\mathbf{P}(f; x^0) : (a, b) \to E$ is differentiable and $\frac{d}{dt}\mathbf{P}(f; x^0)(t)\Big|_{t=\theta} = f(\theta)$ for every $\theta \in (a, b)$.

Proof. To improve legibility (during the course of the proof) I will denote the primitive by $\widetilde{f}$.

Let $\varepsilon > 0$ be given. Since f is uniformly continuous on $[a, b]$, there is a $\delta > 0$ such that if $|h| < \delta$ and $x + h \in [a, b]$, then $||f(x+h) - f(x)||_E < \varepsilon$. Suppose $\theta \in (a, b)$. Choose h so that $(\theta - h, \theta + h) \subset [a, b]$ and $h < \delta$. Then,

$$\begin{aligned} ||\widetilde{f}(\theta + h) - f(\theta) - hf(\theta)||_E &= \left|\left| \int_{x^0}^{\theta+h} f(x)dx - \int_{x^0}^{\theta} f(x)dx - hf(\theta) \right|\right|_E \\ &= \left|\left| \int_{\theta}^{\theta+h} \big(f(x) - f(\theta)\big)dx \right|\right|_E \\ &\leq h\varepsilon. \end{aligned}$$

This shows that $\widetilde{f}(\theta+h) - \widetilde{f}(\theta) - hf(\theta) = o(h)$. This proves the result. □

The next two observations, while not crucial for further progress, are worth keeping in mind.

Remark 4.1.25a. Assuming the differentiability of the primitive of f, here is an elegant method for computing its derivative.

For any continuous linear functional, $\lambda : E \to \mathbb{R}$, define $F_\lambda : [a, b] \to \mathbb{R}$ by setting $F_\lambda(t) = \lambda\big(\widetilde{f}(t)\big)$, where, as before, $\widetilde{f}$ is the primitive of f. Then for any $\theta \in (a, b)$ we have:

$$\begin{aligned} \frac{d}{dt} F_\lambda(t)\Big|_{t=\theta} &= \frac{d}{dt}\lambda(\widetilde{f})\Big|_{t=\theta} \\ &= \frac{d}{dt}\lambda\left(\int_{x^0}^{t} f(x)dx\right)\Big|_{t=\theta} \\ &= \frac{d}{dt}\left(\int_{x^0}^{t} \lambda \circ f(x)dx\right)\Big|_{t=\theta} \\ &= \lambda(f(\theta)), \end{aligned}$$

the last equality follows from the "classical version" of the Fundamental Theorem of Calculus, since $(\lambda \circ f)$ is a continuous $\mathbb{R}$-valued function on $[a,b]$.

Now the chain rule implies that

$$\lambda\big(\tfrac{d\tilde{f}}{dt}(\theta)\big) = F_\lambda'(\theta) = \lambda\big(f(\theta)\big).$$

Since this is true for all continuous linear functionals, the result follows from the Hahn-Banach theorem. □

Remark 4.1.25b. In some situations, the following idea turns out to be useful. One says that a function $f : U \to F$ is *weakly differentiable* or has a *weak derivative* at x if there is a linear transformation $L_f(x) : E \to F$ such that for all continuous linear functionals $\alpha : F \to \mathbb{R}$ the function $\alpha \circ f$ is differentiable at x with derivative $\alpha \circ L_f(x)$. This notion of weak derivative is rather useful in certain situations. For example, it might be difficult to prove that a certain function is differentiable (since this in effect means one is computing the derivative) but on the other hand it might be easier to show that weak derivatives exist. For certain classes of functions (like `holomorphic` functions defined on open subsets of $\mathbb{C}$), weak differentiability (of course, we have to look at $\mathbb{C}$-valued continuous linear functionals,) actually implies differentiability.

Remark 4.1.25c. In the next section, we will compose maps $f : [a,b] \to E$, with a continuous linear functional to reduce an otherwise difficult problem to a simple calculation in the differential Calculus of real-valued functions defined on subintervals of $\mathbb{R}$.

Exercises for § 4.1

The object of the first exercise is to introduce you to a method by which the tools of Differential Calculus can be extended to certain, not necessarily open, subsets of a Banach space. Of course, I do not intend to formally introduce the notion of `differentiable manifolds` in this book. However the computation you are being invited to perform is a classic example of how `differentiable manifolds` are constructed.

4.1.1 The Stereographic Projection Maps

Consider the "unit sphere" in $\mathbb{R}^{n+1}$:

$$\mathbb{S}^n = \{x \in \mathbb{R}^{n+1} \ : \ ||x|| = 1\}.$$

The points, $P_+ = (0, \ldots, 0, 1)$, and $P_- = (0, \ldots, 0, -1)$ are known as the North and South poles respectively. Let $U_+ = \mathbb{S}^n - P_-$ and $U_- = \mathbb{S}^n - P_+$. These two sets constitute an open cover of the unit sphere of $\mathbb{R}^{n+1}$.

Denote by Π the "equatorial plane" $\{(x_1, \ldots, x_n, x_{n+1}) \in \mathbb{R}^{n+1} : x_{n+1} = 0\}$. We will think of Π as $\mathbb{R}^n$ and denote its points by n-tuples of real numbers. $\mathbf{0}$ will denote the common origin of $\mathbb{R}^{n+1}$ and Π. Define

$$h_\pm : U_\pm \to \Pi - \mathbf{0} \subset \mathbb{R}^n$$

as follows:

Given a point $x \in U_\pm$ consider the straight line joining $P_\mp$ to x and define $h_\pm(x)$ to be the point at which this line meets the plane Π.

Let $j_\pm = h_\pm|(U_- \cap U_+)$.

(1.) Show that $h_\pm$ and their inverses are bijections between $U_\pm$ and $\Pi - \{\mathbf{0}\}$.

(2.) Show that the maps $\phi = j_+ \circ (j_-^{-1})$ and $\psi = j_- \circ (j_+^{-1})$ of $\Pi - \mathbf{0}$ to itself are differentiable and compute their derivatives.

4.1.2 Counterexamples

In the following we will be considering functions defined on $\mathbb{R}^2$ whose points will be denoted (x, y) and ρ will stand for $(x^2 + y^2)^{\frac{1}{2}}$.

a. Consider the function $f : \mathbb{R}^2 \to \mathbb{R}$ defined by

$$f(x,y) = \begin{cases} \dfrac{xy}{\rho} & \text{if } \rho \neq 0, \\ 0 & \text{when } (x,y) = (0,0). \end{cases}$$

Show that f is a map, that its partial derivatives exist at every point of $\mathbb{R}^2$ but that f is not differentiable at $(0,0)$.

> You may feel that the failure of differentiability at the origin is because of the fact that we only know that the "directional derivatives" exist only along the x- and y-axes. The next example shows that even the existence of directional derivatives along every vector is not sufficient to ensure differentiability.

b. Consider the function $g : \mathbb{R}^2 \to \mathbb{R}$ defined by

$$g(x,y) = \begin{cases} \dfrac{xy^2}{x^3 + y^2} & \text{if } (x,y) \neq (0,0) \\ 0 & \text{if } (x,y) = (0,0). \end{cases}$$

Show that for any $p \in \mathbb{R}^2$ and $\boldsymbol{v} \in \mathbb{R}^2$ the limit

$$\widetilde{g}(p, \boldsymbol{v}) \doteq \lim_{t \to 0, t \neq 0} \left(\frac{g(p + tv) - g(p)}{t} \right)$$

exists. Show that $\boldsymbol{v} \mapsto \widetilde{g}((0,0), \boldsymbol{v})$ is not a linear function of $\boldsymbol{v}$ and that g is not differentiable at the origin.

The next example indicates further subtleties.

c. Consider the function $h : \mathbb{R}^2 \to \mathbb{R}$ defined by

$$h(x,y) = \begin{cases} \dfrac{x^3 y}{x^4 + y^2} & \text{if } (x,y) \neq (0,0), \\ 0 & \text{if } (x,y) = (0,0). \end{cases}$$

Let $\widetilde{h}(p; \boldsymbol{v}) = \lim_{t \to 0}(h(p + tv) - h(p))/t$ for each $p, \boldsymbol{v} \in \mathbb{R}^2$.

Show that the limit $\widetilde{h}(p, \boldsymbol{v})$ exists for all $p, \boldsymbol{v} \in \mathbb{R}^2$ and that $\boldsymbol{v} \mapsto \widetilde{h}(p, \boldsymbol{v})$ is linear in $\boldsymbol{v}$ for all $p \in \mathbb{R}^2$. Nevertheless, h is not differentiable at the origin. Why?

4.1.3 Differentiability of Norms

1. Show that if E is a Banach space then the function $x \mapsto ||x||_E$ cannot be differentiable at $0 \in E$.

2. For $p \in \mathbb{N}$ let E_p^n be $\mathbb{R}^n$ with the ℓ_p-norm. At what points $x \in \mathbb{R}^n$ are the maps $|| \; ||_p : E_p^n \to \mathbb{R}$ differentiable?

3. Let $\mathbf{c}_0$ be the Banach space of sequences $\mathbf{x} = \{x_n\}_{n>0}$ of real numbers such that $\lim_{n \to \infty} x_n = 0$. (See Exercise 3.3.3.b)

   Prove that the norm in $\mathbf{c}_0$ is differentiable at a point $\mathbf{x}$ iff there is an index $m = m(\mathbf{x})$ such that $|x_m| > |x_n|$ for all $n \neq m(\mathbf{x})$. Compute the derivative of the norm at such points.

   (Hint: Review the computation of the continuous dual of $\mathbf{c}_0$.)

4.1.4 Velocities of Curves

Throughout this exercise $\mathbb{k}$ stands for either $\mathbb{R}$ or $\mathbb{C}$; your proofs (or answers) may need slight modifications to include both these fields. I will always denote the identity matrix of appropriate size.

1. EULER'S THEOREM

Let α be a positive real number. A function $f : E \to F$ between Banach spaces is said to be *homogeneous* of order α if for all $t \in \mathbb{R}_+$ we have the relation $f(t \cdot x) = t^\alpha \cdot f(x)$. Prove Euler's theorem: if f is a homogeneous differentiable map of order α then for every $x \neq 0 \in E$ we have $\mathbf{D}f(x)(\boldsymbol{x}) = \alpha \cdot f(x)$.

2. Let $\mathrm{SL}_n(\Bbbk)$ be the set of $(n \times n)$-matrices whose determinant is equal to 1. Show that if $\gamma : (-1, 1) \to \mathrm{SL}_n(\Bbbk)$ is a differentiable curve such that $\gamma(0) = \mathrm{I}_n$, then the velocity satisfies the condition $\mathrm{Trace}\left(\gamma'(0)\right) = 0$. What can you say about the velocity if $\gamma(0) = \mathrm{A}$?

In contrast, if $\mathrm{M} \in \mathrm{GL}_n(\Bbbk)$ and X any $(n \times n)$-matrix, then there is a curve γ lying entirely in $\mathrm{GL}_n(\Bbbk)$ such that $\gamma(0) = \mathrm{M}$ and having velocity X at $t = 0$.

3. Let $\mathrm{O}(n)$ be the group of $(n \times n)$ orthogonal matrices. Show that if $\gamma : (-1, 1) \to \mathrm{O}(n)$ is a curve such that $\gamma(0) = \mathrm{I}_n$ then $\gamma'(0)$ is a skew-symmetric matrix. What can you say about $\gamma'(0)$ if $\gamma(0)$ is an arbitrary element, $\mathrm{O} \in \mathrm{O}(n)$? Repeat the exercise replacing $\mathrm{O}(n)$ by the *special orthogonal group*: $\mathrm{SO}(n) \doteq \{\mathrm{X} \in \mathrm{O}(n) \ : \ \det \mathrm{X} = 1\}$.

4. Rework the previous problem with $\mathrm{O}(n)$ (resp. $\mathrm{SO}(n)$) replaced by

$$\begin{aligned} \mathrm{U}(n) &\doteq \{\mathrm{X} \in \mathrm{M}_n(\mathbb{C}) \ : \ \mathrm{X} \cdot \mathrm{X}^* = \mathrm{I}\} \\ (\text{resp. } \mathrm{SU}(n) &\doteq \{\mathrm{X} \in \mathrm{U}(n) \ : \ \det \mathrm{X} = 1\}). \end{aligned}$$

where X^* is the conjugate-transpose of X i.e. $\mathrm{x}^*_{ij} = \overline{x}_{ji}$.

5. Let A, B be any two matrices. Let $\phi(t) = \boldsymbol{e}^{-\mathrm{A}t}\mathrm{B}\boldsymbol{e}^{\mathrm{A}t}$.

Show that $\phi'(0) = [\mathrm{B}, \mathrm{A}] \doteq \mathrm{BA} - \mathrm{AB}$, the *commutator* of B and A.

4.1.5 MISCELLANEOUS COMPUTATIONS

1. Let $\boldsymbol{v} \neq 0$ be any vector in $\mathbb{R}^n$ and let
$$U_{\boldsymbol{v}} = \{\boldsymbol{x} \in \mathbb{R}^n \ : \ \langle \boldsymbol{v}, \boldsymbol{x} \rangle \neq 1\}.$$
Show that for each $\boldsymbol{x} \in U_{\boldsymbol{v}}$ the matrix $(\mathrm{I}_n - \boldsymbol{v}^t\boldsymbol{x})$ is invertible. Compute the derivative of the map $f : U_{\boldsymbol{v}} \to \mathrm{GL}_n(\Bbbk)$ defined by

$$f(\boldsymbol{x}) = (\mathrm{I} - \boldsymbol{v}^t\boldsymbol{x})^{-1}.$$

(HINT: Use the "fake Neumann series" explained in Example 3.4.4b; see page 120.)

2. Let $\mathrm{GL}_n^+(\mathbb{R})$ be the set of $(n \times n)$ matrices with determinant $>$ 0. Define $\phi : \mathrm{GL}_n^+(\mathbb{R}) \to \mathbb{R}$ by $\phi(\mathrm{X}) = \log(\det \mathrm{X})$. Show that $\mathbf{D}\phi(\mathrm{X})(\boldsymbol{H}) = \mathrm{Trace}\,(\mathrm{X}^{-1}\mathrm{H})$.

4.1.6 On the Gram-Schmidt Procedure

For this exercise we will be considering vectors of $\mathbb{R}^n$ as column vectors and an $(n \times n)$ matrix will be thought interchangeably, as a matrix or as an ordered n-tuple of elements of $\mathbb{R}^n$.

Let $\mathbb{R}^{n-1}$ be thought of as the set of vectors in $\mathbb{R}^n$ whose last coordinate is zero. Define the map $\phi : \mathbb{R}^{n-1} \to \mathrm{O}(n)$ as follows. Given $\boldsymbol{x} \in \mathbb{R}^{n-1}$ such that $||\boldsymbol{x}|| < 1$, consider the matrix $\mathrm{M}_x = [\boldsymbol{e}_1 + \boldsymbol{x}, \boldsymbol{e}_2, \dots, \boldsymbol{e}_n]$. Show that these columns form a basis of $\mathbb{R}^n$ and let $\phi(x)$ be the orthonormal basis given by the Gram-Schmidt process applied to M_x; $\phi(x)$ being thought of as an orthogonal matrix. Prove that ϕ is a differentiable mapping and show that $\mathbf{D}\phi(0)(\boldsymbol{v})$ is a skew-symmetric matrix which has a block-decomposition

$$\begin{bmatrix} 0 & -\boldsymbol{v}^t \\ \boldsymbol{v} & \mathbf{0} \end{bmatrix}$$

where $\mathbf{0}$ is the zero matrix of order $(n-1) \times (n-1)$.

Formulate and prove the corresponding results for the G–S process in the complex linear space $\mathbb{C}^n$.

(HINT: You will get a differentiable map $\mathbb{C}^{n-1} \to \mathrm{U}(n)$. The derivative of this map, the answer you are being invited to provide, will be a skew-Hermitian matrix.)

4.1.7 Integration by Parts

Let $[a, b]$ be a closed, bounded interval. Suppose that f and g are differentiable maps defined on $[a, b]$ and taking values in Banach spaces F and G respectively. Let $B : F \times G \to H$ be a continuous bilinear function taking values in a Banach space H.

Show that,

$$\int_a^b B(f'(x), g(x))dx = (B(f(b), g(b)) - B(f(a), g(a))) - \int_a^b B(f(x), g'(x))dx.$$

4.2 The Mean-Value Theorem

If you look back upon your earlier courses on Analysis you will find that many of the most important results that you learnt, depend crucially upon the Mean-Value Theorem: Taylor's Formula and the fact that a function is constant if its derivative vanishes everywhere, are two examples at opposite ends on a scale of difficulty. Clearly if we are to make any further progress with our present version of Differential Calculus, we need to have an analogue of the MVT. (I will henceforth use this abbreviation for "Mean-Value Theorem".)

Let us recall the classical result. It asserts that:
If $f : [a, b] \to \mathbb{R}$ is a map which is differentiable on (a, b), then there exists $\xi \in (a, b)$ such that $\big(f(b) - f(a)\big) = (b - a)f'(\xi)$.

The next example warns us to proceed with caution in our quest for a generalization of the MVT.

Example 4.2.1. Consider the function $f : \mathbb{R} \to \mathbb{R}^2$ defined by

$$f(t) = (t^2 - t, t^3 - t).$$

If the MVT were true in the same form as in the case of real-valued differentiable maps defined on closed intervals of $\mathbb{R}$, then there would exist $0 < \xi < 1$ such that $f(1) - f(0) = \mathbf{D}f(\xi)(\mathbf{1} - \mathbf{0})$. (Bold fonts have been used for '0' and '1', as a reminder that the derivative, $\mathbf{D}f(\xi)$, is acting on a vector, which happens to be the difference of two real numbers.)

Using the rule for differentiating vector-valued functions we see that $f'(t) = (2t - 1, 3t^2 - 1)$ and hence $f'(t) \neq 0$ for any $t \in \mathbb{R}$. On the other hand, $f(0) = f(1)$!

So it is clear that MVT for functions taking values in spaces of dimension > 1, *must be somewhat different* from the classical result.

The point to note about the classical result is that it does not give any information about the location of $\xi \in (a,b)$. Furthermore, if you look at the various applications, you will see that all that is ever used is that $\sup\{|f'(\xi)| : \xi \in (a,b)\}$ cannobt be *less than* the absolute value of the average rate of change of f in $[a,b]$ that is:

$$\sup_{\xi\in(a,b)} |f'(\xi)| \geq \left|\frac{f(b)-f(a)}{b-a}\right|.$$

This suggests that what we should look for is a result which will allow us to *estimate* the distance between the values of a differentiable map at two points, say x and y in terms of $||\boldsymbol{x}-\boldsymbol{y}||$ and the magnitude, (i. e. the operator norm,) of the derivative at points lying "between" x and y. Indeed there is such a generalization of the MVT for differentiable maps between open subsets of Banach spaces; this will be the first result of this section.

Theorem 4.2.2 (The Mean-Value Theorem).
Let E, F be Banach spaces, $U \subset E$ an open set and $f : U \to F$ a differentiable map. Suppose that x and y are two points of U such that the line segment

$$L_{x,y} = \{tx + (1-t)y : t \in [0,1]\}$$

joining x and y is entirely contained in U. Then

$$||f(x) - f(y)||_F \leq ||\boldsymbol{x}-\boldsymbol{y}||_E \cdot \sup_{\xi\in L_{x,y}} ||\mathbf{D}f(\xi)||_{\text{op}}.$$

Proof. Let $\lambda : F \to \mathbb{R}$ be any continuous linear functional and consider the function $\phi_\lambda : [0,1] \to \mathbb{R}$ defined by

$$\phi_\lambda(t) = \lambda(f(tx + (1-t)y)).$$

The map, $t \mapsto (tx + (1-t)y)$, is differentiable and its velocity (i. e. derivative) at each $t \in \mathbf{I}$ is $(\boldsymbol{x}-\boldsymbol{y})$. Since the maps f and λ are also differentiable so is the map ϕ_λ and by the chain rule:

$$\phi'_\lambda(t) = \lambda\big(\mathbf{D}f(tx + (1-t)y)(\boldsymbol{x}-\boldsymbol{y})\big). \tag{4.2.1}$$

So, by the MVT for real-valued differentiable maps on $[0,1]$ there exists $\theta_\lambda \in (0,1)$, depending on λ, such that

$$\lambda(f(x) - f(y)) = \phi_\lambda(1) - \phi_\lambda(0) = \phi'_\lambda(\theta_\lambda). \tag{4.2.2}$$

We now appeal to the Hahn-Banach Theorem and choose $\mu \in F'$ such that $||\mu|| = 1$ and $\mu(f(x) - f(y)) = ||f(x) - f(y)||_F$. I will denote by ξ the point $\theta_\mu x + (1 - \theta_\mu)y \in E$, where θ_μ is a point in $[0, 1]$ at which the equality 4.2.2 holds with $\lambda = \mu$.

Now if we take norms, on the right-hand side of 4.2.2 (of course, with $\lambda = \mu$), we get:

$$\begin{aligned} ||f(x) - f(y)||_F &= |\phi'(\theta_\mu)| \\ &= |\mathbf{D}\phi(\theta_\mu)(1)| \\ &\leq |\mu \circ \mathbf{D}f(\xi)(\boldsymbol{x} - \boldsymbol{y})| \\ &\leq ||\mu|| \cdot ||\mathbf{D}f(\xi)||_{\text{op}} ||\boldsymbol{x} - \boldsymbol{y}||_E \\ &\leq \sup_{\xi \in L_{x,y}} ||\mathbf{D}f(\xi)||_{\text{op}} \cdot ||\boldsymbol{x} - \boldsymbol{y}||_E \end{aligned}$$

as required. □

An easy corollary is the following fundamental (and familiar) result.

Corollary 4.2.2 (i). *Let U be a connected, open set in a Banach space E and suppose that $f : U \to F$ is a differentiable map into a Banach space F. If $\mathbf{D}f(x) = 0$ for all $x \in U$ then f is a constant.*

Proof. Given any $x^0 \in U$ and an $r > 0$ such that $\mathbb{B}_r(x^0) \subset U$ it is clear that for any $x, y \in \mathbb{B}_r(x^0)$ the line segment joining x and y lies in U. Hence, by the MVT proved above, $f(x) = f(y)$.

This actually shows that the set

$$U' = \{x \in U \,:\, f(x) = f(x^0)\}$$

is open in U. But this set is also closed, since f is continuous. Since U is connected the only open-and-closed subsets of U are U itself and $\emptyset$. Since U' is nonempty, ($x^0 \in U'$), $U' = U$. Hence f is constant. □

In future applications, the following variant of the MVT proves extremely useful; it is generally known as the "Second Mean-Value Theorem".

Theorem 4.2.3 (The second Mean-Value Theorem).
Let a, b be points in a Banach space E and U an open set containing the line segment L joining a and b. If $f : U \to F$ is a differentiable map into a Banach space F, then for any $x^0 \in U$ we have the relation:
$||f(a) - f(b) - \mathbf{D}f(x^0)(\boldsymbol{a} - \boldsymbol{b})|| \leq ||\boldsymbol{a} - \boldsymbol{b}|| \cdot \sup_{\xi \in L} ||\mathbf{D}f(\xi) - \mathbf{D}f(x^0)||_{\text{op}}$.

Proof. Consider the map $g : U \to F$ defined by $g(x) = f(x) - \mathbf{D}f(x^0)(\boldsymbol{x})$. Clearly $\mathbf{D}g(x) = \mathbf{D}f(x) - \mathbf{D}f(x^0)$ and the result follows from the Theorem 4.2.2. □

Here is a more substantial application of the MVT. The result establishes conditions, on a sequence of differentiable functions which, when fulfilled, allow us to interchange the operations of taking limits and differentiation.

Theorem 4.2.4.
Let U be an open, connected subset in a Banach space E. Suppose that:

1. *$\{f_n\}$ is a sequence of differentiable mappings of U into a Banach space F;*
2. *there is a point $x^0 \in U$ such that $\{f_n(x^0)\}$ converges in F;*
3. *for each $a \in U$ there is a ball $\mathbb{B}_r(a) \subset U$ such that in this ball the sequence $\mathbf{D}f_n(x)$ converges uniformly.*

Then for each $a \in U$, the sequence f_n is uniformly convergent in $\mathbb{B}_r(a)$. Moreover, if for each $x \in U$, we set

$$f(x) = \lim_{n\to\infty} f_n(x), \quad \textit{and } \Gamma(x) = \lim_{n\to\infty} \mathbf{D}f_n(x).$$

then $\Gamma(x) = \mathbf{D}f(x)$ for each $x \in U$.

In other words, $\lim_{n\to\infty} \big(\mathbf{D}f_n(x)\big) = \mathbf{D}\Big(\lim_{n\to\infty} f_n\Big)(x)$ for each $x \in U$.

Proof. By the second MVT, for any $m, n \in \mathbb{N}$ and $x \in \mathbb{B}_r(a)$ we have:

$$\begin{aligned} ||\big(f_n(x) - f_m(x)\big) - \big(f_n(a) - f_m(a)\big)|| \\ \leq ||\boldsymbol{x} - \boldsymbol{a}|| \cdot \sup_{z\in\mathbb{B}_r(a)} ||\mathbf{D}f_n(z) - \mathbf{D}f_m(z)|| \\ \leq r \cdot \sup_{z\in\mathbb{B}_r(a)} ||\mathbf{D}f_n(z) - \mathbf{D}f_m(z)|| \end{aligned}$$

Hence,

$$||\big(f_n(x) - f_m(x)\big) - \big(f_n(a) - f_m(a)\big)|| \leq r \cdot \sup_{z\in\mathbb{B}_r(a)} ||\mathbf{D}f_n(z) - \mathbf{D}f_m(z)|| \quad (*)$$

As the sequence $\{\mathbf{D}f_n(x)\}$ is uniformly convergent in $\mathbb{B}_r(a)$ and since F is complete, this proves that if the sequence $\{f_n(x)\}$ is convergent at some point of $\mathbb{B}_r(a)$, then it is also convergent (in fact, uniformly convergent) at *every* point of $\mathbb{B}_r(a)$ This shows that the set

$$V \doteq \{x \in A \, : \, \{f_n(x)\} \text{ is a convergent sequence}\}$$

is open. The same estimate also implies that the complement of V is open. Since $x^0 \in V$, $V \neq \emptyset$. Hence the connectedness of U implies $V = U$.

I will now show that $\mathbf{D}f(x) = \Gamma(x)$. Given $\varepsilon > 0$, by (c), there is an integer N_0 such that for $m, n \geq N_0$, $||\mathbf{D}f_m(z) - \mathbf{D}f_n(z)|| \leq \varepsilon/r$ for every $z \in \mathbb{B}_r(a)$, and $||\Gamma(a) - \mathbf{D}f_n(a)|| \leq \varepsilon$. Letting m tend to ∞ in $(*)$, we see that, for $n > N_0$ and $x \in \mathbb{B}_r(a)$, we have

$$(\alpha) \qquad ||(f(x) - f(a)) - (f_n(x) - f_n(a))|| \leq \varepsilon \cdot ||x - a||.$$

On the other hand for any $n > N_0$ there is $r' \leq r$ such that for $||x - a|| < r'$ we have

$$(\beta) \qquad ||f_n(x) - f_n(a) - \mathbf{D}f_n(a)(\boldsymbol{x} - \boldsymbol{a})|| \leq \varepsilon \cdot ||\boldsymbol{x} - \boldsymbol{a}||.$$

Also, if $||x - a|| \leq r'$, then we have,

$$(\gamma) \qquad ||\mathbf{D}f_n(a)(x - a) - g(a)(x - a)|| \leq \varepsilon ||x - a||.$$

From the inequalities α, β, γ and the triangle inequality, we get

$$||f(x) - f(a) - g(a)(\boldsymbol{x} - \boldsymbol{a})|| \leq 3 \cdot \varepsilon ||\boldsymbol{x} - \boldsymbol{a}||$$

which shows that f is differentiable at a and that $\mathbf{D}f(a) = \Gamma(a)$. □

You should make the easy adaptation of the above result and produce a theorem about the differentiability properties of convergent *series* of differentiable functions.

Example 4.2.5. For any Banach space E the function $\boldsymbol{e} : \mathbb{L}(E) \to \mathbf{GL}(E)$ which takes a bounded operator A to its exponential $\boldsymbol{e}^A$ is differentiable. The computation of the derivative $\mathbf{De}(A)(B)$ is quite subtle. See the last exercise at the end of this section.

One of the most important applications of the MVT is in studying the differentiability properties of "`functions of several variables`".

My objective now is to obtain a result which will relate the differentiability of a function which depends on more than one variable to the properties of the "partial derivatives" of the function.

It is necessary to be careful in setting up the notation and nomenclature. Let $E_1, \ldots, E_n$ be Banach spaces and let $E = E_1 \oplus \cdots \oplus E_n$ be their direct sum equipped with the ℓ_∞-norm:

$$||(\boldsymbol{v}_1, \ldots, \boldsymbol{v}_n)||_E = \max\{||\boldsymbol{v}_1||_{E_1}, \ldots, ||\boldsymbol{v}_n||_{E_n}\}.$$

Since I wish to take the view that functions on E and its subsets depend on the n components of the vectors in $E = E_1 \oplus \cdots E_n$, I will denote E by $E_1 \times \cdots \times E_n$ which, after all, is the underlying set of E. Let $U \subset E = E_1 \times \cdots \times E_n$ be an open set and let $\phi : U \to F$ be a map taking values in a Banach space F. Such a function will be said to be a *function of n variables.* Let $p = (p_1, \ldots, p_n) \in U \subset E$ and choose $\delta > 0$ so that the ball $\mathbb{B}_\delta(p) \subset U$. Then for each $j = 1, \ldots, n$ we have a function $\phi_j^p : \mathbb{B}_\delta(p_j) \to F$ defined by:

$$\phi_j^p(t_j) = \phi_j^p(p_1, \ldots, p_{j-1}, t_j, p_{j+1}, \ldots, p_n).$$

I will say that the jth partial derivative of ϕ exists at the point $p \in U$ if ϕ_j^p is differentiable at $p_j \in E_j$. The derivative, $\mathbf{D}\phi_j^p(p_j) : E_j \to F$ will be called the jth *partial derivative* of ϕ at p and denoted $\mathbf{D}_j\phi(p)$.

You have already seen that the existence of partial derivatives does not imply the differentiability of ϕ. The next theorem is a very beautiful and useful result relating the derivative of a function and its partial derivatives. But first I have to introduce a new idea; one which had been hinted at, as early as the preamble to chapter 2. Let $U \subset E$ be an open subset of a Banach space and suppose $f : U \to F$ is a differentiable map taking values in another Banach space F. f is said to be *continuously differentiable* at $u \in U$, if the function $\mathbf{D}f : U \to \mathbb{L}(E, F)$ defined by $x \mapsto \mathbf{D}f(x)$ is continuous at u. If f is continuously differentiable at each point of U f is said to be of class C^1 or, more simply, f is a C^1-*map* from U to F.

Theorem 4.2.6 (C^1-maps and partial derivatives).
Let $E_1, \ldots, E_n$ and F be Banach spaces and let $f : U \to F$ be a map defined on an open set $U \subset E = E_1 \times \cdots \times E_n$. Then f is a C^1-map if and only if *all the partial derivatives of f exist at each point in U and if each of the partial derivative functions, $\mathbf{D}_j f : U \to \mathbb{L}(E_j, F)$ are continuous.*

Proof. We begin by proving the necessity of the condition.

For $j = 1, \ldots, n$ let $\alpha_j : E_j \to E$ be the natural injections defined by:

$$\boldsymbol{v} \mapsto (\delta_{1j}\boldsymbol{v}, \ldots, \delta_{nj}\boldsymbol{v}),$$

where it is understood that when $m \neq j$, $(\delta_{mj} \cdot \boldsymbol{v})$ is the zero vector of E_m. Clearly the α_j's are continuous linear transformations.

Fix any point $p \in U$ and an index $1 \leq j \leq n$. Let $\mathbb{B}_{\delta,j}$ be the ball of radius δ centred at the origin of E_j and let $\alpha_j(\mathbb{B}_{\delta,j}) \subset E_j$ be denoted $\mathbb{B}^j_\delta$. Then if $\delta > 0$, is sufficiently small, the function $E_j \supset \mathbb{B}_\delta(0) \xrightarrow{f^p_j} F$ defined by $v \mapsto f(p_1, \ldots, p_j + v, \ldots, p_n)$ is a composition of functions as indicated in the following commutative diagram.

$$\begin{array}{ccc}
\mathbb{B}_{\delta,j}(0) & \xrightarrow{\alpha_j} & \mathbb{B}^j_\delta(0) \\
{\scriptstyle f^p_j}\downarrow & & \downarrow{\scriptstyle +p} \\
F & \xleftarrow[f]{} & U
\end{array}$$

The derivative of the top arrow is obviously α_j at all points, hence $\mathbf{D}\alpha_j$ is continuous. The derivative of the vertical arrow on the right is 1_E, again a constant and therefore continuous. $\mathbf{D}f : U \to \mathbb{L}(E, F)$ is continuous, by hypothesis. So by the chain rule $\mathbf{D}_j f(p) \doteq \mathbf{D}\big(f \circ (+p) \circ \alpha_j\big)(p) = \mathbf{D}f(p) \circ 1_E \circ \alpha_j$. Clearly, $p \mapsto \mathbf{D}_j f(p)$ is continuous for all $1 \leq j \leq n$.

I will now establish the sufficiency of the condition. I will write out the proof for $n = 2$ to simplify the notation; no new idea is needed to treat the general case.

So let us suppose that $x = (x_1, x_2) \in U \subset E = E_1 \times E_2$ and that all the partial derivatives of f exist and are continuous at all points of U. If $\boldsymbol{t} = (\boldsymbol{t}_1, \boldsymbol{t}_2) \in E$ is sufficiently small in norm, then $x+t = (x_1+t_1, x_2+t_2)$ will be a point in U. To show the differentiability of f at x (and, in fact, compute this derivative) I will show that,

$$\text{(4.2.3)} \quad \big\|\big(f(x_1 + h_1, x_2 + h_2) - f(x_1, x_2)\big) - \mathbf{D}_1 f(x)(\boldsymbol{h}_1) - \mathbf{D}_2 f(x)(\boldsymbol{h}_2)\big\| = o(\boldsymbol{h})$$

where $\boldsymbol{h} = (\boldsymbol{h}_1, \boldsymbol{h}_2)$.

To this end, let $\varepsilon > 0$ be arbitrarily fixed. We apply the Second MVT to the left-hand side of (4.2.3) in the following way. We write the term inside the norm, in (4.2.3) as a sum of:

$$(\dagger) \qquad f(x_1 + h_1, x_2 + h_2) - f(x_1 + h_1, x_2) - \mathbf{D}_2 f(x)(\boldsymbol{h}_2) \text{ and}$$

$$(\ddagger) \qquad f(x_1 + h_1, x_2) - f(x_1, x_2) - \mathbf{D}_1 f(x)(\boldsymbol{h}_1).$$

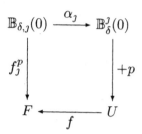

Applying the Second MVT to (†) we conclude that

$$(1)\qquad \begin{aligned}&||f(x_1+h_1,x_2+h_2)-f(x_1+h_1,x_2)-\mathbf{D}_2f(x)(\boldsymbol{h}_2)||\\ &\le \sup_{||\boldsymbol{v}||<||\boldsymbol{h}_2||} ||\mathbf{D}_2f(x_1+h_1,x+v)-\mathbf{D}_2f(x_1+h_1,x_2)||.\end{aligned}$$

Now from the definition of the derivative applied to (‡) we get that

$$(2)\qquad ||f(x_1+h_1,x_2)-f(x_1,x_2)-\mathbf{D}_1f(x)(\boldsymbol{h}_1)|| \le \varepsilon||\boldsymbol{h}_1||$$

provided that $||\boldsymbol{h}_1||<\delta$ and $\delta=\delta(\varepsilon)>0$, is sufficiently small. Now the continuity of $\mathbf{D}_2f$ implies that if δ is small enough, then

$$(3)\qquad \sup_{||\boldsymbol{v}||<||\boldsymbol{h}_2||} ||\mathbf{D}_2f(x_1+h_1,x_2+v)-\mathbf{D}_2f(x_1+h_1,x_2)|| < \varepsilon.$$

From 1, 2 and 3 we see that if $\delta>0$ is suitably small and if $0<||h||<\delta$ then the left-hand side of (4.2.3) is $< 2\varepsilon\sup(||\boldsymbol{h}_1||,||\boldsymbol{h}_2||)=2\varepsilon||\boldsymbol{h}||$. This establishes (4.2.3). Hence $\mathbf{D}f(x_1,x_2)$ exists and is given by

$$(4.2.4)\qquad \mathbf{D}f(x_1,x_2)=\mathbf{D}_1f(x_1,x_2)\circ\pi_1+\mathbf{D}_2f(x_1,x_2)\circ\pi_2$$

where $\pi_j : E\to E_j$ are the jth projections. Since the π_j and the $\mathbf{D}_jf$'s are continuous, so is $\mathbf{D}f$. □

Remark. The proof for functions of n variables, $n>2$, is being omitted because it would be pointless to obscure the straightforward application of the Second MVT in a mass of subscripts and summation signs which would be the only new ingredients needed to prove the more general case.

The basic idea is to write $f(x_1+h_1,\ldots,x_n+h_n)-f(x_1,\ldots,x_n)$ as a sum of terms of the form $f(x)-f(y)$ where $x,y\in E$ differ only in one place, say the jth place. Then this term can be estimated by the jth partial derivative of f acting on a suitable vector. Now one applies the second MVT to $f(x)-f(y)-\mathbf{D}_jf(y)(\boldsymbol{h}_j)$ as we did in the case of 2 variables.

Remarks 4.2.6 (on Differentiability and Continuous Differentiability)

Remark 4.2.6a. Notice that the theorem tells us more than differentiability: it assures us continuity of the derivative as well. It would be nice if there was a simple necessary and sufficient condition for the existence of $\mathbf{D}f$ in terms of the properties of the partial derivatives; in the exercises some results in this direction are described.

Remark 4.2.6b. In most applications one requires not just C^1-maps but, in fact, that even the higher derivatives (which I will introduce in the next section) exist and are also continuous. So the above theorem is really one of the fundamental results of the subject.

I should point out one fairly easy extension of this result; but first a recap of notation.

Let $(E_1, \ldots, E_n)$ and $(F_1, \ldots, F_m)$ be ordered tuples of Banach spaces and let $E = E_1 \times \cdots \times E_n$ and $F = F_1 \times \cdots \times F_n$ be their Cartesian products regarded as Banach spaces with the linear space structure of the appropriate direct sum and the ℓ_∞-norm.

I will regard elements of E and F as column vectors and use the space-saving notation. Thus if $x \in E$ and $y \in F$, I will write these points in the forms:

$$x = \begin{bmatrix} x_1 & \ldots & x_n \end{bmatrix}^{\boldsymbol{t}} \equiv (x_1, \ldots, x_n) \quad \text{and}$$
$$y = \begin{bmatrix} y_1 & \ldots & y_m \end{bmatrix}^{\boldsymbol{t}} \equiv (y_1, \ldots, y_m) \quad \text{respectively.}$$

Now if $T : E \to F$ is a continuous linear transformation, there are unique linear transformations, $T_{pq} \in \mathbb{L}(E_p, F_q)$ such that:

$$T(\boldsymbol{x}_1, \ldots, \boldsymbol{x}_n) = \begin{bmatrix} \sum_{\imath=1}^{n} T_{1\imath}(\boldsymbol{x}_\imath) \\ \vdots \\ \sum_{\imath=1}^{n} T_{m\imath}(\boldsymbol{x}_\imath) \end{bmatrix}.$$

where $1 \leq p \leq n$ and $1 \leq q \leq m$. (Note that I have reverted to column vector notation, since when each component is a sum of n terms, the "space-saving notation" ceases to save space!)

Thus we can write $T \in \mathbb{L}(E, F)$ as an $(m \times n)$ matrix of continuous linear transformations:

$$T = \begin{bmatrix} T_{11} & \cdots & T_{1n} \\ \vdots & T_{pq} & \vdots \\ T_{m1} & \cdots & T_{mn} \end{bmatrix}.$$

This is the "partitioned form" of T corresponding to the direct sum decompositions of E and F.

The next result follows from Theorem 4.2.6 and Proposition 4.1.7.

Proposition 4.2.7 (The "Big" Jacobian Formula).
Let $(E_1, \ldots, E_n), (F_1, \ldots, F_m), E, F$ be as above. Suppose that $U \subset E$ is an open set and that $\phi : U \to F$ is a map.

If $\phi(x) = (\phi_1(x), \ldots, \phi_m(x)) \in F_1 \times \cdots \times F_m$, *then* ϕ *is a* C^1*-map iff each* $\mathbf{D}_p\phi_q : U \to \mathbb{L}(E_q, F_p)$*is continuous. Furthermore,* $\mathbf{D}\phi(x) \in \mathbb{L}(E, F)$ *has the partitioned form:*

$$\mathbf{D}\phi(x) = \begin{bmatrix} \mathbf{D}_1\phi_1(x) & \cdots & \mathbf{D}_n\phi_1(x) \\ \vdots & \mathbf{D}_p\phi_q(x) & \mathbf{D}_p\phi_q(x) \\ \mathbf{D}_1\phi_m(x) & \cdots & \mathbf{D}_n\phi_m(x) \end{bmatrix}.$$

This is sometimes called the Jacobian of ϕ. □

Exercises for § 4.2

4.2.1 Existence of Continuous Derivatives

Let E, F be Banach spaces, $U \subset E$ an open set and $f : U \to F$ a map. Suppose that for each $x \in U$ there is a continuous linear transformation $u(x) \in \mathbb{L}(E, F)$ such that for every $\boldsymbol{v} \in E$ the limit

$$\lim_{t \to 0, t \neq 0} \frac{f(x + tv) - f(x)}{t}$$

exists and is equal to $u(x)(\boldsymbol{v})$,

Suppose that in addition the function $x \mapsto u(x)$ is a map of U into $\mathbb{L}(E, F)$, Show that f is then a C^1-map of U into F.

(Hint: Apply the MVT to the function $[0, 1] \to F$ defined by $t \mapsto f(x + ty) - f(x)$.)

4.2.2 Continuity Properties of the Derivative

Let E, F be Banach spaces, $U \subset E$ an open set and $f : U \to F$ a differentiable map.

1. Show that for $\mathbf{D}f$ to be continuous at $p \in U$ it is necessary and sufficient that for any $\varepsilon > 0$ there exist a $\delta = \delta(\varepsilon) > 0$ such that whenever $\boldsymbol{h}, \boldsymbol{k} \in E$ have norm less than δ we have:

$$||f(p + h) - f(p + k) - \mathbf{D}f(p)(\boldsymbol{h} - \boldsymbol{k})|| \leq \varepsilon||h - k||.$$

2. Show that for $\mathbf{D}f$ to be *uniformly continuous* in U a necessary and sufficient condition is that for every $\varepsilon > 0$ there exist a $\delta = \delta(\varepsilon) > 0$ such that the following holds:

whenever $p \in U$ and $\boldsymbol{v} \in E$ are such that $p+tv \in U$ for all $t \in [0,1]$ and $||\boldsymbol{v}|| < \delta$ then

$$||f(p+v) - f(p) - \mathbf{D}f(p)(\boldsymbol{v})|| \leq \varepsilon ||\boldsymbol{v}||.$$

4.2.3 Derivatives of Functions of Several Variables

Let E_1, E_2 be Banach spaces, $U \subset E_1 \times E_2$ an open set and $f : U \to F$ a map into a Banach space F. Show that for f to be differentiable at $(x, y) \in U$ a necessary and sufficient condition is that:

a. the partial derivatives of f exist at (x, y);

b. for any $\varepsilon > 0$ there exists a $\delta = \delta(\varepsilon) > 0$ such that if $(\boldsymbol{u}, \boldsymbol{v}) \in E_1 \times E_2$ and $||(\boldsymbol{u}, \boldsymbol{v})|| < \delta$, then

$$||f(x+u, y+v) - f(x+u, y) - f(x, y+v) + f(x, y)|| \leq \varepsilon(||u|| + ||v||).$$

Show that this second condition is satisfied if $\mathbf{D}_1 f(x, y)$ exists and in some open set $V \subset E_1 \times E_2$ containing (x, y) the partial derivative $\mathbf{D}_2 f$ exists and $(u, v) \mapsto \mathbf{D}_2 f(u, v)$ is a continuous map from V to $\mathbb{L}(E_2, F)$.

4.2.4 Integrals depending on a Parameter

a. Let $U \subset E$ be an open set in a Banach space, and $f : [\alpha, \beta] \times U \to F$ a map into a Banach space F. Show that the function $g : U \to F$ defined by

$$g(x) = \int_\alpha^\beta f(t, x)dt$$

is continuous.

Suppose that the partial derivative $\mathbf{D}_2 f$ exists and defines a map from $[\alpha, \beta] \times E$ to $\mathbb{L}(E, F)$. Show that g is a C^1-map and that

$$\mathbf{D}g(x) = \int_\alpha^\beta \mathbf{D}_2 f(t, x)dt$$

b. Continuing with the same notation and hypothesis as above, suppose $\xi, \eta : U \to \mathbb{R}$ are two differentiable maps. Define $h : U \to F$ by

$$h(x) = \int_{\xi(x)}^{\eta(x)} f(t, x)dt.$$

Show that h is continuously differentiable in U and that its derivative at the point $z \in U$ is the linear transformation

$$\boldsymbol{v} \mapsto \int_{\xi(z)}^{\eta(z)} \mathbf{D}_2 f(t,z)(\boldsymbol{v})dt + \langle \mathbf{d}\beta(z), \boldsymbol{v}\rangle \cdot f(\beta(z),z) - \langle \mathbf{d}\alpha(z), \boldsymbol{v}\rangle \cdot f(\alpha(z),z).$$

4.2.5 Improper Integrals and Applications

a. Let $f : [0,\infty) \to E$ be a map into a Banach space. Imitate the procedure for real-valued functions and define the improper integral $\displaystyle\int_0^\infty f(x)dx$.

b. Derive the rule for "*differentiation under the integral sign*" as stated below:

Differentiating under the integral sign.
Let $U \subset E$ be an open set in a Banach space and $f : [0,\infty) \times U \to F$ be a map such that $\mathbf{D}_2 f(t,x)$ exists for all $(t,x) \in [0,\infty) \times U$. Suppose the integral $\left(\int_0^N f(t,x)dt\right)$ converges to a limit, $\phi(x)$, uniformly in $\xi \in U$ as $N \to \infty$. Then, $\phi : U \to F$ is a differentiable map and

$$\mathbf{D}\phi(x^0) = \int_0^\infty \mathbf{D}_2 f(t,x^0)dt$$

for all $x^0 \in U$.

c. The Lyapunov Equation.
Show that if A is a matrix such that for every eigenvalue, λ_{A}, of A, the real part of λ_{A} is strictly positive then $\int_0^\infty e^{-t\mathrm{A}}dt$ exists.

d. Show that for any *positive-definite* matrix, A

$$\int_0^\infty e^{-t\mathrm{A}} S e^{-t\mathrm{A}} dt$$

is a solution of the matrix equation:

$$(*) \qquad AX + XA = S.$$

Indeed the above is the unique solution of the Lyapunov equation (*).

4.2.6 THE DERIVATIVE OF THE EXPONENTIAL MAP, $e : A \mapsto e^A$

a. By using induction, or otherwise, show that if A, B are $(p \times p)$-matrices, then

$$(1) \qquad (\mathrm{A}+\mathrm{B})^n - \mathrm{A}^n = \sum_{k=0}^{n-1} (\mathrm{A}+\mathrm{B})^k \mathrm{B} \mathrm{A}^{n-(k+1)}.$$

b. Verify the formula:

$$(2) \qquad \int_0^1 t^k (1-t)^m \, dt = \frac{k!m!}{(k+m+1)!}.$$

Using (1) and (2), prove "Dyson's expansion":

$$\boldsymbol{e}^{(\mathrm{A}+\mathrm{B})} - \boldsymbol{e}^{\mathrm{A}} = \int_0^1 \boldsymbol{e}^{t(\mathrm{A}+\mathrm{B})} \mathrm{B} \boldsymbol{e}^{(1-t)\mathrm{A}} dt$$

c. Now show that

$$(\star) \quad ||\boldsymbol{e}^{(\mathrm{A}+\mathrm{B})} - \boldsymbol{e}^{\mathrm{A}} - \int_0^1 \boldsymbol{e}^{t\mathrm{A}} \mathrm{B} \boldsymbol{e}^{(1-t)\mathrm{A}} dt|| \leq ||\mathrm{B}|| e^{||\mathrm{A}||} \cdot \left|\left| \int_0^1 \boldsymbol{e}^{t(\mathrm{A}+\mathrm{B})} \mathrm{B} \boldsymbol{e}^{(1-t)\mathrm{A}} dt \right|\right|.$$

Show that the integral on the right hand side tends to 0 as $||\mathrm{B}|| \to 0$. Conclude that

$$\mathbf{D}\boldsymbol{e}(\mathrm{A})(\boldsymbol{B}) = \int_0^1 \boldsymbol{e}^{t\mathrm{A}} \mathrm{B} \boldsymbol{e}^{(1-t)\mathrm{A}} dt.$$

4.3 Higher Derivatives

I will begin by "fine tuning" my conventions for handling multilinear functions. This will simplify the notation used in this chapter.

Let E, F be Banach spaces. As before, $\mathbb{M}_r(E; F)$ will denote the Banach space of continuous r-linear functions from the r-fold Cartesian product of E to F.

If $\boldsymbol{v} \in E$ and $\phi \in \mathbb{M}_r(E; F)$ then $\phi(\boldsymbol{v}^{[r]})$ will denote the value of ϕ on the r-tuple all of whose entries are equal to $\boldsymbol{v}$.

Similarly, if $\phi \in \mathbb{L}_p(E;F)$ and $\boldsymbol{v} \in E$, for $k < p$ $\phi(\boldsymbol{v}^{(k)})$ will be the element $\phi(\boldsymbol{v} : \overset{k \text{ terms}}{\cdots\cdots} : \boldsymbol{v}) \in \mathbb{L}_{p-k}(E;F)$ obtained by successively applying ϕ. That is in the first step we get $\phi(\boldsymbol{v}) \in \mathbb{L}_{p-1}(E;F)$, then $\phi(\boldsymbol{v} : \boldsymbol{v}) \doteq \phi(\boldsymbol{v})(\boldsymbol{v}) \in \mathbb{L}_{p-2}(E;F)$ and so on.

Let U be an open set in a Banach space E and $f : U \to F$ a differentiable map. You have already seen that studying the continuity properties of the function $\mathbf{D}f : U \to \mathbb{L}(E,F)$ is fruitful. If this function is a map defined on U, then it is natural to ask the question, 'is this map differentiable?"

If $\mathbf{D}f : U \to \mathbb{L}(E,F)$ is differentiable at $x \in U$, then the derivative at x is called the *second derivative* of f at x. By definition, it is an element of $\mathbb{L}_2(E;F)$ and is denoted $\mathbf{D}^2 f(x)$. Using the Law of Exponents this yields a continuous bilinear transformation of $E \times E$ into F; that is, an element of $\mathbb{M}_2(E;F)$. *This bilinear transformation will also be called the second derivative of f at x and it will also be denoted* $\mathbf{D}^2 f(x)$. We inductively define the higher derivatives and C^p-maps as follows. Suppose for some $p \geq 2$ and all $1 \leq k < p$ the kth derivative has been defined; the kth derivative of f at x being an element $\mathbf{D}^k f(x) \in \mathbb{L}_k(E;F)$. The associated k-linear transformation in $\mathbb{M}_k(E;F)$ will also be denoted $\mathbf{D}^k f(x)$. Let $f : U \to F$ be a map. If f is $(p-1)$-times differentiable in an open set V_x containing a point $x \in U$ and if the function $\mathbf{D}^{p-1} f : V_x \to \mathbb{L}_{p-1}(E;F)$ is differentiable at x, then f is said to be p-times differentiable at x and the derivative of $\mathbf{D}^{p-1} f$ (which will be an element of $\mathbb{L}_p(E;F)$) is the pth derivative of f at x. The corresponding element of $\mathbb{M}_p(E;F)$ is also called the pth derivative of f.

Let us now return to the second derivative. Because of the peculiar way in which the order of the arguments appear in $\mathbb{L}_2(E,F)$ and $\mathbb{M}_2(E,F)$ I am reiterating that these two ways of looking at the second derivative of a map are related in the following manner. If $\boldsymbol{v}_1, \boldsymbol{v}_2 \in E$ then

$$\mathbf{D}^2 f(p)(\boldsymbol{v}_2 : \boldsymbol{v}_1) = \mathbf{D}^2 f(p)(\boldsymbol{v}_1, \boldsymbol{v}_2).$$

Notice that we can always tell which *avatar*, (bilinear transformation or linear 2-function) we are dealing with by looking at how the argument is written: separated by ',' for bilinear functions, separated by ':' for linear 2-functions.

Analogous relations hold between the two *avatars* of the higher derivatives, $\mathbf{D}^k f(p)$.

A function $f : U \to F$ is said to be C^∞ or *infinitely differentiable* if it is of class C^n for all $n \in \mathbb{N}$. These are also known as *smooth maps.*

We will soon see that, in fact, the order in which we write the arguments of the higher derivatives of a C^p-map is irrelevant.

I now give some basic examples of the computation of higher derivatives.

Example 4.3.1a. If $A : E \to F$ is a continuous linear transformation then the derivative $\mathbf{D}A \in \mathbb{L}(E, F)$ is the constant map equal to A; consequently $\mathbf{D}^p A = 0$ for all $p \geq 2$. □

Example 4.3.1b. Let $B : E \times E \to F$ be a continuous bilinear function and define $f : E \to F$ by $f(x) = B(x, x)$. Then

$$\begin{aligned} \mathbf{D}f(x)(\boldsymbol{h}) &= B(x, h) + B(h, x) \\ \text{and hence} \qquad \mathbf{D}^2 f(x)(\boldsymbol{k} : \boldsymbol{h}) &= B(k, h) + B(h, k). \end{aligned}$$

So in terms of continuous multilinear transformations, we have

$$\mathbf{D}^2 f(x)(\boldsymbol{h}, k) = B(\boldsymbol{h}, \boldsymbol{k}) + B(\boldsymbol{k}, \boldsymbol{h})$$ □

Notice that the second derivative is symmetric in its two arguments.

Example 4.3.1c. This is really a continuation of the previous example. I wish to compute the higher derivatives of the continuous bilinear map

$$B : E \times E \to F \text{ given by } (x, y) \mapsto B(\boldsymbol{x}, \boldsymbol{y}).$$

One can easily see that $\mathbf{D}^2 B$ is independent of (x, y) and hence the derivatives of B of order ≥ 3 are zero. □

Notice that in these examples the higher derivatives trivially satisfied the condition $\mathbf{D}^k f(x)(\boldsymbol{v}_1, \ldots, \boldsymbol{v}_k) = \mathbf{D}^k f(x)(\boldsymbol{v}_{1'}, \ldots, \boldsymbol{v}_{k'})$ for all choices of permutations $\left(\begin{smallmatrix} 1 & 2 & \cdots & k \\ 1' & 2' & \cdots & k' \end{smallmatrix}\right)$ of $\{1, \ldots, k\}$.

This symmetry property of the second and higher derivatives in these particular cases, in fact, is true *in general* for *all* C^p-maps. This is a fundamental result and will be the first major theorem of this section.

I begin with a few simple but useful, lemmata.

Lemma 4.3.2. *Let E, F be Banach spaces, $U \subset E$ an open set and $f : U \to F$ a twice-differentiable map: this means that the second derivative of f exists at each $p \in U$. Then, for any $\boldsymbol{v} \in E$, the function $f_{\boldsymbol{v}} : U \to F$ defined by*

$$f_{\boldsymbol{v}}(x) = \mathbf{D}f(x)(\boldsymbol{v})$$

is differentiable and

$$\mathbf{D}f_{\boldsymbol{v}}(x)(\boldsymbol{w}) = \mathbf{D}^2 f(x)(w : v) \equiv \mathbf{D}^2 f(x)(v, w) \tag{4.3.1}$$

Proof. Let $\phi : U \times E \to \mathbb{L}(E,F) \times E$ be the function $(x, \boldsymbol{v}) \mapsto (\mathbf{D}f(x), \boldsymbol{v})$. and define $\boldsymbol{ev}$, as usual, to be the evaluation function: $\boldsymbol{ev} : \mathbb{L}(E,F) \times E \to F$ defined as $\boldsymbol{ev}(\alpha, \boldsymbol{u}) = \alpha(\boldsymbol{u})$. Clearly $f_{\boldsymbol{v}} = (\boldsymbol{ev} \circ \phi)$. Since $\boldsymbol{ev}$ is a continuous bilinear function, it is differentiable and we can compute the derivative of $f_{\boldsymbol{v}}$ using the chain rule and the product rule as follows.

$$\begin{aligned}\mathbf{D}f_{\boldsymbol{v}}(x)(\boldsymbol{w}) &= \boldsymbol{ev}\left(\ \mathbf{D}^2 f(x)(\boldsymbol{w}), \boldsymbol{v}\ \right) + \boldsymbol{ev}(\mathbf{D}f(x), 0) \\ &= \mathbf{D}^2 f(x)(\boldsymbol{w} : \boldsymbol{v}) \\ &= \mathbf{D}^2 f(x)(\boldsymbol{v}, \boldsymbol{w}).\end{aligned}$$

□

Lemma 4.3.3. *Let E, F be Banach spaces, $U \subset E$ an open set and $f : U \to F$ a C^p-map. If $\lambda : F \to \mathbb{R}$ is any continuous linear functional on F then the composite map $\psi = \lambda \circ f$ is also p-times differentiable and $\mathbf{D}^p\psi(x) = \lambda \circ \mathbf{D}^p f(x)$.*

Proof. Example 4.3.1a shows that λ is p-times differentiable for all p and that its derivatives of order ≥ 2 are zero. The result follows upon noting that $x \mapsto \psi(x)$ is the evaluation of the continuous bilinear transformation $\boldsymbol{ev}$ on $(\lambda, f(x))$ and using the product rule. □

We now prove the symmetry property of the higher derivatives. For reasons that will be explained later this result is referred to, in classical textbooks, as the "`equality of mixed partial derivatives`".

Theorem 4.3.4 (Symmetry of Higher Derivatives).
Let U be an open set in a Banach space E and suppose $f : U \to F$ is a C^p-map into a Banach space F. Then for all $x \in U$ and for any p vectors $\boldsymbol{v}_1, \ldots, \boldsymbol{v}_p \in E$, and for all permutation σ of $\{1, \ldots, p\}$. we have

$$\mathbf{D}^p f(x)(\boldsymbol{v}_1, \ldots, \boldsymbol{v}_p) = \mathbf{D}^p f(x)(\boldsymbol{v}_{\sigma(1)}, \ldots, \boldsymbol{v}_{\sigma(p)}).$$

Proof. The proof proceeds by induction on p. The case $p = 1$ is trivial. Unfortunately, one cannot start the induction at this stage: some hard work is necessary to establish the case $p = 2$ after which one can use induction. So suppose that f is a C^2-map; we must show that for each point $x \in U$ and all $\boldsymbol{v}, \boldsymbol{w} \in E$ we have the relation:

$$\mathbf{D}^2 f(x)(\boldsymbol{v}, \boldsymbol{w}) = \mathbf{D}^2 f(x)(\boldsymbol{w}, \boldsymbol{v}).$$

(In order to avoid unnecessary cluttering of notations, I will not use subscripts for the various norms that will arise during our calculations: the reader should keep track of these.)

First choose $r > 0$ so that $\mathbb{B}_r(x) \subset U$ and then choose $\boldsymbol{v}, \boldsymbol{w} \in E$ so that $||\boldsymbol{v}||, ||\boldsymbol{w}|| < \frac{r}{2}$. Observe that for all $t \in [0,1]$ the points $(x + tv + w)$ and $(x + tv)$ are in U. We define a map $g : [0,1] \to F$ by setting

$$g(t) = f(x + tv + w) - f(x + tv).$$

This g is clearly differentiable and the second MVT implies that

$$(\star) \qquad ||g(1) - g(0) - g'(0)|| \leq \sup_{\theta \in (0,1)} ||g'(\theta) - g'(0)||.$$

Using the chain rule, we get

$$\begin{aligned} g'(\theta) &= \Big(\mathbf{D}f(x + \theta\boldsymbol{v} + \boldsymbol{w}) - \mathbf{D}f(x + \theta\boldsymbol{v})\Big)(\boldsymbol{v}) \\ (\alpha) \qquad &= \Big(\mathbf{D}f(x + \theta\boldsymbol{v} + \boldsymbol{w}) - \mathbf{D}f(x)\Big)(\boldsymbol{v}) \\ (\beta) \qquad &\qquad - \Big(\mathbf{D}f(x + \theta\boldsymbol{v}) - \mathbf{D}f(x)\Big)(\boldsymbol{v}). \end{aligned}$$

We estimate the two terms on the right-hand side of (α) and (β) separately.

Since f is twice differentiable $\mathbf{D}f$ is differentiable and so we can choose $0 < r' < r$ so that for $||\boldsymbol{v}||, ||\boldsymbol{w}|| < r'$ and for all $t \in [0,1]$ we have

$$||\mathbf{D}f(x + \theta v + w) - \mathbf{D}f(x) - \mathbf{D}^2 f(x)(\theta\boldsymbol{v} + \boldsymbol{w})|| \leq \varepsilon(||\boldsymbol{v}|| + ||\boldsymbol{w}||), \text{ and}$$

$$(1) \qquad ||\mathbf{D}f(x + \theta v) - \mathbf{D}f(x) - \mathbf{D}^2 f(x)(\theta\boldsymbol{v})|| \leq \varepsilon.||\boldsymbol{v}||.$$

Now $\mathbf{D}^2 f(x)$ being bilinear,

$$\mathbf{D}^2 f(x)(\boldsymbol{v}, \theta\boldsymbol{v} + \boldsymbol{w}) - \mathbf{D}^2 f(x)(\boldsymbol{v}, \theta\boldsymbol{v}) = \mathbf{D}^2 f(x)(\boldsymbol{v}, \boldsymbol{w})$$

and hence it follows that for $\theta \in [0,1]$,

$$\begin{aligned} (2) \qquad & ||g'(\theta) - \mathbf{D}^2 f(x)(\boldsymbol{v}, \boldsymbol{w})|| \\ & \leq ||(\mathbf{D}f(x + \theta\boldsymbol{v}) - \mathbf{D}f(x))(\boldsymbol{v}) - \mathbf{D}^2 f(x)(\boldsymbol{v}, \theta\boldsymbol{v})|| \\ & + ||\left(\mathbf{D}f(x + \theta\boldsymbol{v} + \boldsymbol{w}) - \mathbf{D}f(x)\right)(\boldsymbol{v}) - \mathbf{D}^2 f(x)(\boldsymbol{v}, \theta\boldsymbol{v} + \boldsymbol{w})||. \end{aligned}$$

Combining (1) and (2) we get

$$(3) \qquad ||g'(\theta) - \mathbf{D}^2 f(x)(\boldsymbol{v}, \boldsymbol{w})|| \leq 2\varepsilon ||\boldsymbol{v}||(||\boldsymbol{v}|| + ||\boldsymbol{w}||).$$

Since the right-hand side of (3) is independent of θ we conclude also that

$$||g'(t) - g'(0)|| \leq ||g'(t) - \mathbf{D}^2 f(x)(\boldsymbol{v}, \boldsymbol{w})|| + ||g'(0) - \mathbf{D}^2 f(x)(\boldsymbol{v}, \boldsymbol{w})||. \quad \text{Hence,}$$

$$(\star\star) \qquad ||g'(t) - g'(0)|| \leq 4\varepsilon ||\boldsymbol{v}||(||\boldsymbol{v}|| + ||\boldsymbol{w}||).$$

Let $\theta_0 \in [0, 1]$ be a point at which $||g'(\theta) - g'(0)||$ attains its supremum. Then we have

$$\begin{aligned} ||g(1) - g(0) - \mathbf{D}^2 f(x)(\boldsymbol{v}, \boldsymbol{w})|| &\leq ||g(1) - g(0) - g'(0)|| + ||g'(0) - g'(\theta_0)|| \\ &\quad + ||g'(\theta_0) - \mathbf{D}^2 f(x)(\boldsymbol{v}, \boldsymbol{w})|| \\ &\leq 2||g'(\theta_0) - g'(0)|| \\ &\quad + ||g'(\theta_0) - \mathbf{D}^2 f(x)(\boldsymbol{v}, \boldsymbol{w})|| \end{aligned}$$

Hence, using (3) and ($\star\star$), we get

$$(\P) \qquad ||g(1) - g(0) - \mathbf{D}^2 f(x)(\boldsymbol{v}, \boldsymbol{w})|| \leq 6\varepsilon ||\boldsymbol{v}||(||\boldsymbol{v}|| + ||\boldsymbol{w}||).$$

Now $g(1) - g(0) = f(x+v+w) - f(x+v) - f(x+w) + f(x)$ is symmetrical with respect to $\boldsymbol{v}$ and $\boldsymbol{w}$. Hence interchanging $\boldsymbol{v}$ and $\boldsymbol{w}$ we get another inequality similar to ¶ but with $\boldsymbol{v}, \boldsymbol{w}$ interchanged. Using these two we get

$$(\dagger) \qquad ||\mathbf{D}^2 f(x)(\boldsymbol{w}, \boldsymbol{v}) - \mathbf{D}^2 f(x)(\boldsymbol{v}, \boldsymbol{w})|| \leq 6\varepsilon(||\boldsymbol{v}|| + ||\boldsymbol{w}||)^2.$$

Now (†) holds for all $\boldsymbol{v}, \boldsymbol{w} \in E$ having norm $\leq r'$. Since both sides of this inequality are homogeneous functions of order 2 the homogeneity implies that this inequality will continue to hold for all v, w and hence $\mathbf{D}^2 f(x)$ is a symmetric bilinear form. This completes the proof of the case $p = 2$.

To establish the result for arbitrary $p > 2$, we need a fact about the group of permutations which you should be able to prove easily:

(A property of $\mathfrak{S}_p$.)
Any permutation σ of $\{1, 2, \ldots, p\}$ can be obtained by composing permutations of $\{1, \ldots, (p-1)\}$ with transpositions of the index "p" and the indices $\{1, 2, \ldots, (p-1)\}$.

Suppose now that the result holds for the $(p-1)^{\text{th}}$ derivatives of arbitrary C^{p-1}-maps. Given the C^p-map $f : U \to F$ and $x \in U$, fix p vectors, $\boldsymbol{v}_1, \boldsymbol{v}_2, \ldots, \boldsymbol{v}_p$ of E and consider the function $g : U \to F$ defined by

$$g(x) = \mathbf{D}^{p-2} f(x)(\boldsymbol{v}_1, \ldots, \boldsymbol{v}_{p-2}).$$

An argument similar to Lemma 4.3.2 shows that g is twice differentiable and the second derivative of g is the bilinear transformation

$$\mathbf{D}^2 g(x) : (\boldsymbol{v}_{p-1}, \boldsymbol{v}_p) \mapsto \mathbf{D}^p f(x)(\boldsymbol{v}_1, \ldots, \boldsymbol{v}_p).$$

Since the left side is a second derivative acting on v_{p-1}, v_p, by the symmetry property of $\mathbf{D}^2$ already proved, we have

$$\mathbf{D}^p f(x)(\boldsymbol{v}_1, \ldots, \boldsymbol{v}_p) = \mathbf{D}^p f(x)(\boldsymbol{v}_1, \ldots, \boldsymbol{v}_{p-2}, \boldsymbol{v}_p, \boldsymbol{v}_{p-1}).$$

Now by the induction hypothesis if τ is a permutation of $\{1, \ldots, p-1\}$ then

$$\mathbf{D}^{p-1} f(x)(\boldsymbol{v}_1, \ldots, \boldsymbol{v}_{p-1}) = \mathbf{D}^{p-1} f(x)(\boldsymbol{v}_{\tau(1)}, \ldots, \boldsymbol{v}_{\tau(p-1)}).$$

Taking the first derivative of both sides and applying the derivative to $\boldsymbol{v}_p$ we get

$$\mathbf{D}^p f(x)(\boldsymbol{v}_1, \boldsymbol{v}_2, \ldots, \boldsymbol{v}_p) = \mathbf{D}^p f(x)(\boldsymbol{v}_{\tau(1)}, \ldots, \boldsymbol{v}_{\tau(p-1)}, \boldsymbol{v}_p).$$

The result now follows from the property of permutations cited above. □

We now discuss an important special case. Suppose $U \subset E$, is an open set in a Euclidean space. Let $\{\boldsymbol{e}_1, \ldots, \boldsymbol{e}_n\}$ be an orthonormal basis and $\{x_i\}_1^n$ the associated coordinates. If a function $f : U \to \mathbb{R}$ is differentiable p-times on U then for $2 \leq k \leq p$ one uses the notation:

$$\frac{\partial^k f}{\partial x_{i_1} \cdots \partial x_{i_k}}(x) \quad \text{or} \quad \left.\frac{\partial^k f}{\partial x_{i_1} \cdots \partial x_{i_k}}\right|_x \quad \text{for } \mathbf{D}^k f(x)(\boldsymbol{e}_{i_1}, \ldots, \boldsymbol{e}_{i_k}).$$

These are called the *mixed or higher partial derivatives* of f. Notice the notation is in consonance with our earlier notation for partial derivatives of real-valued functions.

The classical version of Theorem 4.3.4 is:

Theorem 4.3.5 (The equality of Mixed partial Derivatives).
Let E be a Euclidean space, $U \subset E$ an open set and $f : U \to \mathbb{R}$ a C^p-map.

Let $x_1, \ldots, x_n$ be the coordinates associated with an orthonormal basis of E. Then for all $p \in U$, $2 \leq k \leq p$ and permutations σ of $\{1, \ldots, k\}$ we have

$$\left.\frac{\partial^k f}{\partial x_1 \cdots \partial x_k}\right|_p = \left.\frac{\partial^k f}{\partial x_{\sigma(1)} \cdots \partial x_{\sigma(k)}}\right|_p . \qquad \square$$

The next few results are easily obtained by induction on p; indeed you may get the impression that this whole section is one long exercise in the use of mathematical induction!

Proposition 4.3.6. *Let E, F be Banach spaces, $U \subset E$ an open subset and suppose that $f : U \to F$ is a continuous map. If f is $(p+q)$-times differentiable then $\mathbf{D}^p f$ is differentiable q-times and the qth derivative of $\mathbf{D}^p f$ is $\mathbf{D}^{p+q} f$.*

Proof. If $q = 1$, this is the definition of the $(p+1)^{\text{st}}$ derivative. The easy induction step is proved using Lemma 4.3.2. $\square$

Combining with the criteria of continuous differentiability, we get:

Proposition 4.3.7. *For $i = 1, \ldots, m$. let $f_i : U \to F_i$ be maps of an open set U of a Banach space E into Banach spaces F_i. Then f is a C^p-map iff all the f_i's are C^p-maps and conversely. Moreover, if we denote the function f by a column vector, $f = (f_1, \ldots, f_m)$ then the pth derivative of f is given by $\mathbf{D}^p f(x) = (\mathbf{D}^p f_1, \ldots, \mathbf{D}^p f_m)$.*

The next result is, of course, nothing surprising, but the proof is, by necessity rather more subtle than you may have expected.

Theorem 4.3.8 (Compositions of C^p-maps).
Let E, F and G be three Banach spaces and suppose that:

1. *$U \subset E$ and $V \subset F$ are open sets;*

2. *$f : U \to F$ and $g : V \to G$ are C^p-maps and that $f(U) \subset V$.*

Then the composition, $(g \circ f) : U \to G$ is a C^p-map.
In particular, if f, g are smooth maps, so is gf.

Proof. For $p = 1$ the chain rule tells us that gf is differentiable at each point $x \in U$ and that the derivative of gf is the composition of the derivative of g with that of f. More precisely:

$$(*) \qquad \mathbf{D}gf(x) = \mathbf{D}g\big(f(x)\big)\mathbf{D}f(x).$$

Now, $x \mapsto \mathbf{D}g(f(x))$ is the composition

$$U \xrightarrow{f} f(U) \xrightarrow{\mathbf{D}g} \mathbb{L}(E,G)$$

and hence is continuous, since g is, by hypothesis C^1.

Let $\phi : U \to \mathbb{L}(E,F)\times\mathbb{L}(F,G)$ be the function $x \mapsto \big(\mathbf{D}f(x), \mathbf{D}g(f(x))\big)$. From (*) $\mathbf{D}gf(x)$ is the composition:

$$U \xrightarrow{\phi} \mathbb{L}(E,F) \times \mathbb{L}(F,G) \xrightarrow{\Gamma} \mathbb{L}(E,G)$$

where $\Gamma(\alpha,\beta) \doteq \beta\circ\alpha$. Since all the functions involved are continuous, so is $x \mapsto \mathbf{D}gf(x)$. (The continuity of Γ was discussed in Example 3.3.10; see page 114.) Hence the composition of C^1 maps is C^1.

Now assume the result is true for C^k-maps for all $k < p$ and suppose f and g are C^p-maps. I will write h for the composed function gf.

Then $\mathbf{D}h(x)$ is the composition of ϕ and Γ. Now Γ being a continuous bilinear function is C^∞. So it will suffice to show that ϕ is C^{p-1}. We have $\phi : x \mapsto \mathbf{D}g(f(x))$ is a composition

$$U \xrightarrow{f} V \xrightarrow{\mathbf{D}g} G.$$

Here f is C^p and hence *a fortiori* C^{p-1}, and $\mathbf{D}g$ is C^{p-1} since g is C^p. So by the induction hypothesis $x \mapsto \mathbf{D}g(f(x))$ is a C^{p-1} map. The map $\mathbf{D}h : x \mapsto \mathbf{D}(gf)(x)$ is C^{p-1} since f is C^p. Both components of

$$x \mapsto \phi(x) = (\mathbf{D}g(f(x)), \mathbf{D}f(x)).$$

being C^{p-1} so is ϕ and the proof is complete. □

We now come to the last major result of this section.

Theorem 4.3.9 (Taylor's Formula).
Let E, F be Banach spaces, $U \subset E$ an open set and $f : U \to F$ a C^p-map. If $x \in U$ and $y \in E$ are points such that the line segment joining x and $(x+y)$ is contained in U then

$$(4.3.2) \quad f(x+y) = f(x) + \mathbf{D}f(x)(\boldsymbol{y}) + \frac{1}{2}\mathbf{D}^2 f(x)(\boldsymbol{y}^{[2]}) + \cdots$$
$$\cdots + \frac{1}{(p-1)!}\mathbf{D}^{p-1} f(x)(\boldsymbol{y}^{[p-1]})$$
$$+ \int_0^1 \frac{(1-t)^{p-1}}{(p-1)!}\mathbf{D}^p f(x+ty)(\boldsymbol{y}^{[p]})dt.$$

Proof. By the Hahn-Banach theorem it suffices to show that if $\Lambda : F \to \mathbb{R}$ is any continuous linear functional then Λ takes the same value on both sides of the equation ($\star$). Fix a $\Lambda \in F'$ and define $\psi : [0,1] \to \mathbb{R}$ by $\psi(t) = \Lambda\big(f(x+ty)\big)$. Then from Lemma 4.3.3 we see that for all $t_0 \in (0,1)$, we have

$$(\dagger) \qquad \frac{d^k\psi}{dt^k}(t_0) = \mathbf{D}^k\psi(t_0)(1^{[k]}) = \Lambda\big(\mathbf{D}^k f(x+t_0y)(\boldsymbol{y}^{[k]})\big).$$

(Here $\mathbf{D}^k f(*)$ is a k-linear function and $(\boldsymbol{y}^{[k]})$ is the k-tuple $(\boldsymbol{y},\ldots,\boldsymbol{y})$.) The result now follows from the classical Taylor's formula, but since the proof is so simple and short we reproduce it.

Consider the function:

$$g(x) = \frac{(1-t)^{(p-1)}}{(p-1)!}\psi^{\{p\}}(x+ty),$$

where $\psi^{\{k\}}$ denotes the kth derivative of ψ. We compute the integral $\int_0^1 g(t)dt$ by successive integration by parts.

$$\begin{aligned}\int_0^1 \frac{(1-t)^{(p-1)}}{(p-1)!}\psi^{\{p\}}(x+ty)dt &= -\frac{1}{(p-1)!}\psi^{\{p-1\}}(x)\\ &\quad+\int_0^1 \frac{(1-t)^{(p-2)}}{(p-2)!}\psi^{\{p-1\}}(x+ty)dt\\ &\ \ \vdots \quad \cdots \quad \vdots \quad \cdots \quad \vdots \quad \cdots\\ &= -\psi(x) - \psi'(x) - \frac{1}{2}\psi''(x) - \cdots\\ &\quad \cdots - \frac{1}{(p-1)!}\psi^{\{p-1\}}(x).\end{aligned}$$

The result follows from ($\dagger$) since for any continuous function ϕ from $\mathbf{I}$ to F, we have

$$\Lambda\left(\int_0^1 \phi(t)dt\right) = \int_0^1 (\Lambda\circ\phi)(t)dt.$$

$\square$

The "remainder term" (that is, the last term) in Taylor's formula is clearly an unwieldy mess and it would be nice to have a more manageable expression, even if it were somewhat weaker in mathematical content.

If we impose a slightly stronger hypothesis on f we get a very simple remainder term, usually known as Lagrange's form of the remainder. On the other hand we can no longer assert an equality between two functions but only estimate the difference between $f(x+y)$ and the first p terms of the Taylor formula.

Theorem 4.3.10 (Taylor's Formula: Lagrange's Form).
Let U, E, F, f, x and y be as in Theorem 4.3.9.

Suppose, in addition, that $||\mathbf{D}^p f(x)|| < M$ for all $x \in U$. Then we have

$$(4.3.3) \quad ||f(x+y) - f(x) - \mathbf{D}f(x)(\boldsymbol{y}) - \frac{1}{2}\mathbf{D}^2 f(x)(\boldsymbol{y}^{[2]}) + \ldots$$
$$\cdots - \frac{1}{(p-1)!}\mathbf{D}^{p-1} f(x)(\boldsymbol{y}^{[p-1]})|| \le \frac{M}{p!}||\boldsymbol{y}||^p.$$

The proof is a simple exercise in estimating the integral in the earlier Taylor's formula. □

There is yet another form of this result which is simple to use and does not require any additional hypotheses like the ones required for the validity of Lagrange's form.

Proposition 4.3.11. *Let U, E, F, f, x and y be as in Theorem 4.3.9. Then*

$$\left|\left|f(x+y) - f(x) - \sum_{k=1}^{p} \frac{1}{k!}\mathbf{D}^k f(x)([\boldsymbol{y}]^k)\right|\right| = o(||\boldsymbol{y}||^p).$$

Proof. For $p = 1$ the assertion is just the definition of differentiability. So we will proceed by induction; assume that the result holds for all $1 \le k \le m < p$. Then

$$||f(x+y) - f(x) - \sum_{l=1}^{m} \frac{1}{l!}\mathbf{D}^l f(x)(\boldsymbol{y}^{[l]})|| = o(||\boldsymbol{y}||^m)$$

Define a new function of y by setting

$$\phi(y) = f(x+y) - f(x) - \sum_{l=1}^{m} \frac{1}{l!}\mathbf{D}^l f(x)(\boldsymbol{y}^{[l]}).$$

Now note that the typical term inside the summation sign is of the form:

$$\mathbf{D}\phi_l(\boldsymbol{y})(\boldsymbol{k}) = l\mathbf{D}^l f(x)(\boldsymbol{y}, \ldots, \boldsymbol{y}, \boldsymbol{k}).$$

So a convenient way of expressing this is that for $l \geq 2$,

$$\mathbf{D}\phi_l(y) = l\mathbf{D}^l f(x)(y^{(l-1)}).$$

Note that since $\mathbf{D}^l f(x)$ is being evaluated on $y^{(l-1)}$, the outcome is a linear transformation from E to F. Thus

$$\mathbf{D}\phi(y) = \mathbf{D}f(x+y) - \mathbf{D}f(x) - \sum_{l=1}^{m} \frac{1}{l!}\mathbf{D}^l f(x)(y^{(l-1)}).$$

If $m < p$ then $\mathbf{D}f$ is differentiable m times and we can apply the induction hypothesis to it and the above expression leads to $||\mathbf{D}\phi(y)|| = o(||y||^m)$, which means that for any $\varepsilon > 0$ there is a $\delta > 0$ such that if $||\boldsymbol{y}|| < \delta$ then $||\mathbf{D}\phi(y)|| < \varepsilon||\boldsymbol{y}||^m$. Since $\phi(0) = 0$ using the the mean value theorem we get

$$||\phi(y)|| = ||\phi(y) - \phi(0)|| < \varepsilon||\boldsymbol{y}||^{m+1}$$

whenever $||\boldsymbol{y}|| < \delta$. This completes the induction step. □

In consonance with the terminology of linear algebra, we will say that a continuous, symmetric bilinear transformation $\beta : E \times E \to \mathbb{R}$ is *positive definite* (resp. *negative definite*) if for all nonzero vectors $\boldsymbol{v} \in E$, $\beta(\boldsymbol{v}, \boldsymbol{v}) > 0$ (resp. < 0.)

The next result should not come as a surprise. The hypotheses look a little more complicated than what you might have expected but the additional hypothesis ensures the validity of the result for functions which are defined on open subsets of infinite-dimensional Banach spaces.

Proposition 4.3.12. *Let E be a Banach space, U an open subset of E and $f : U \to \mathbb{R}$ a C^2-function. Suppose that $x^0 \in U$ and the following conditions hold:*

1. $\mathbf{D}f(x^0) = 0$;
2. *there is a constant $K > 0$ such that for all vectors $\boldsymbol{v} \in E$ of we have $\mathbf{D}^2 f(x^0)(\boldsymbol{v}, \boldsymbol{v}) < -K$.*

Then there is an $r > 0$ such that $\mathbb{B}_r(x^0) \subset U$ and $f(x) < f(x^0)$ for every $x \neq x^0 \in \mathbb{B}_r(x^0)$.

Proof. For any $r > 0$ such that $\mathbb{B}_r(x^0) \subset U$ and $x \in B_r(x^0)$, (because $\mathbf{D}f(x^0) = 0$) the Taylor formula estimate given by Proposition 4.3.11 (with $p = 2$) reduces to

$$f(x) = f(x^0) + \tfrac{1}{2}\mathbf{D}^2 f(x^0)(\boldsymbol{v}_x, \boldsymbol{v}_x) + o_2(x),$$

where $\boldsymbol{x} = (\boldsymbol{x}^0 + \boldsymbol{v}_x)$ and $o_2(x) = o(||\boldsymbol{v}_x||^2)$.

Choose $\varepsilon = K/2$ and r so that $|o_2(x)| < \varepsilon||\boldsymbol{v}_x||^2$ whenever $x \in B_r(x^0)$. Since $\mathbf{D}f^2(x^0)([x - x^0]^2) < -K||x - x^0||^2$ the result follows. □

Exercises for § 4.3

4.3.1 Some computations

(a). Let $U \subset E$ and $V \subset F$ be open sets of the Banach spaces E and F and suppose that $f : U \to V$ and $g : V \to G$ are two maps of class C^3. Let $x^0 \in U$ and for $k = 1, 2, 3$ denote the derivatives $\mathbf{D}^k f(x^0)$ (resp. $\mathbf{D}^k g(f(x^0))$) by f', f'', f''' (resp. g', g'', g''') for brevity. Show that if $h = (g \circ f)$ then

$$\begin{aligned}\mathbf{D}^3 h(x^0)(\boldsymbol{u}, \boldsymbol{v}, \boldsymbol{w}) &= g'(f'''(u, v, w)) + g''(f'(\boldsymbol{u}), f''(\boldsymbol{v}, \boldsymbol{w})) \\ &\quad + g''(f'(\boldsymbol{v}), f''(\boldsymbol{w}, \boldsymbol{u})) + g''(f'(\boldsymbol{w}), f''(\boldsymbol{u}, \boldsymbol{v})) \\ &\quad + g'''(f'(\boldsymbol{u}), f'(\boldsymbol{v}), f'(\boldsymbol{w})).\end{aligned}$$

Perhaps now you realize why the proof of Proposition 4.3.8, that compositions of C^p-maps are C^p was complicated: there is no nice formula for the higher derivatives of a composite function.

(b). Let E be a Hilbert space. Show that the map $f : E_\times \to \mathbb{R}$ defined by $f(x) = \boldsymbol{x}/||\boldsymbol{x}||$ is a C^∞-map.

4.3.2 Products and the Leibniz Formula

Let $f, g : U \to F$ be C^p-maps of an open subset of a Banach space E into a Banach space F and $B : F \times F \to F$ a symmetric and continuous bilinear function. Let $\phi(x) = B(f(x), g(x))$ for each $x \in U$. Obtain an expression for $\mathbf{D}^n\phi$ in terms of derivatives of order $\leq n$ of f and g modelled on the classical *Leibniz's Formula*.

4.4 Tangent Vectors and all that

Throughout this section I will be looking at functions defined on finite-dimensional Banach spaces. So, without loss of generality, I will assume the functions are defined on *Euclidean spaces*. Moreover we will look only at functions of class C^∞. These will be referred to as *smooth functions*.

The object of this section is to review the process of differentiation from a slightly different perspective. This is useful in further extending the scope of Differential Calculus to the realm of Differentiable manifolds. These ideas are also useful in defining, in a suitably flexible manner, the important concepts of `differential equations` and `differential operators`.

Let E, F be Banach spaces, $U \subset E$ an open set and $f : U \to F$ a smooth map. We have seen that while there is no systematic procedure for computing the derivative of f at a point p, given a $\boldsymbol{v} \in E$, $\mathbf{D}f(p)(\boldsymbol{v}) \in F$ is determined by the 'formula':

$$\mathbf{D}f(p)(\boldsymbol{v}) = \lim_{t\to 0} \frac{f(p+tv) - f(p)}{t}. \tag{4.4.1}$$

Recall, that in Definition 4.1.12, I introduced the notion of tangent vectors at p, represented by differentiable curves satisfying the conditions **TV 1** – **TV 3** (see page 148). Since, we are working with smooth, and hence, C^1-functions the warnings sounded in Remark 4.1.17a about not assuming that the directional derivatives can ensure the existence of the derivative (or determine the linear transformation $\mathbf{D}f(p)$) do not apply. So the set

$$\left\{\lim_{t\to 0} \frac{f(p+tv) - f(p)}{t}\right\}_{\boldsymbol{v}\in T_pE}$$

completely determines the derivative of f at p.

What I now intend to do is to investigate the nature of the operation which takes the function f to $\mathbf{D}f(p)(\boldsymbol{v})$.

So far we have regarded $\boldsymbol{v}$ as an element of E, but really speaking we were thinking of $\boldsymbol{v}$ as a vector in the way one thinks of a vector like 'force' in physics: it has a point of application, a direction and a magnitude. We have been implicitly visualizing a tangent vector $\boldsymbol{v}$ at $p \in E$ like an arrow starting at p: look at the first term in the numerator of Equation 4.4.1: one can think of $f(p+tv)$ as the value of f as one moves a small distance from p (the point of application) along the vector $\boldsymbol{v}$. This was the motivation behind the definition of tangent vectors and the tangent space, T_pE at p. The derivative of f at p acts on these vectors. So there are two ways of thinking about the argument of $\mathbf{D}f(p)$:

1. it is a vector $\boldsymbol{v}$ of E; except that the origin is shifted to the point p.

2. It is the velocity of a curve $\alpha_{\boldsymbol{v}} : (-\varepsilon, \varepsilon) \to U$ such that $\alpha_{\boldsymbol{v}}(0) = p$ and $\alpha'_{\boldsymbol{v}}(0) = \boldsymbol{v}$.

(At this point, perhaps you should go back to page 149 and review Proposition 4.1.13.)

The second definition is preferable because it clarifies the nature of $\mathbf{D}f(p)(\boldsymbol{v})$. According to the first definition, this is a vector belonging to F, and not a "tangent vector" unless we drag the base of the arrow to the point $f(p)$. But since $f \circ \alpha_{\boldsymbol{v}}$ is a differentiable curve passing through $f(p)$ at time $t = 0$ and $\mathbf{D}f(p)(\boldsymbol{v})$ is the limit:

$$\mathbf{D}f(p)(\boldsymbol{v}) = \lim_{t \to 0} \frac{f \circ \alpha_{\boldsymbol{v}}(t) - f \circ \alpha_{\boldsymbol{v}}(0)}{t}$$

it is clear that $\mathbf{D}f(p)(\boldsymbol{v})$ is a tangent vector of F at $f(p)$. *So the derivative of f at p takes tangent vectors at p to tangent vectors at $f(p)$.* The derivative can now be thought of as a linear transformation $\mathbf{D}f(p) : T_pU \to T_{f(p)}F$.

The only difficulty with T_pU is that the linear space structure cannot be determined without reference to the linear space structure of E.

A third way of thinking about tangent vectors is that it is simply something which "computes directional derivatives" of smooth functions. The advantages of this approach will become clear soon.

I will begin by looking at directional derivatives of smooth $\mathbb{R}$-valued functions $f : U \to \mathbb{R}$. Since, now we are thinking of tangent vectors at p to be entities which compute directional derivatives of functions, I introduce a slightly different, but more suggestive, notation. If $p \in U$ and $\boldsymbol{v} \in E$ then the tangent vector $\boldsymbol{v}_p$ at p corresponding to $\boldsymbol{v}$ takes a smooth function $f : U \to \mathbb{R}$ to the real number, $\mathbf{D}f(p)(\boldsymbol{v})$. To highlight the fact that $\boldsymbol{v}_p$ is regarded as something which is operating upon f, I will write '$\boldsymbol{v}_p[f]$' for $\mathbf{D}f(p)(\boldsymbol{v})$.

If $f, g : U \to \mathbb{R}$ are two smooth functions then their pointwise product, that is the function, $x \mapsto (f(x\cdot)g(x))$ is also smooth and, by the product rule, we have the relation:

(Derivation Condition) $$\boldsymbol{v}_p[f \cdot g] = f(p)\boldsymbol{v}_p[g] + g(p)\boldsymbol{v}_p[f].$$

Now observe that to determine $\boldsymbol{v}_p[f]$ it is necessary only to know the values taken by f in *any* open set containing p. If two functions g_1, g_2 agree on some open set containing p, then $\boldsymbol{v}_p[g_1] = \boldsymbol{v}_p[g_2]$. We now define for each $p \in U$ an equivalence relation $\sim_p$ on the set of smooth R-valued functions defined on U.

> If $g, h : U \to \mathbb{R}$ are smooth functions and $p \in U$, then $g \sim_p h$ if there is some $\delta > 0$ such that for all $q \in \mathbb{B}_\delta(p) \cap U$, $g(q) = h(q)$.

An equivalence class of smooth functions under this relation is called a *germ at p*; the germ at p corresponding to a smooth function f will be denoted $[f]_p$ and the set of all germs at p will be denoted $\mathcal{G}_p$.

We now look at some of the formal properties of the set of germs at a point.

Let $C^\infty(U)$ be the ring of smooth functions from U to $\mathbb{R}$. Then $\mathcal{G}_p$ satisfies the following properties.

1. *$\mathcal{G}_p$ is an $\mathbb{R}$-linear space.*
   If $[g]_p, [h]_p \in \mathcal{G}_p$ and $r \in \mathbb{R}$ define

   (Addition) $$[g]_p + [h]_p \doteq [g + h]_p$$
   and
   (Scalar Multiplication) $$r[g]_p \doteq [r \cdot g]_p.$$

   It is easy to see that these operations are well-defined and yield a structure of linear space on $\mathcal{G}_p$.

2. *$\mathcal{G}_p$ is a ring.*
   We just have to define the product of $[g]_p$ and $[h]_p$ to be $[(g \cdot h)]_p$.

3. There is a function $m : C^\infty(U) \times \mathcal{G}_p \to \mathcal{G}_p$ defined by $m(\phi, [g]_p) = [(\phi \cdot \gamma)]_p$, where γ is any smooth function on U such that $g \sim_p h$.

   It is clear that m is well-defined and satisfies the following properties: if $\phi, \psi \in C^\infty(U)$ and $[g]_p, [h]_p \in \mathcal{G}_p$, then

   (a) $$m(\phi, [g]_p + [h]_p) = m(\phi, [g]_p) + m(\phi, [h]_p)$$
   (b) $$m(\phi, m(\psi, [g]_p) = m(\phi \cdot \psi, [g]_p).$$

   In algebraic jargon, $\mathcal{G}_p$ is a *module* over the ring $C^\infty(U)$, of smooth real-valued functions defined on U.

Consider the dual space, $\mathcal{G}_p^*$, of the linear space of germs at p. (Note that there is no norm defined on $\mathcal{G}_p$. So $\mathcal{G}_p^*$ is simply the set of $\mathbb{R}$-linear transformations $\mathcal{G}_p \to \mathbb{R}$.)

Let $\mathcal{T}_p \subset \mathcal{G}_p^*$ be those linear functionals which also satisfy the following "derivation condition".

Definition 4.4.1 (Derivation).
An element $\Lambda \in \mathcal{G}_p^*$ is said to be a *derivation* if for any two germs, $[g]_p, [h]_p \in \mathcal{G}_p$, we have the relation,

$$\Lambda([g]_p \cdot [h]_p) = g(p)\Lambda([h]_p) + h(p)\Lambda([g]_p)$$

where the each of the two terms on the right side is a product of two real numbers.

$\mathcal{T}_pU$ will be called the **analytic tangent space** to U at the point p. We will see that $\mathcal{T}_pU$ and $T_pU \equiv E$ are isomorphic.

I begin by studying the formal properties of $\mathcal{T}_pU$.

Let us suppose that E is n-dimensional with an onb, $\mathcal{E} = (\boldsymbol{e}_1, \ldots, \boldsymbol{e}_n)$ and let $(\epsilon_1, \ldots, \epsilon_n)$ be the dual basis. Then we can (and will) use the basis, $\mathcal{E}$, to identify E with $\mathbb{R}^n$.

We will regard the restrictions (to U) of the $\epsilon_\imath$'s, as real-valued functions on U. Then $x_\imath \doteq \epsilon_\imath|U : U \to \mathbb{R}$ are obviously smooth functions. $x_\imath$ is called the ith *coordinatfe function* of U, corresponding to the basis $\mathcal{E}$.

Now corresponding to each $\boldsymbol{e}_\imath \in \mathcal{E}$, there is a $\delta_\imath > 0$ such that the curves $\alpha_\imath : (-\delta_\imath, \delta_\imath) \to E$ given by $\alpha_\imath(t) = (\boldsymbol{p} + t \cdot \boldsymbol{e}_\imath)$, take values in U. These curves represent the standard basis vectors as tangent vectors. The tangent vectors they represent will be denoted, $\boldsymbol{e}_{\imath,p} \in T_pE$. Since $\alpha_\imath(t) = (p_1, \ldots, p_n) + (0, \ldots, tx_\imath, 0, \ldots, 0)$,

$$\boldsymbol{e}_{\imath,p}[f] = \left.\frac{\partial f}{\partial x_\imath}\right|_p \quad \text{for each } i = 1, \ldots, n.$$

But since f is a smooth function, so are its partial derivatives and so we will, henceforth, write: $\left.\dfrac{\partial f}{\partial x_\imath}\right|_p$ as $\dfrac{\partial f}{\partial x_\imath}(p)$.

The right-hand side of the equation $\boldsymbol{e}_{\imath,p}[f] = \frac{\partial f}{\partial x_\imath}(p)$ will not change if f is replaced by a function g such that $f \sim_p g$. So we can define an operation of the $\boldsymbol{e}_{\imath,p}$'s on *germs* by setting:

$$L_p(\boldsymbol{e}_{\imath,p})([\phi]_p) = \frac{\partial \phi}{\partial x_\imath}(p) \quad \text{for } i = 1, \ldots, n.$$

Since taking partial derivatives is obviously a derivation, $L_p(\boldsymbol{e}_{\imath,p})$ is a derivation of $\mathcal{G}_p$. Extending by linearity, we get a linear transformation, $L_p : T_pU \to \mathfrak{T}_pU$.

We will see that L_p is an isomorphism.

We first check that elements of the form $L_p(\boldsymbol{v})$ do indeed "compute directional derivatives", which tangent vectors ought to do.

Lemma 4.4.2. *For any germ* $[f]_p \in \mathcal{G}_p$ *and* $\boldsymbol{v} \in E$: $L_p(\boldsymbol{v})([f]_p) = \mathbf{D}f(p)(\boldsymbol{v})$.

Proof. Suppose $\boldsymbol{v} = \sum_{\imath=1}^{n} v_\imath \boldsymbol{e}_\imath$. Then from the definition of L_p we get:

$$\begin{aligned} L_p(\boldsymbol{v})([f]_p) &= \sum_{\imath=1}^{n} v_\imath L_p(\boldsymbol{e}_{\imath,p})[f]_p \\ &= \sum_{\imath=1}^{n} v_\imath \frac{\partial f}{\partial x_\imath}(p) \\ &= \boldsymbol{v}_p[f]. \end{aligned}$$

Thus L_p takes each tangent vector to the "operator" which computes directional derivatives of smooth functions along the tangent vector. □

Let $\Lambda \in \mathfrak{T}_pU$. Then Λ has the following important properties.

1. *Λ takes germs of constant functions to* 0.

   Proof. Consider the constant function $\widetilde{1} : U \to \mathbb{R}$, which always takes the value 1. Clearly $\widetilde{1}\cdot\widetilde{1} = \widetilde{1}$. Then using the Derivation Condition we get:
   $$\Lambda\big([\widetilde{1}\cdot\widetilde{1}]_p\big) = 1\cdot\Lambda\big([\widetilde{1}]_p\big) + \Lambda\big([\widetilde{1}]_p\big)\cdot 1,$$
   showing that $\Lambda\big([\widetilde{1}]_p\big) = 2\cdot\Lambda\big([\widetilde{1}]_p\big)$. Hence, $\Lambda\big([\widetilde{1}]_p\big) = 0$. Since any constant function is a scalar multiple of $\widetilde{1}$, and Λ is a linear functional the result follows. □

2. *Suppose $p = 0$, the origin of E. If $B : E \times E \to \mathbb{R}$ is a bilinear function and $\bar{B}(\boldsymbol{v}) = B(\boldsymbol{v},\boldsymbol{v})$, then $\Lambda([\bar{B}]_0) = 0$.* ($\bar{B} : E \to \mathbb{R}$ is called the *quadratic form* associated with B.)

   Proof. If $B : E\times E \to \mathbb{R}$ is a bilinear function it will be continuous since E is finite-dimensional. From Example 4.3.1b we know that B is smooth.

As before let $x_\imath : E \to \mathbb{R}$ be the coordinate functions, that is the linear functionals which constitute the basis dual to $\mathcal{E}$. Then if $\boldsymbol{v} \in E$ is written as $\boldsymbol{v} = \sum_{\imath=1}^{n} x_\imath \boldsymbol{e}_\imath$, then we get

$$\bar{B}(\boldsymbol{v}) = \sum_{\imath,\jmath=1}^{n} x_\imath x_\jmath B(\boldsymbol{e}_\imath, \boldsymbol{e}_\jmath).$$

Hence, upon going to the germ at 0, we get

$$[\bar{B}]_0 = \sum_{\imath,\jmath=1}^{n} m_{\imath\jmath}[x_\imath]_0 \cdot [x_\jmath]_0,$$

where $m_{\imath\jmath} = B(\boldsymbol{e}_\imath, \boldsymbol{e}_\jmath)$. Now using the derivation condition we see that:

$$\Lambda\big([\bar{B}]_0\big) = \sum_{\imath,\jmath=1}^{n} m_{\imath\jmath}\big(x_\imath(0) \cdot \Lambda([x_\jmath]) + \Lambda([x_\imath]_0) \cdot x_\jmath(0)\big) = 0$$

since $x_k(0) = 0$ for all $1 \le k \le n$. □

We are now in a position to determine how aan element $\Lambda \in \mathcal{T}_pU$ acts on the germ of an arbitrary smooth function defined on U.

If $f : U \to \mathbb{R}$ is a smooth function then for all $q \in U$ suitably near p there is a vector $\boldsymbol{v} \in E$ such that $\boldsymbol{q} = \boldsymbol{p} + \boldsymbol{v}$ and the line segment $\{(p+tv) : t \in [0,1[\}$ is contained in U. So using Taylor's formula (4.3.2) with $p = 2$, we get:

$$f(q) - f(p) = \mathbf{d}f(p)(\boldsymbol{v}) + \int_0^1 \mathbf{D}^2 f(p+tv)(\boldsymbol{v}, \boldsymbol{v})dt. \tag{4.4.2}$$

Setting $R_f^2 = \int_0^1 \mathbf{D}^2 f(p+tv)(\boldsymbol{v}, \boldsymbol{v})dt$ and regarding f as a function of v, we get:

$$f(q) = f(p+tv) = f(p) + \mathbf{d}f(p)(\boldsymbol{v}) + \bar{R}_f^2(\boldsymbol{v}). \tag{4.4.3}$$

So taking germs at p, i. e., at $\boldsymbol{v} = 0$ we get:

$$[f]_p = [f]_0 + [\mathbf{d}f(p)]_0 + [\bar{R}_f^2]_0. \tag{4.4.4}$$

Notice that the last term is the germ of a quadratic form at 0. Hence, for any $\Lambda \in \mathcal{T}_pU$,

$$\Lambda([f]_p) = \Lambda([\mathbf{d}f(p)]_0)$$

I have substituted $\mathbf{D}f(p)$ by $\mathbf{d}f(p)$ since f is a real-valued function and $\mathbf{d}f(p) : E \to \mathbb{R}$ is being thought of as a smooth function when we write $[df(p)]_0$. Observe, also, that in terms of the coordinate functions on U, (4.4.3) has the form

$$(4.4.4^\star) \qquad [f]_p = [f(p)]_p + \sum_{i=1}^{n} \frac{\partial f}{\partial x_i}(p)[x_i]_p + \sum_{i,j=1}^{n} Q_{ij}[x_i]_p[x_j]_p$$

where Q_{ij} is the symmetric matrix representing R_f^2.

Equation (4.4.3) shows that the elements of $\mathcal{T}_pU$ are completely determined by their values on differentials of smooth functions which means that $\Lambda \in \mathcal{T}_pU$ is determined by the values it takes on E^*: clearly suggesting that $\mathcal{T}_pU$ is related to E^{**} which is, of course, identified with E since we are working with an Euclidean space.

We can now write down an explicit inverse of $L_p : T_pU \to \mathcal{T}_pU$.

Theorem 4.4.3 (The analytic definition of Tangent vectors).
Given $\Lambda \in \mathcal{T}_pU$ *define*

$$\mathcal{L}_p(\Lambda) = \sum_{i=1}^{n} \lambda_i \boldsymbol{e}_{i,p}.$$

where $\lambda_i = \Lambda([x_i]_p)$. *Then* $\mathcal{L}_p$ *is an inverse of the linear transformation* $L_p : T_pU \to \mathcal{T}_pU$ *which was defined on page* 196.

Remark. I am calling this an "analytic" definition because it relies only on the notion of differentiability. The other point of view, is *geometric*. A little later, we will arrive at an "analytic definition" of the derivative of smooth functions. This will be a way of formulating the notion of derivative based upon the present analytic definition of tangent vectors.

Proof. Clearly $\mathcal{L}_p$ is a linear transformation. We first show that it is a left inverse of $L_p : T_pU \to \mathcal{T}_pU$. Recall that in Lemma 4.4.2 we had shown that:

$$L_p(\boldsymbol{v}_p)([f]_p) = \mathbf{d}f(p)(\boldsymbol{v}).$$

First observe that, $\frac{\partial}{\partial x_i}[x_j]_0 = \delta_{ij}$. This implies that for each $1 \leq i \leq n$, $\mathcal{L}_p \circ L_p(\boldsymbol{e}_{i,p}) = \boldsymbol{e}_{i,p}$. Since the collection, $\{\boldsymbol{e}_{1,p}, \dots, \boldsymbol{e}_{n,p}\}$, is a basis of T_pU, this shows that $\mathcal{L}_p \circ L_p = 1_{T_pU}$.

It remains then to establish that $L_p \circ \mathcal{L}_p = 1_{\mathcal{T}_pU}$.

To this end, note that by definition, $\Big(L_p(\mathcal{L}_p(\Lambda))\Big)[f]_p = \sum_{i=1}^{n} \lambda_i \frac{\partial f}{\partial x_i}(p)$.

The result now follows from equation 4.4.4*. ∎

REMARKS 4.4.4 (ON TANGENT VECTORS)

I am going to talk about tangent spaces in a broader context than open sets of Euclidean spaces. But you have already computed the tangent spaces of $\mathrm{SL}_n(R), \mathrm{O}_n$ (see the exercises under Exercise 4.1.4) and should not feel intimidated by this.

Remark 4.4.4a (The linear space structure of $\mathcal{T}_p$). I had mentioned that one difficulty with defining tangent vectors using the velocity of curves is that if $\boldsymbol{v}, \boldsymbol{w}$ are represented by the curves $\alpha_{\boldsymbol{v}}, \alpha_{\boldsymbol{w}}$ respectively, then the only way to get a curve whose velocity will represent $\boldsymbol{v}+\boldsymbol{w}$ at p is to define $\alpha_{\boldsymbol{v}+\boldsymbol{w}} = \alpha_{\boldsymbol{v}} + \alpha_{\boldsymbol{w}} - \boldsymbol{p}$, *using the linear structure of* E.

This is quite all right for dealing with tangent spaces of open sets but obviously will cause problems if we are trying to define the tangent spaces, for instance, of the unit sphere in a Euclidean space. A simple differentiation shows that if $\alpha : (-\delta, \delta) \to \mathbb{S}_1 \subset \mathbb{R}^n$ is a C^1 curve passing through $p \in \mathbb{S}_1$ then $\langle \boldsymbol{p}, \alpha'(0) \rangle = 0$. With a little more effort one can show that the tangent space at p is given by

$$T_p(\mathbb{S}_1) = \{\boldsymbol{x} \in E \,:\, \langle \boldsymbol{x}, \boldsymbol{p} \rangle = 0\}.$$

which is the “right answer” intuitively. But what is the linear space structure in this situation? If one adds curves taking values in $\mathbb{S}_1 \subset \mathbb{R}^n$ using the linear space structure of E, then we will not have a curve taking values in $\mathbb{S}_1$.

In the analytic definition this problem does not arise since $\mathcal{T}_p$ is, *by definition,* a subspace of the dual of the linear space $\mathcal{G}_p$. There is a fairly straight-forward way to define smooth functions on $\mathbb{S}_1$. A function $f : \mathbb{S}_1$ is smooth if there is an open set $U \supset \mathbb{S}_1$ and a smooth function $\bar{f} : U \to \mathbb{R}$, such that $\bar{f}|\mathbb{S}_1 = f$. Germs of smooth functions can now be defined in the obvious way and the analytic definition can be used in such a situation, to define the tangent space of $\mathbb{S}_1$ at any point of $\mathbb{S}_1$.

Remark 4.4.4b. The open set U had played, virtually, no role in our discussion because any smooth function on an open set of E can be extended to a smooth function on all of E. (This is a fairly technical result: its proof is quite different in spirit from the topics we are discussing and is being omitted.) This result implies that:

For any open set U in E, and any point $p \in U$, $T_pU = T_pE$.

This is false for arbitrary sets.

Remark 4.4.4c (On germs).
It is time to confess that the definition of germs of smooth functions that I have used is a (harmless) simplification of the usual definition of germs. Normally germs are defined as follows.

Fix a point $p \in \mathbb{R}^n$ and consider the set of all ordered pairs, (V, ϕ) where V is an open set containing p and $\phi : V \to \mathbb{R}$ is a smooth function. Two such pairs (V, ϕ) and (W, ψ) are said to be equivalent if $\phi|(V \cap W) = \psi|(V \cap W)$.

The equivalence classes are known as germs of smooth functions at p. In view of the previous remark, it is clear that this definition is equivalent to the one I have used.

The set of germs at p will be henceforth denoted $\mathcal{G}_p$. I will use these modified notation of germs when I discuss the derivative of smooth functions between Euclidean spaces.

Remark 4.4.4d. One point should be noted. *There is no natural way* of constructing isomorphisms between $\mathcal{T}_pU$ and $\mathcal{T}_qU$ at different points because they are subspaces of the duals of different linear spaces. This behaviour is only to be expected of a notion of tangent spaces which is more generally applicable than to open subsets of Banach spaces. (Review Remark 4.4.4a where the tangent spaces of $\mathbb{S}_1 \subset \mathbb{R}^n$ are described.)

This has considerable implications in physics! For instance using the analytic notion of tangent vectors, one cannot define the concept of *acceleration* because $\alpha'(t_0)$ and $\alpha'(t_0 + t)$ are in different linear spaces and we cannot form the "differential quotient" $(\alpha'(t_0 + t) - \alpha'(t_0))/t$. Acceleration is not an analytic notion: it requires that the spaces we are dealing with have some additional *geometric structure* known as a `connection`. Indeed, this is one of the reasons why Einstein's Theory of General Relativity is formulated in the language of `Differential geometry`.

I will now reformulate, in terms of the analytic definition of tangent spaces, the definition of the derivative of a smooth map between open subsets of Banach spaces. This can be done satisfactorily only for Hilbert spaces and we will content ourselves by treating only the finite-dimensional situation.

For Euclidean spaces, the procedure is both simple and elegant.

Let E, F be Euclidean spaces, $U \subset E$ and $V \subset F$ open subsets. Let $f : U \to V$ be a smooth map. Then f gives rise to a linear transformation $f_\sharp : C^\infty(V) \to C^\infty(U)$ defined by $f_\sharp(\phi) = (\phi \circ f)$. This induces, in an obvious fashion (by taking equivalence classes with respect to $\sim_p$ and $\sim_{f(p)}$) a linear transformation, $f_\sharp^p : \mathcal{G}_{f(p)} \to \mathcal{G}_p$. Explicitly, this linear transformation acts as folllows:

$$\text{For any } [\gamma]_{f(p)}, \quad f_\sharp^p\big([\gamma]_{f(p)}\big) = [(\gamma \circ f)]_p.$$

Now the 'analytic tangent spaces", $\mathfrak{T}$'s, were subspaces of the duals of the spaces of germs. So it is reasonable to speculate that

$$f_\sharp^{p*} : \mathcal{G}_p^* \to \mathcal{G}_{f(p)}^*$$

is related to $\mathbf{D}f(p) : E \to F$.

The first step is to check that $f_\sharp^{p*}$ takes a tangent vector to tangent vector; that is a derivation to a derivation. In view of Remark 4.4.4b I will write $\mathfrak{T}_x$ for the tangent space at any point x in a Euclidean space.

Lemma 4.4.5. *Let $f : U \to V$ be a smooth map between open subsets of Euclidean spaces. Suppose $p \in U$ and $q = f(p)$.*

Let $\Lambda \in \mathfrak{T}_p$ and set $f_\sharp^{p*}(\Lambda) = \Omega \in \mathfrak{T}_q$ Then for any two germs $[\phi]_q, [\psi]_q \in \mathcal{G}_q$, we have:

$$\Omega\big([\phi]_q \cdot [\psi]_q\big) = \phi(q) \cdot \Omega\big([\psi]_q\big) + \psi(q) \cdot \Omega\big([\phi]_q\big).$$

Thus, $f_\sharp^{p*}$ maps $\mathfrak{T}_p$ into $\mathfrak{T}_q$.

Proof. By the definition of duals of linear transformations, we have:

$$\begin{aligned}\Omega\big([\phi]_q \cdot [\psi]_q\big) &= \Lambda\Big(f_\sharp^p\big([\phi]_q \cdot [\psi]_q\big)\Big), \\ &= \Lambda\big([(\phi \circ f)]_p \cdot [(\psi \circ f)]_p\big).\end{aligned}$$

Since Λ is a derivation of $\mathcal{G}_p$, we get:

$$\begin{aligned}\Omega\big([\phi]_q \cdot [\psi]_q\big) &= \Lambda\big([(\phi \circ f)]_p \cdot [(\psi \circ f)]_p\big), \\ &= \phi(q) \cdot \Lambda\big(f_\sharp([\psi]_q)\big) + \psi(q) \cdot \Lambda\big(f_\sharp([\phi]_q)\big), \\ &= \phi(q) \cdot \Omega\big([\psi]_q\big) + \psi(q) \cdot \Omega\big([\phi]_q\big)\end{aligned}$$

and this proves the lemma. □

We can now show that $f_\sharp^{p*}$ is indeed another *avatar* of $\mathbf{D}f(p) : E \to F$.

Theorem 4.4.6 (The analytic definition of the derivative).
Let E, F be Euclidean spaces, $U \subset E$ and $V \subset F$ two open subsets and $f : U \to V$ a smooth map. Then for each $p \in U$ there is a commutative diagram:

$$\begin{array}{ccc} E & \xrightarrow{\mathbf{D}f(p)} & F \\ {\scriptstyle L_p^E}\big\downarrow & & \big\downarrow{\scriptstyle L_{f(p)}^F} \\ \mathcal{T}_p & \xrightarrow[f_\sharp^{p*}]{} & \mathcal{T}_{f(p)} \end{array}$$

Proof. As before, to simplify the notation, I will write "q" for "$f(p)$". The linear transformations which identify E and F with the analytic tangent spaces, at p and q will be denoted By $L_p^E : E \to \mathcal{T}_p$ and $L_q^F : F \to \mathcal{T}_q$ respectively.

Let $\boldsymbol{v} \in E$ and $[u]_q \in \mathcal{G}_q$.

Then
$$\begin{aligned} L_q^F\Big(\mathbf{D}f(p)(\boldsymbol{v})\Big)\big([u]_q\big) &= \mathbf{D}u(q)\big(\mathbf{D}f(p)(\boldsymbol{v})\big) \quad \text{(From Lemma 4.4.2)} \\ &= \mathbf{D}(u \circ f)(p)(\boldsymbol{v}) \quad \text{(using the chain rule)} \\ &= f_\sharp^{p*}(L_p^E(\boldsymbol{v}))\big([u]_q\big), \end{aligned}$$
where in the last step we are again using Lemma 4.4.2.

This establishes the commutativity of the diagram. □

Remark 4.4.7. Notice that although $\mathcal{T}_p$ depends only on smooth functions taking values in $\mathbb{R}$, it contains enough structure to compute derivatives of smooth maps taking values in any Euclidean space *without invoking coordinates.*

$f_\sharp^{p*}$ is obviously independent of coordinates! In fact, everything we have done would go through without significant change to the case when E, F are infinite dimensional Hilbert spaces.

The new notion of "tangent spaces" attached to each point of open sets, $U \subset E$ has some other implications which I discuss briefly.

Vector Fields: Suppose that $U \subset \mathbb{R}^n$ is an open set. A *vector field*, $\boldsymbol{V}$, on U is an assignment of a tangent vector $\boldsymbol{V}(p) \in \mathcal{T}_p$ to each point $p \in U$. Clearly we can always find functions $v_1, \dots, v_n : U \to \mathbb{R}$ such that

$$\boldsymbol{V}(p) = \sum_{i=1}^{n} v_i(p)\left(\frac{\partial}{\partial x_i}\right)_p. \tag{4.4.5}$$

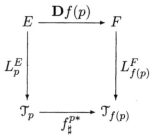

If all the $v_\imath$'s are smooth, then $\boldsymbol{V}$ is called a *smooth vector field* on U. From Equation 4.4.5 it is clear that the pointwise sum of two smooth vector fields is again a smooth vector field. Further if $\phi \in C^\infty(U)$ is a smooth function and $\boldsymbol{V}$ a smooth vector field, then the assignment $p \mapsto \phi(p) \cdot \boldsymbol{V}(p)$ is a smooth vector field on U.

Differential Operators: Let $\boldsymbol{V}$ be a vector field on U. Then one can associate to this vector field an "operation", $\widetilde{\boldsymbol{V}} : C^\infty(U) \to C^\infty(U)$ by setting :

$$(\star)\ldots\ldots\ldots\ldots\ldots \quad \big(\widetilde{\boldsymbol{V}}(\phi)\big)(p) = \sum_{\imath=1}^{n} v_\imath(p)\frac{\partial \phi}{\partial x_\imath}(p),$$

for any smooth function $\phi : U \to \mathbb{R}$. Since ϕ is smooth so are its partial derivatives. Thus $p \mapsto \widetilde{\boldsymbol{V}}\phi$ is also smooth. This is why smooth vector fields are sometimes called first order "differential operators".

Now $\widetilde{V} : C^\infty(U) \to C^\infty(U)$ is clearly a linear endomorphism of $C^\infty(U)$. So such endomorphisms can be composed and added to yield new endomorphisms of $C^\infty(U)$. All these endomorphisms are known as partial differential operators.

We will look at one example.

The differential operators:

$$p \mapsto \left(\frac{\partial}{\partial x_\imath}\right)_p \qquad (i = 1, \ldots, n)$$

are called "basic operators". This is because any smooth vector field can be written as a linear combination of these with elements of $C^\infty(U)$ as coefficients.

Example 4.4.8 (The Laplace Operator). The operator

$$\text{(The Laplacian)} \qquad \nabla^2 \doteq \boldsymbol{X}_1^2 + \cdots + \boldsymbol{X}_n^2$$

is known as the *Laplacian* or Laplace operator in $\mathbb{R}^n$ and is a fundamental object of study in higher Analysis and Physics.

Generalized coordinates

So far whenever I have used the word "coordinate(s)" I have meant the function(s) defined on a Euclidean space or an open subset thereof by the dual vectors corresponding to an orthonormal basis.

However the word "coordinate" has a somewhat wider connotation with which you ought to be familiar.

Definition 4.4.9 (Generalized coordinate systems).
Let $U, V \subset \mathbb{R}^n$ be open sets. A C^∞ map $\Gamma : U \to V$ is called a *coordinate system* on U if:
1. Γ is a bijection, and
2. for each $p \in U$, $\mathbf{D}\Gamma(p) : \mathbb{R}^n \to \mathbb{R}^n$ is a nonsingular linear transformation.

If $\Gamma : U \to V$ is a coordinate system, then the `Inverse Function Theorem`, which we will prove in the next chapter, implies that $\Gamma^{-1} : V \to U$ is a coordinate system on V. (In fact, this theorem also implies that $\Gamma(U) \subset \mathbb{R}^n$ is open and so we need not have required the set V to be open.)

Γ is usually thought of as an ordered n-tuple, $(\gamma_1, \dots, \gamma_n)$, of smooth functions defined on U. The γ_i are called *coordinates* for the open set U.

Coordinate systems are particularly useful in tackling problems of physics and geometry. Some illustrations follow: note however I am not defining the coordinate systems precisely except for polar coordinates on the plane. The object is to point out situations where generalized coordinate systems are useful and roughly describe the idea behind three of the most useful ones and the context in which they appear.

1 Planetary motion in the plane and polar coordinates
It is convenient when dealing with the motion of planets under the sole influence of (Newtonian) gravitation in a *planar*, heliocentric system to choose the origin of $\mathbb{R}^2$ at the centre of mass of the 'sun' and describe points $(x, y) \in \mathbb{R}^2_\times$ by two numbers $[r, \theta]$ where, $r > 0$, is known as the *radial* coordinate and is equal to the distance from the origin (that is $r = \left(x^2 + y^2\right)^{\frac{1}{2}}$) and θ, known as the *azimuth* or angular coordinate, is determined by the conditions:

$$x = r\cos\theta,\ y = \sin\theta \text{ and } \pi < \theta \leq \pi.$$

Of course this way of describing the location of points in $\mathbb{R}^2$ does not yield a coordinate system. Since I will be using the associated coordinate systems shortly, I explain below, in a precise fashion, how polar coordinates give rise to coordinate systems on $\mathbb{R}^2_\times$.

For $k = 0, 1, 2, 3$, let

$$\begin{aligned} U_0 &\doteq \{(x, y) \in \mathbb{R}^2 \ : \ y > 0\}; \\ U_1 &\doteq \{(x, y) \in \mathbb{R}^2 \ : \ x < 0\}; \\ U_2 &\doteq \{(x, y) \in \mathbb{R}^2 \ : \ y < 0\}; \\ U_3 &\doteq \{(x, y) \in \mathbb{R}^2 \ : \ x > 0\}. \end{aligned}$$

I will define the polar coordinate system on these open sets, U_k, as follows. Let $\mathbb{R}^2 \supset V_k \doteq \mathbb{R}_+ \times (\frac{k\pi}{2}, \pi + \frac{k\pi}{2})$ where $k = 0, 1, 2, 3$. (Points in V_k will be denoted $[r, \theta]$ instead of the usual parentheses to distinguish them from points of U_k.) Define $\Gamma_k : V_k \to U_k$ by:

$$\Gamma_k([r, \theta]) = (r \cos \theta, r \sin \theta).$$

Note that although the right-hand side of the equation which defines the four functions, $\gamma_0, \ldots, \Gamma_3$ are the same, the domains of these functions are distinct.

The functions Γ may be thought of as defined by a pair of functions:

$$\begin{aligned} x &= r \cos \theta \\ \text{and } y &= r \sin \theta \end{aligned}$$

and one thinks of (x, y) as the standard coordinates of U_k being defined in terms of polar coordinates, by the functions $\Gamma : V_k \to U_k$.

2 Planetary motion in space and spherical polar coordinates

If one is interested in planetary motion in three dimensions, then one has to introduce spherical polar coordinates as described below.

A point $P \in \mathbb{R}^3_\times$ whose standard coordinates are (x, y, z) is represented by (r, θ, ϕ). The *radial coordinate*, $r = \left(x^2 + y^2 + z^2\right)^{\frac{1}{2}}$, is the distance of P from the origin, O. The other two coordinates ϕ and θ known as the *polar angle* and *azimuth* respectively may be described as follows. (We refer throughout to the configuration shown in Figure 4.4.1.)

If P' is the orthogonal projection of P onto the plane $z = 0$ then the azimuth, θ, of P ranges over $[0, 2\pi)$ and is the azimuth or angular coordinate of P' with respect to planar polar coordinates in the (x, y)-plane. The polar angle, ϕ, ranges over $[0, \pi]$ and measures the angle between the line OP and the z-axis.

3 Cylindrical polar coordinates:

In studying the motion in $\mathbb{R}^3$ of charged particles in the electromagnetic field due to a current flowing along a straight wire, (of infinite length,) it is convenient to choose a standard basis in which the wire is the line $L \doteq \{(x, y, z) \in \mathbb{R}^3 \ : \ x = y = 0\}$ and then use "cylindrical polar coordinates" to represent points of $\mathbb{R}^3 - L$.

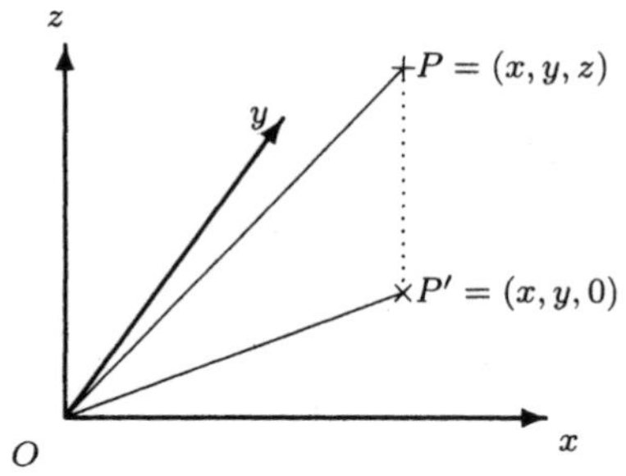

FIGURE 4.4.1. Spherical Polar Coordinates

In this system, the height above the (x, y)-plane augments the information supplied by polar coordinates on the (x, y)-plane and this third coordinate is denoted by z. Again referring to Figure 4.4.1, the cylindrical polar coordinates (r, θ, z) of the point $P = (x, y, z)$ are simply the polar coordinates of $P' = (x, y, 0)$ and the coordinate z.

Remark 4.4.10. While one can, in a straight-forward manner, use the coordinate system associated with plane polar coordinates to define coordinate systems which cover $\mathbb{R}^3 - \{(x, y, z) \in \mathbb{R}^3 : x = y = 0\}$, the corresponding exercise for spherical polar coordinates is not entirely trivial.

I now illustrate how the idea of (generalized) coordinates can be used to simplify problems by discussing the Laplace operator in $\mathbb{R}^2_\times$.

The Laplacian in $\mathbb{R}^2_\times$ and planar polar coordinates

Suppose one wants to find a solution to the Laplace equation (in $\mathbb{R}^2_\times$) and the "physical conditions" imply that the solution we seek will have some symmetry properties with respect to rotations of the plane, (for example that the solution depends only on the distance from the origin) then it is not clear how to exploit this information in solving:

$$\nabla^2 u \doteq \frac{\partial^2 u}{\partial x^2} + \frac{\partial^2 u}{\partial y^2} = 0.$$

In such a situation it would be more convenient if only differentiation with respect to r and θ were involved. Notice that here x, y refer to the standard coordinates: the dual of $\boldsymbol{e}_1$ and $\boldsymbol{e}_2$ respectively. So we wish to find out:

The exact combination of $\frac{\partial}{\partial r}$ and $\frac{\partial}{\partial \theta}$ which applied upon a smooth function, u, defined on the set $U_k$will have the same effect as the Laplacian.

The obvious first step is to "express u in terms of r and θ"; what this means mathematically is that we choose $k = 0, \ldots, 3$ such that $P \in U_k$, and then consider the function, $u \circ \Gamma = \Gamma_\sharp(u) : V_k \to \mathbb{R}$.

Now on the sets V_k, r, θ are indeed restrictions of the dual vectors of an orthonormal basis. Fix a point $P = (x_0, y_0) \in \mathbb{R}^2_\times$. Let $\Gamma^{-1}(x_0, y_0) = [r_0, \theta_0]$ and let us denote the tangent vectors to V_k at $[r_0, \theta_0]$ represented by the curves $\gamma_r(t) = [r_0 + t, \theta_0]$ and $\gamma_\theta = [r_0, \theta_0 + t]$ by $\boldsymbol{e}_r$ and $\boldsymbol{e}_\theta$ respectively. These are the tangent vectors which give rise to the operators $\frac{\partial}{\partial r}$ and $\frac{\partial}{\partial \theta}$.

In the computation which follows, to simplify the notation, I am going to omit the point (x_0, y_0) or $[r_0, \theta_0]$ at which the tangent vectors are located. For any $\boldsymbol{X} \in T_{[r_0,\theta_0]}V_k$, we have

$$\begin{aligned}
\boldsymbol{X}\left(\Gamma_\sharp(u)\right) &= \langle \boldsymbol{X}, \Gamma_\sharp(u) \rangle \\
&\quad \text{(Recall, that } \boldsymbol{X} \text{ is a linear functional on smooth functions,)} \\
&= \langle \Gamma_\sharp^*(\boldsymbol{X}), u \rangle \\
&= \left(\mathbf{D}\Gamma([r_0, \theta_0])\boldsymbol{X} \right)[u] \quad \text{(by Theorem 4.4.6.)}
\end{aligned}$$

This tells us how to convert the action of tangent vectors in V_k to the action of tangent vectors on U_k.

By the Jacobian Formula (Theorem 4.1.16) the matrix of $\mathbf{D}\Gamma(r_0, \theta_0)$ with respect to $\{\boldsymbol{e}_r, \boldsymbol{e}_\theta\}$ of $T_{[r_0,\theta_0]}V_k$ and $\{\boldsymbol{e}_x, \boldsymbol{e}_y\}$ of T_pU_k is

$$\frac{\partial(x, y)}{\partial(r, \theta)} \doteq \begin{bmatrix} \cos\theta_0 & -r_0 \sin\theta_0 \\ \sin\theta_0 & r_0 \cos\theta_0 \end{bmatrix}$$

This matrix tells us the effect of $\mathbf{D}\Gamma$ on the tangent vectors of V_k. But what we need is a description of the opposite kind: How are $\boldsymbol{e}_x, \boldsymbol{e}_y$ affected by $\mathbf{D}\Gamma^{-1}$?

I have already mentioned that by virtue of the Inverse Function Theorem, Γ^{-1} is C^∞. Since $\Gamma \circ \Gamma^{-1} = 1_{\mathbb{R}^2}$, the chain rule implies that $\mathbf{D}\Gamma^{-1} = (\mathbf{D}\Gamma)^{-1}$. Inverting the matrix $\frac{\partial(x,y)}{\partial(r,\theta)}$ we get:

$$\left(\frac{\partial(x, y)}{\partial(r, \theta)} \right)^{-1} = \begin{bmatrix} \cos\theta_0 & -\sin\theta_0 \\ \frac{1}{r_0}\sin\theta_0 & \frac{1}{r_0}\cos\theta_0 \end{bmatrix}$$

This implies that under Γ^{-1},

$$\begin{aligned} \boldsymbol{e}_x &\mapsto \cos\theta \boldsymbol{e}_r - \frac{1}{r}\sin\theta \boldsymbol{e}_\theta \\ \boldsymbol{e}_y &\mapsto \sin\theta \boldsymbol{e}_r + \frac{1}{r}\cos\theta \boldsymbol{e}_\theta \end{aligned}$$

Now we have a straightforward computation ahead of us.

$$\begin{aligned} \nabla^2 u &\doteq \boldsymbol{e}_x[\boldsymbol{e}_x[u]] + \boldsymbol{e}_y[\boldsymbol{e}_y[u]] \\ &= \boldsymbol{e}_x\left[\cos\theta\frac{\partial u}{\partial r} - \frac{1}{r}\sin\theta\frac{\partial u}{\partial \theta}\right] + \boldsymbol{e}_y\left[\sin\theta\frac{\partial u}{\partial r} + \frac{1}{r}\cos\theta\frac{\partial u}{\partial \theta},\right] \end{aligned}$$

and *carefully* repeating the operation of $\boldsymbol{e}_x$ and $\boldsymbol{e}_y$, we get

$$\nabla^2 u = \frac{\partial^2 u}{\partial r^2} + \frac{2}{r}\frac{\partial u}{\partial r} + \frac{1}{r^2}\frac{\partial^2 u}{\partial \theta^2}$$

which is Laplace's equation, on the "punctured plane", in polar coordinates.

Remark. If we are seeking a solution of $\nabla^2 u = 0$ in $\mathbb{R}^2_\times$, which depends only on the distance from the origin, then we only need to solve:

$$\nabla^2 u = \frac{\partial^2 u}{\partial r^2} + \frac{2}{r}\frac{\partial u}{\partial r}.$$

Elementary manipulations now yield the two linearly independent solutions:

$$u_1(r) = \frac{\text{constant}}{r} \text{ and } u_2(r) = \text{constant}.$$

5

Existence Theorems

This chapter is devoted to establishing three basic existence theorems which lie at the foundations of the "three great differential theories" of Modern Mathematics: Differential Topology, Differential Geometry and Differential Equations.

All the three results affirm the existence of a function having certain desirable properties provided appropriate hypotheses are satisfied. The optimal method of proving all three results uses an existence theorem of a different kind: one that asserts the existence of a `fixed point` for a map of a space into itself. This result, which is proved in § 5.1, will probably be your first encounter with a geometric or `topological` result with profound consequences in Analysis.

§ 5.2 is devoted to a brief account of the existence and uniqueness of solutions of ordinary differential equations. The treatment is less formal in manner than the earlier sections of this book because it is not my aim to give an introduction to the theory of ordinary differential equations. I am simply trying to make you aware of how different the modern theory of differential equations is from the classical approach to the subject, (With the latter, most of you would have possibly had some previous encounter.) It is appropriate to do this in a book devoted to Calculus in normed linear spaces since most of the essential features of the modern theory get obscured when one is solely interested in the 1-dimensional context.

In § 5.3 I establish the Inverse Function Theorem which I have already invoked while discussing generalized coordinates. This (together with the Implicit Function Theorem) constitute the foundation stones of Differential Topology and Geometry. After proving the Inverse Function Theorem, as a nice application of the result, I prove the "grand-daddy" of all existence theorems of Analysis: the Fundamental Theorem of Algebra.

In the final section I deal with the Implicit Function Theorem. One part of this result is equivalent to the Inverse Function Theorem. But the uniqueness of the implicitly defined function is quite a delicate matter and the major emphasis is on this. (Since this is used, only at one point in the remainder of the book, it may be omitted on a first reading.)

5.1 The Contraction Mapping Principle

As I have already indicated, this section introduces you to a new kind of theorem; I would like to reiterate that it does not require the notions of Calculus to state or prove this result.

Theorem 5.1.1 (The Contraction Mapping Principle).
Let (X,d) be a complete metric space and $f : X \to X$ a function. Suppose that there is a constant, $0 < K < 1$, such that for all distinct points, $x, y \in X$ we have $d(f(x), f(y)) \leq Kd(x,y)$. (This, of course, implies that f is continuous and such functions will be called *contraction maps.*) *Then there is a unique point $\xi \in X$ such that $f(\xi) = \xi$ or, in other words,* a contraction map has a unique fixed point.

Proof. Choose any point $x_0 \in X$ and for $n \geq 1$, inductively define the sequence of points, $x_n = f(x_{n-1})$.

Let $p > q \geq 1$. Then, by repeated use of the "contraction condition" we get:

$$d(x_p, x_q) = d\big(f(x_{p-1}), f(x_{q-1})\big) \leq K \cdot d\big(f(x_{p-1}, x_{q-1})\big) \leq \cdots \leq K^q d(x_{p-q}, x_0).$$

So by repeated application of the triangle inequality we see that

$$d(x_m, x_{m+n}) \leq K^{m-1}\big(1+K+\cdots+K^n\big)d(x_1, x_0) \leq \frac{K^{m-1}}{1-K} d(x_m, x_0) \to 0,$$

as $m \to \infty$ which means that $\{x_n\}$ is a Cauchy sequence. Since (X,d) is a complete metric space, there is a unique point $\xi \in X$ such that

$\lim_{n\to\infty} x_n = \xi$. Since $f : X \to X$ is continuous,

$$f(\xi) = f\Big(\lim_{n\to\infty} x_n\Big) = \lim_{n\to\infty} f(x_n) = \lim_{n\to\infty} x_{n+1} = \xi$$

that is ξ is a fixed point of f.

If $\xi \neq \eta$ are distinct fixed points of f we have from the contraction condition that,

$$d(\xi,\eta) = d\big(f(\xi), f(\eta)\big) \leq Kd(\xi,\eta) < d(\xi,\eta)$$

which is a contradiction. Hence f cannot possess more than one fixed point. The proof of the CMP is complete. □

Remark 5.1.2 (Caveat!).
One word of caution; it is not sufficient to assume that the distance of any two distinct points is greater than the distance between their images under f; that is, $d(f(x), f(y)) < d(x, y)$. for all distinct points $x, y \in X$. I would like to thank S. M. Srivastava for showing me the following example which illustrates this.

A COUNTER-EXAMPLE

We consider $\mathbb{R}$ as a union of subintervals $J_m = [a_m, b_m]$ where m ranges over *all* integers. These intervals are defined as follows. For $m \geq 0$ we define $J_0 = [0, 1]$, that is, $a_0 = 0, b_0 = 1$. Inductively define $a_n = b_{n-1}$ and $b_n = a_n + \frac{1}{n+1}$. The divergence of the harmonic series ensures that $\bigcup_{m\geq 0} J_m$ is the set of all non-negative real numbers. Now for $m > 0$ we define $J_{-m} = [a_{-m}, b_{-m}]$ as follows. $b_{-1} = 0$ and $a_{-1} = -2$. If a_m, b_m have been defined for all $m \geq -p$, we define $b_{-p-1} = a_{-p}$ and $a_{-p-1} = b_{-p-1} - (p+2)$. Clearly the $J_\imath$'s constitute a covering of $\mathbb{R}$ by a countable family of closed intervals such that $J_m \cap J_n \neq \emptyset$ iff $|m-n| = 1$. Moreover J_m and $J_{m\pm 1}$ have exactly one end-point in common.

Define a map $\phi : \mathbb{R} \to \mathbb{R}$ as follows: ϕ takes the points of the interval J_i to those of J_{i+1} taking a_i to a_{i+1} and b_i to b_{i+1} and is linear on each of these intervals. Clearly ϕ does not have a fixed point and it is obvious (and easy to check formally) that if $x, y \in \mathbb{R}$ are distinct points then $d(\phi(x), \phi(y)) < d(x, y)$. □

The following points concerning the CMP are also worth noting.

Remark 5.1.2a. Notice that the choice of the initial point x_0 is irrelevant: all points in X converge to the unique fixed point under repeated application of the map f.

Remark 5.1.2b. In a "real-world situation" this repeated application of the function f can be used to produce approximate fixed points.

Exercises for § 5.1

5.1.1 A FIXED-POINT THEOREM FOR COMPACT METRIC SPACES

a. Let X be a compact metric space and $f : X \to X$ a *shrinking* function; more precisely:

for every $x, y \in X$, $d\big(f(x), f(y)\big) < d(x,y)$ *if* $x \neq y$.

Show that there is a unique point $\xi \in X$ such that $f(\xi) = \xi$.

b. Let $\mathbf{R} = \{(x,y) \mid 0 \leq x, y \leq 1\}$ and suppose that f, g are C^1-functions defined on some open set containing $\mathbf{R}$, satisfying the conditions:

1. $0 \leq f(p), g(p) \leq 1$ for all $p \in \mathbf{R}$, and
2. $(\star)$ $\quad 0 \leq \dfrac{\partial f}{\partial x}(p), \dfrac{\partial f}{\partial y}(p), \dfrac{\partial g}{\partial x}(p), \dfrac{\partial g}{\partial y}(p) \leq \dfrac{1}{2}$ for every $p \in \mathbf{R}$.

Define $\Phi : \mathbf{R} \to \mathbf{R}$ by $\Phi(x,y) = \big(f(x,y), g(x,y)\big)$.

Show that there is a unique $(\xi, \eta) \in \mathbf{R}$ such that $\Phi(\xi, \eta) = (\xi, \eta)$

(HINT: Use the MVT.)

c. Will this conclusion be valid if in $(\star)$ we replace '$<$' by '$\leq$'?

d. Obtain a generalization for the unit hypercube in $\mathbb{R}^n$.

5.2 Ordinary Differential Equations

This section is devoted to a *discussion* of Ordinary Differential Equations and is somewhat different in spirit and content from the rest of the book. The subject is too vast and varied to treat in less than a separate volume and, fortunately, a wonderful introductory text ([HS74]) already exists! My objective here is to acquaint you with the basic *definitions* and terminology of the theory of ordinary differential equations in normed linear spaces and (hopefully) impress upon you how different it is from the traditional accounts of the subject.

I will first state, without explanations the classical form of the basic existence result.

THE PROBLEM: Suppose I, J are open subintervals of $\mathbb{R}$ and $f : I \times J \to \mathbb{R}$ a map. Then the *first-order differential equation* corresponding to this map is written

$(*)$ .... $x' = f(t, x(t)).$

If $(p, \xi) \in I \times J$,

the *initial-value problem* corresponding to $(*)$ and the condition $x(p) = \xi$ requires one to find a function $\phi : I \to \mathbb{R}$ such that for all $t \in I$, $(t, \phi(t)) \in I \times J$, $\phi(p) = \xi$ and $\phi'(t) = x\big(t, \phi(t)\big)$.

Let me begin by quoting the *classical form* of the fundamental result of the subject of ordinary differential equations:

Theorem (Cauchy–Picard).

Let $I, J \subset \mathbb{R}$ be open intervals and $f : I \times J \to \mathbb{R}$ a map satisfying the following "Lipschitz condition":

for all $(t, x_1), (t, x_2) \in I \times J$,

$$|f(t, x_1) - f(t, x_2)| \leq C|x_1 - x_2| \text{ for some constant } C > 0.$$

Then, for any $p = (t_0, x_0)$ in $I \times J$ there exists $\varepsilon_p > 0$ (which may depend on p,) and a differentiable map $y_p : (t_0 - \varepsilon_p, t_0 + \varepsilon_p) \to \mathbb{R}$ such that:

$$y_p'(t) = f\big(t, y_p(t)\big) \text{ and } y_p(t_0) = x_0.$$

There are other refinements to the result which I have not included in the above statement: these will be included in the more general result I will state and prove a little later. My objective now is to highlight the difference between the points of view in the classical treatment and that which we are going to adopt.

I begin with some general remarks upon the question:

What is a differential equation?

Let us look at a very simple example.

Consider the problem:

$$(*)\ldots\ldots\ldots\ldots\ldots\ldots \qquad \frac{dy}{dx} = x^2 y \quad \text{and } y(0) = 2.$$

The first equation in $(*)$ yields:

$$\begin{aligned} \frac{dy}{y} &= x^2. \text{ Integration gives:} \\ \log y &= \tfrac{1}{3}x^3 + c \\ \text{or, } y &= c \cdot e^{\left(\frac{1}{3}x^3\right)}. \end{aligned}$$

and using the initial condition, we get $y = 2e^{\left(\frac{1}{3}x^3\right)}$.

So in this simple situation, solving an initial-value problem seems to proceed in two parts:

Step 1 Evaluating an indefinite integral;

Step 2 Using the initial condition to determine the constant of integration.

In fact, one might get the impression that an initial value problem for an ordinary differential equation of first order is merely an alternative way of setting up a problem of "integral calculus": for instance, the first set of exercises in the classic textbook of Ince ([Inc56]) begins with the invitation: "Integrate the following differential equations."

Without further ado, I now turn to the formulation of the problem of ordinary differential equations in the setting of Banach spaces.

The first point to observe is that the classical formulation is in terms of two open intervals, $U, V \subset \mathbb{R}$, and a function $f : U \times V \to \mathbb{R}$. The solution of the differential equation is a real-valued function $\alpha : U_0 \to \mathbb{R}$ of a subinterval $U_0 \subset U$, whose derivative is "pre-assigned" by the function f.

In the general setting of Banach spaces, we will be concerned with an open set $U \subset E$ and the solution of the yet to be defined notion of "differential equation" is to be a function $\alpha : (a, b) \to U$ from an open interval into U, whose derivative is "pre-assigned". So the role of the function "$f : U \times V \to \mathbb{R}$" now *must* be replaced by an assignment of tangent vectors at points of U which may depend on $t \in J$ where $J \subset \mathbb{R}$ is an open interval.

This is described by saying $\boldsymbol{f} : I \times U \to E$ is a time-dependent vector field on U.The important thing to note is that $\boldsymbol{f}(t, x)$ is to be thought of as a *tangent vector* belonging to $T_x E$.

Definition (Ordinary Differential Equations, integral curves).
Let U be an open subset of a Banach space E and $J \subset \mathbb{R}$ an open interval

containing 0. An *ordinary differential equation* (sometimes called a *time-dependent vector field*), is a C^k-map ($k \geq 0$) $\boldsymbol{f} : J \times U \to E$.

The *initial-value problem* corresponding to the point $(0, x_0) \in J \times U$ is to find an open subinterval $J_0 \subset J$ containing 0 and a curve $\alpha : J_0 \to U$ such that

($\star$) for all $t \in J_0$, $\alpha'(t) = \boldsymbol{f}\big(t, \alpha(t)\big)$ and $\alpha(0) = x_0$.

Such a curve $\alpha : J_0 \to U$ is called an *integral curve* of the vector field (or differential equation) *starting at* x_0. *This is also called a solution* of the initial value problem.

(The choice of 0 is not significant: it is merely chosen so that the "initial point" will be determined by $t = 0$.)

REMARKS 5.2.2 (ON THE NOTION OF A DIFFERENTIAL EQUATION)

Remark 5.2.2a. Notice that on the right-hand side of ($\star$), I have used "$\boldsymbol{f}(t,x)$" rather than "$f(t,x)$". This is to emphasize that f is thought of as an assignment of a tangent vector to each point of U, which, possibly, depends on 'time' represented by t.

Remark 5.2.2b (Linear Differential Equations).
One particularly interesting class of vector fields are those of the form:

$$\boldsymbol{f}(t,x) = A(t)(\boldsymbol{x}) + b(t)$$

where $A : J \to \mathbb{L}(E)$ and $b : J \to E$ are, say, continuous and bounded maps. The associated problems are referred to as *linear differential equations* of order 1. The differential equation is called *homogeneous* if $b(t) = 0$ for all t.

Before proceeding further, it might be instructive to write down in *long-hand* what a linear differential equation looks like. Let E be a finite-dimensional linear space with an onb $\{\boldsymbol{e}_1, \ldots, \boldsymbol{e}_n\}$. If $x_1, \ldots, x_n$ are the coordinates corresponding to this basis then a linear differential equation in E looks like (using matrix notation):

$$\begin{bmatrix} x_1' \\ x_2' \\ \vdots \\ x_n' \end{bmatrix} = \begin{bmatrix} a_{11}(t) & a_{12}(t) & \ldots & a_{1n}(t) \\ a_{21}(t) & a_{22}(t) & \ldots & a_{2n}(t) \\ \vdots & \vdots & \ldots & \vdots \\ a_{n1}(t) & a_{n2}(t) & \ldots & a_{nn}(t) \end{bmatrix} \begin{bmatrix} x_1(t) \\ x_2(t) \\ \vdots \\ x_n(t) \end{bmatrix} + \begin{bmatrix} b_1(t) \\ b_2(t) \\ \vdots \\ b_n(t) \end{bmatrix}$$

Linear Differential Equations are generally written in the form

$$\frac{d\boldsymbol{x}}{dt} + A(t)(\boldsymbol{x}) = b(t).$$

Remark 5.2.2c (The geometric viewpoint). Since the right-hand side of a differential equation $x'(t) = \boldsymbol{f}(t,x)$ is being regarded as a vector field, the initial value problem has a very striking "picturization" which is exemplified below.

Let us suppose for simplicity that $E = \mathbb{R}^2$ and $U \subset E$ is an open ball centred at the origin. Then for each point $p \in U$ we have a (time-dependent) tangent vector $\boldsymbol{f}(t,p)$ assigned. The initial value problem $y(0) = \xi$ is the problem of finding a curve $y : J \to U$ starting at ξ, (that is, $y(0) = \xi$,) whose velocity at any 'time' $t \in J$, is the value of the vector field at the "current position and time" i. e. $y'(t) = \boldsymbol{f}\big(t, y(t)\big)$.

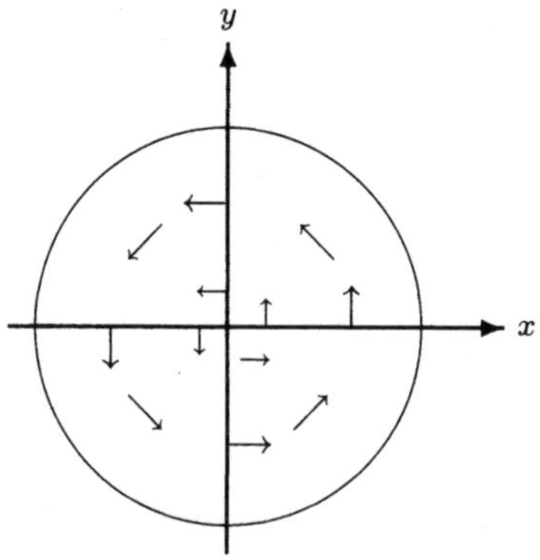

FIGURE 5.2.1. The vector field $(x,y) \mapsto (y,-x)$

For instance, Figure 5.2.1 gives a crude and approximate representation of the (time-independent) vector field:

$$\boldsymbol{f}(x,y) = (y,-x) \quad \text{defined on } \mathbb{R}^2. \tag{5.2.1}$$

It might be helpful to think of this in terms of the motion of a fluid. The vector field at each point represents the direction and strength of the current and the integral curves describe motion of a particle being carried along by the fluid.

It is immediately clear that one can now see questions which one would not have thought of asking before. For instance: "Is there an integral curve defined on $[a,b]$ such that the set $\{\gamma(t) : t \in [a,b]\}$ is a closed curve?" This is a property of the vector field and the answer is dependent on the domain on which the vector field is defined. This question will not be pursued here.

Remark 5.2.2d (Concerning the initial condition).
In the classical formulation the role of the initial condition was to enable us to "evaluate the constant of integration"; here it determines the starting point of the trajectory and influences how large an interval J can support an integral curve. The vector field $\boldsymbol{f}$ defined by 5.2.1 illustrates this.

It is easy to check that if $(x_0, y_0) = (r_0 \cos\theta_0, r_0 \sin\theta_0)$ then,
$\alpha : \mathbb{R} \to \mathbb{R}^2$ defined by:

$$\alpha(t) = (r_0 \cos(t + \theta_0), r_0 \sin(t + \theta_0))$$

is an integral curve starting at (x_0, y_0). Indeed it is an example of a closed integral curve: I had mentioned this point earlier.

It is clear that if we choose open rectangles, (with sides parallel to the axes and centred at the origin,) and restrict the vector field to this region and try to construct integral curves, they will stray out of the domain unless r_0 is less than half the length of the shortest edge. Thus the initial conditions determine the 'length' of the integral curve; that is, the size of the interval on which an integral curve may be defined.

Remark 5.2.2e (On finding integral curves).
Given any continuous curve $\beta : J \to U$, consider the function $F_\beta(t)$ defined by

$$F_\beta(t) = x_0 + \int_0^t \boldsymbol{f}(u, \beta(u))du,$$

where $\boldsymbol{f}$ is a C^p-vector field. Clearly, F_β is C^1. In fact, an obvious induction argument shows that if β is C^p, then F_β is C^{p+1}. F_β satisfies the relations:

$$(\dagger) \quad \ldots\ldots\ldots\ldots\ldots \quad F'_\beta(t) = \boldsymbol{f}\big(t, \beta(t)\big) \text{ and } F(0) = x_0.$$

Since $x_0 \in U$, if $|t|$ is small enough then F_β will be a C^1 curve defined on $(-|t|, |t|)$ and taking values in U.

If we regard the transformation $\beta \Longrightarrow F_\beta$ as some sort of a "function", ϕ, whose domain and range is the set, $\mathcal{P}$, of maps of J into U (that is, the set of continuous curves taking values in U), then the relations $(\dagger)$ can be interpreted as saying that $\phi \in \mathcal{P}$ is a fixed point of this ϕ *iff it is an integral curve of f starting at x_0*. So, perhaps the CMP can be invoked to prove the existence of integral curves. Of course, before this makes any sense, we have to endow the set, $\mathcal{P}$, or some suitable subset thereof, with the structure of a complete metric space.

We do this in two steps.

1. For each $x \in U$, we look at the following restricted set of paths. Let $r > 0$ be such that $\overline{\mathbb{B}}_r(x) \subset U$. Denote by $\mathcal{P}_x$ the set of continuous maps defined on an interval $[-a, a] \subset J$ and taking values in $\overline{\mathbb{B}}_r(x)$. For $u, v \in \mathcal{P}_x$ define the metric, ρ, on $\mathcal{P}_x$ by setting

$$\rho(u, v) = \sup_{|t|<a} ||u(t) - v(t)||.$$

This makes $\mathcal{P}_x$ a complete metric space since a sequence of paths $\{u_n\}$ is Cauchy with respect to this metric iff it is a uniformly convergent sequence of paths.

2. If further hypotheses are imposed upon the vector field $\boldsymbol{f}$, then indeed $\phi : \mathcal{P}_x \to \mathcal{P}_x$ defined by $\phi(\beta) = F_\beta$, will turn out to be a contraction mapping of $\mathcal{P}_x$.

This is the strategy we will adopt in proving the existence of solutions to the initial value problem.

I will now set up suitable hypotheses to enable us to apply the CMP as indicated in the preceding remarks. Let U and J be as before.

Definition 5.2.3 (The Lipschitz Condition).
A map $\boldsymbol{f} : J \times U \to U$ is said to satisfy a *Lipschitz condition* on U, if for each $t \in J$ there is a constant $K_t > 0$ such that:

$$||\boldsymbol{f}(t, x_1) - \boldsymbol{f}(t, x_2)|| \le K_t \cdot ||x_1 - x_2||$$

for all $x_1, x_2 \in U$. If a constant $K > 0$ can be found so that (**Lip**) holds for all $t \in J$, then we say f *is uniformly Lipschitz on* U (or that f is a Lipschitz map on $J \times U$ which is uniformly Lipschitz with respect to J). K is usually referred to as a Lipschitz constant for f.

Example 5.2.4. If $U \subset E$ is a convex open set and $\boldsymbol{f} : J \times U \to E$ a C^1-map which satisfies the condition that $\mathbf{D}_2\boldsymbol{f}(t, x)$ exists and is bounded on U then $\boldsymbol{f}$ is Lipschitz, for we can then choose

$$K_t = \sup_{x \in U} ||\mathbf{D}_2\boldsymbol{f}(t, x)||.$$

The MVT will then ensure that $||f(t, x_1) - f(t, x_2)|| \le K_t||x_1 - x_2||$.

At this point I could simply state and prove the basic theorem giving conditions under which integral curves starting at x_0 exist. But since I have already announced that this section is of a somewhat informal nature, I am now going to work backwards and see under what conditions the function $\Phi : \mathcal{P} \to \mathcal{P}$ will be a contraction mapping. The role of the Lipschitz condition defined above will then become transparent.

For any fixed differential equation, $\boldsymbol{f} : J \times U \to E$ and any two curves $u, v : [-a, a] \to \overline{\mathbb{B}}_r(x_0)$, we estimate the distance between the curves, $\Phi(u)$ and $\Phi(v)$ as follows:

$$\begin{aligned}\rho\big(\Phi(u), \Phi(v)\big) &= \sup_{|t|\le a} \left\| \int_0^t \big(\boldsymbol{f}(s, u(s)) - \boldsymbol{f}(s, v(s))\big) ds \right\| \\ &\le \sup_{|t|\le a} \left| \int_0^t ||\boldsymbol{f}(s, u(s)) - \boldsymbol{f}(s, v(s))|| ds \right| \\ &\le a \cdot \sup_{|s|\le a} ||\boldsymbol{f}(s, u(s)) - \boldsymbol{f}(s, v(s))||.\end{aligned}$$

If f happens to be uniformly Lipschitz with constant $K > 0$ then we get

$$\rho\big(\Phi(u), \Phi(v)\big) \le aK \cdot \rho(u, v).$$

So if there is a constant, c, such that $aK \le c < 1$, we can apply the CMP and deduce that there is an integral curve for $\boldsymbol{f}$ starting from x_0. (Of course, we must also check that Φ indeed maps a path in $\overline{\mathbb{B}}_r(x_0)$ into another such path.)

It will now be easy for you to appreciate the need for the various technical conditions laid down in the following fundamental result on the existence of integral curves.

Theorem 5.2.5 (Cauchy–Picard: the Existence of integral curves). *Let J be an open interval containing 0 and suppose that $U \subset E$ is an open set in a Banach space. Let $x_0 \in U$ and suppose that $0 < a < 1$ is such that $\mathbb{B}_{3a}(x_0) \subset U$.*

Suppose that $\boldsymbol{f} : J \times U \to E$ is a map which satisfies the following conditions.

1. *$\boldsymbol{f}$ is bounded on $J \times U$; we will assume that $||f(t, x)|| < M$ for all $(t, x) \in J \times U$ where $M \ge 1$.*

2. *$\boldsymbol{f}$ is uniformly Lipschitz on U with Lipschitz constant $K \ge 1$.*

Then for each $x \in B_a(x_0)$ and each constant, b, satisfying the inequality $0 < b < \frac{a}{KM}$, there is an integral curve $\alpha_x : (-b, b) \to U$ starting at x. If $\boldsymbol{f}$ is C^p, then so is the integral curve α_x.

Proof. Let $x \in \bar{\mathbb{B}}_a(x_0)$ be any point and let $\mathcal{P}_x$ be the set of maps from $[-b, b]$ into $\overline{\mathbb{B}}_{2a}(x)$ and, as before, let us define $\Phi : \mathcal{P}_x \to \mathcal{P}_x$ by setting:

$$\Phi(u)(t) = x + \int_0^t \boldsymbol{f}\big(s, u(s)\big)ds.$$

Then the distance of $\Phi(u)(t)$ from x is bounded by the norm of the integral in the formula for Φ which is, in turn, less than or equal to $\int_0^t ||\boldsymbol{f}\big(s, u(s)\big)||ds \leq bM < a$. Hence, Φ is, indeed, a self-map of the complete metric space $\mathcal{P}_x$.

The calculations just before this theorem shows that $\rho\big(\Phi(u), \Phi(v)\big) \leq bK\rho(u, v)$ and so by the choice of the constants K, M and a, $\Phi : \mathcal{P}_x \to \mathcal{P}_x$ satisfies the conditions of the CMP. Hence Φ has a unique fixed point which, by my earlier remarks, is an integral curve of $\boldsymbol{f}$ starting at x. □

Now that you have seen the power of the CMP, I must make you aware of how useful this result is and how much more could have been proved with a little more effort.

REMARKS 5.2.5 (ON THE CAUCHY–PICARD THEOREM)

Remark 5.2.5a (Uniqueness of solutions).
In fact, the integral curve starting at x is *unique*; but I will not go into this in any detail. The only sense in which the curve we obtained is non-unique is that there may be another interval $(-b', b')$ on which one can define an integral curve. One can show that for each x there is a unique 'maximal' integral curve through x.

Remark 5.2.5b (Necessity of the Lipschitz condition).
The following example indicates how important the Lipschitz condition really is. Consider the time-independent vector field on $\mathbb{R}$:

$$\boldsymbol{f}(t, x) = \begin{cases} +\sqrt{x} & \text{if } x > 0 \\ -\sqrt{|x|} & \text{if } x < 0. \end{cases}$$

and the initial-value problem of finding integral curves starting at $(0, 0)$. We can 'integrate' this, to use the old-fashioned term, easily enough.

It is simple to check that the map $\alpha : \mathbb{R} \to \mathbb{R}$ defined by $\alpha(t) = \frac{t|t|}{4}$ is an integral curve starting at $(0, 0)$. But clearly $\alpha(t) = 0$ for all t also is a solution of this initial value problem! This spectacular breakdown of the uniqueness of integral curves should convince you of the importance of the Lipschitz condition.

Before ending this cursory (and totally inadequate) treatment of differential equations, I want to make miscellaneous remarks on linear differential equations, mainly to impress upon you the flexibility and power of the approach to differential equations that I have adopted.

Remarks on Linear Differential Equations

I have already introduced differential equations of the form:

$$\frac{d\boldsymbol{x}}{dt} = A(t)(\boldsymbol{x}) + \boldsymbol{b}(t)$$

where $\boldsymbol{x}, \boldsymbol{b}$ are functions defined on an open interval (containing 0) and taking values in a Banach space E and A takes values in $\mathbb{L}(E)$.

Now guided by the simplest situation, when $E = \mathbb{R}$, it is not difficult to see that when $\boldsymbol{b}(t) = 0$, (the homogeneous case,) given the initial condition, $x(0) = \boldsymbol{v}$, the function, $x(t) = \big(\mathbf{A}(t)\big)(\boldsymbol{v})$ is a solution for all $t \in J$, where

$$\mathbf{A}(t) = \mathbf{e}^{\left(\int_0^t A(s)ds\right)}.$$

I want to point out some of the ramifications of this simple observation.

Remark 5.2.6a (Reduction of order).
A homogeneous linear differential equation of *order* n has the form:

$$\frac{d^n\boldsymbol{x}}{dt^n} + a_1(t)\frac{d^{n-1}\boldsymbol{x}}{dt^{n-1}} + \cdots + a_{n-1}(t)\frac{d\boldsymbol{x}}{dt} + a_n(t)\boldsymbol{x} = 0,$$

the $a_k(t)$ being continuous $\mathbb{L}(E)$-valued functions defined on some interval J which contains 0. Suppose we want to find in a subinterval $I \subset J$ a solution $x : I \to E$ satisfying the initial conditions

$$\boldsymbol{x}(0) = \boldsymbol{v}_0, \frac{d\boldsymbol{x}}{dt}(0) = \boldsymbol{v}_1, \ldots, \frac{d^{n-1}\boldsymbol{x}}{dt^{n-1}}(0) = \boldsymbol{v}_{n-1}.$$

We can replace this n^{th} order differential equation by an equivalent problem which involves an initial-value problem associated with a new first-order linear differential equation in the Banach space, $E^{[n]} \doteq E\oplus\cdots\oplus E$, ($n$ summands).

We define the $E^{[n]}$-valued function:

$$\boldsymbol{Y}(t) = \big(\boldsymbol{y}_1(t), \boldsymbol{y}_2(t), \ldots, \boldsymbol{y}_n(t)\big)$$

and require that $\boldsymbol{y}_1(t) = \boldsymbol{x}(t)$ and $\boldsymbol{y}_{k+1}(t) = \boldsymbol{y}_k'(t)$. Then,

$$\boldsymbol{Y}(t) = \left(\boldsymbol{x}(t), \frac{d\boldsymbol{x}}{dt}(t), \ldots, \frac{d^{n-1}\boldsymbol{x}}{dt^{n-1}}(t)\right).$$

We now have an initial-value problem associated with the first-order linear equation, $\frac{d\boldsymbol{Y}}{dt} = A(t)(\boldsymbol{Y})$ where

$$A(t) = \begin{bmatrix} 0 & 1_E & 0 & \ldots & 0 \\ 0 & 0 & 1_E & \ldots & 0 \\ \vdots & \vdots & & \ddots & \vdots \\ 0 & 0 & \ldots 0 & & 1_E \\ -a_n(t) & -a_{n-1}(t) & \cdots\cdot & & -a_1(t) \end{bmatrix}.$$

The initial condition now is:

$$\boldsymbol{y}(0) = \boldsymbol{v} \doteq (\boldsymbol{v}_0, \ldots, \boldsymbol{v}_{n-1})^{\boldsymbol{t}}.$$

Notice that the initial-value problem must involve the data concerning y and its first $(n-1)$ derivatives at the initial point. I have used block matrices (as explained in Chapter 2) to describe $A(t)$.

To clarify the situation obtaining, I now specialize further to the case when $E = \mathbb{R}$ in the above setup.

Remark 5.2.6b. Suppose we are looking at a linear differential equation of order n:

$$\mathcal{L}(y) = a_0 \frac{d^n y}{dt^n} + \cdots + a_n y = 0 \quad \text{where } y : (a,b) \to \mathbb{R}.$$

and the corresponding initial-value problem.

(Recall that a special case of the Laplace equation on $\mathbb{R}^2_\times$, which we encountered in the previous chapter, page 208, was a second-order equation of this form.)

As mentioned earlier the initial data is a vector

$$\boldsymbol{v} = \left(y(0), \frac{dy}{dt}(0), \ldots, \frac{d^{n-1}y}{dt^n}(0) \right) \in \mathbb{R}^n.$$

Clearly the set of functions $y : (a,b) \to \mathbb{R}$ which are differentiable n times and satisfy $\mathcal{L}(y) = 0$ is a linear space, say $\mathrm{sol}(\mathcal{L})$. Then obviously, the map

$$\mathrm{sol}(\mathcal{L}) \to \mathbb{R}^n \text{ defined by } y \mapsto \left(y(0), \frac{dy}{dt}(0), \ldots, \frac{d^{n-1}y}{dt^n}(0) \right)$$

is a linear transformation. But then existence and uniqueness of solutions of such equations (Cauchy–Picard) implies that this map is, in fact, *a bijection!*

Hence an nth-order linear equation has exactly n linearly independent solutions.

Example 5.2.7 (The harmonic oscillator).
Perhaps the most important/celebrated differential equation is the *harmonic oscillator*:

$$\frac{d^2y}{dt^2} + y = 0,$$

which represents oscillations or vibrations of a constant frequency. By the previous remark the set of solutions is isomorphic to $\mathbb{R}^2$. The solutions arising from initial conditions corresponding to the standard basis $\boldsymbol{e}_1 = (1,0)$ and $\boldsymbol{e}_2 = (0,1)$ are $\cos t$ and $\sin t$ respectively.

5.3 The Inverse Function Theorem

If $f : (a, b) \to \mathbb{R}$ is a continuously differentiable function and its derivative is nonzero at $\theta \in (a, b)$, say $f'(\theta) > 0$, then in an interval $(\theta - \delta, \theta + \delta) \subset (a, b)$ the function will be strictly monotonic increasing and hence a bijection onto its image. So $f : (\theta - \delta, \theta + \delta) \to (f(\theta - \delta), f(\theta + \delta))$ will have an inverse function. The inverse function theorem generalizes this result to C^p-maps defined on open subsets of Banach spaces. What the theorem says is that if at some point, p, in its domain, the derivative of a C^1-map has a (continuous) inverse, then the restriction of the function to a suitable open set containing p has a C^1-inverse.

In a sense this is a validation of the way we have approached Calculus: the derivative was defined as the "best linear approximation" of f at a point. And the Inverse Function Theorem affirms that if the "best local linear approximation" (the derivative) is invertible at some point p, then the function is itself invertible in a neighbourhood of p.

Since I have already advertised the fact that this result has far-reaching consequences, you should expect that its proof will not be as trivial as the observation which I made at the beginning of this section.

Before we begin, let me remind you that if E, F are Banach spaces, then a continuous linear transformation $\alpha : E \to F$ is an *isomorphism* if there is a *continuous*, linear transformation $\beta : F \to E$ such that $\beta\alpha = 1_E$ and $\alpha\beta = 1_F$.

Theorem 5.3.1 (The Inverse Function Theorem).
Let E, F be Banach spaces and $U \subset E$ an open set containing the point x_0. Suppose that $f : U \to F$ is a C^p-map whose derivative at x_0, $\mathbf{D}f(x_0) : E \to F$, is an isomorphism. Then there are open sets $V \subset E$, $W \subset F$ and a function $\phi : W \to V$ satisfying the conditions:

1. $x_0 \in V$, $f(x_0) \in W$;

2. $f_V\phi = 1_W$, $\phi f_V = 1_V$, *where* $f_V = f|V$;

3. $\phi : W \to V \subset F$ *is a* C^p-*map, and*

4. *for each* $w \in W$, $\mathbf{D}\phi(w) = \left(\mathbf{D}f(\phi(w))\right)^{-1}$

An ordered triple (V, ϕ, W) satisfying the conditions (a) and (b) above is called a **local inverse of f at $f(x_0)$**.

Proof. I will begin by assuming that $E = F$, $x_0 = f(x_0) = 0_E$ and that $\mathbf{D}f(0) = 1_E$. As we will see, this will, in fact, imply the seemingly more general statement enunciated above.

Consider the function $g : U \to E$ defined by $g(x) = x - f(x)$. Clearly, $\mathbf{D}g(x) = 1_E - \mathbf{D}f(x)$ and hence $\mathbf{D}g(0) = 0$. Since $\mathbf{D}g$ is continuous we can choose $r > 0$ so that $\overline{\mathbb{B}}_r(0) \subset U$, and for $||x|| \leq r$, $||\mathbf{D}g(x)|| < \frac{1}{2}$. Then by the MVT, $||g(x)|| < \frac{||x||}{2}$. For any $x \in \overline{\mathbb{B}}_r(0)$ and each $y \in \mathbb{B}_{\frac{r}{2}}(0)$, consider the function $g_y(x) \doteq y + g(x) = y + x - f(x)$. Now, by the MVT,

$$||g(x)|| \leq \sup_{t \in \mathbb{B}_r(0_E)} ||\mathbf{D}g(t)|| \cdot ||x||,$$

$$\text{and hence, } ||g_y(x)|| \leq ||y|| + ||g(x)|| \leq ||y|| + \frac{||x||}{2} r \leq r.$$

So, g_y maps $\overline{\mathbb{B}}_r(0)$ into itself. Further, for $x_1, x_2 \in \overline{\mathbb{B}}_r(0)$, we have:

$$\begin{aligned} ||g_y(x_1) - g_y(x_2)|| &= ||g(x_1) - g(x_2)|| \\ &\leq \tfrac{1}{2} \cdot ||x_1 - x_2||. \end{aligned}$$

So g_y is a contraction mapping of the complete metric space, $\overline{\mathbb{B}}_r(0)$ into itself. Hence by the CMP it has a unique fixed point which we denote by $\phi(y)$.

This defines a function from $W \doteq \mathbb{B}_{r/2}(0)$ to E. Define V to be the set $f^{-1}(W) \cap \mathbb{B}_r(0)$.

Obviously, if $t \in V$, then $g_{f(t)}(t) = t - f(t) + f(t) = t$, and thus t is the unique fixed point of $g_{f(t)}$ Hence, $\phi(f(t)) = t$, that is, $\phi f_V = 1_V$.

On the other hand, if $y \in W$, then

$$\begin{aligned} g_y(\phi(y)) &= y + \phi(y) - f(\phi(y)) \\ \text{or,} \quad \phi(y) &= y + \phi(y) - f(\phi(y)). \quad \text{Hence,} \\ f(\phi(y)) &= y. \quad \text{Hence, } f_V \phi = 1_W. \end{aligned}$$

This establishes the fact that there exists a local inverse of f at 0_E. I will now show step by step that $\phi : W \to E$ is a C^p-map.

To avoid repeated explanations of the same (and obvious notation), I will write $x, x', x_\imath$ for points of V and $y, y', y_\imath$ for the images of these points under f. (Similarly if $y \in W$ then $\phi(y)$ will be denoted x.) This convention will be used for the remainder of the proof.

Continuity of ϕ: If $y_1, y_2 \in W$, then:

(1) $$x_1 = g_{y_1}(x_1) = y_1 + g(x_1)$$

(2) $$\text{and } x_2 = g_{y_2}(x_2) = y_2 + g(x_2).$$

Subtracting (2) from (1) and taking norms, we get:

$$||x_1 - x_2|| \leq ||y_1 - y_2|| + \frac{1}{2}||x_1 - x_2||$$

$$\text{and hence, } \frac{1}{2}||x_1 - x_2|| \leq ||y_1 - y_2||$$
$$\text{or, } ||\phi(y_1) - \phi(y_2)|| \leq 2||y_1 - y_2||.$$

This establishes the continuity of $\phi : W \to V$. □

Differentiability of ϕ: Let $y, y' \in \mathbb{B}_{\frac{r}{2}}(0)$. Then, the differentiability of f at $x = \phi(y)$ implies that

$$y' - y - \mathbf{D}f(x)(\boldsymbol{x}' - \boldsymbol{x}) = o(\boldsymbol{x}' - \boldsymbol{x}).$$

By our choice of r, $(\mathbf{D}f(x))^{-1}$ exists in $\mathbb{B}_r(0)$. Hence applying the continuous linear transformation $\varPhi_y \doteq (\mathbf{D}f(x))^{-1}$ on both sides we get:

$$\varPhi_y(\boldsymbol{y}' - \boldsymbol{y}) - \varPhi_y(\phi(y') - \phi(y)) = o(\boldsymbol{x}' - \boldsymbol{x})$$

$$\text{(5.3.1)} \qquad \text{Hence, } ||\phi(y') - \phi(y) - \varPhi_y(\boldsymbol{y}' - \boldsymbol{y})|| = o(\boldsymbol{y}' - \boldsymbol{y})$$

From equation (5.3.1), it follows that, ϕ is differentiable for all $y \in W$ and the derivative of ϕ at y is $\varPhi_y$, as stated in the theorem. □

Smoothness of ϕ: Recall that the "inversion map" $\mathbf{GL}(E) \to \mathbf{GL}(E) \subset \mathbb{L}(E)$ is C^∞. Hence, if f is C^p, then the map $x \mapsto (\mathbf{D}f(x))^{-1}$ is C^{p-1}. So the map $y \mapsto \mathbf{D}\phi(y) = \big(\mathbf{D}f(\phi(x))\big)$ is C^{p-1}. This proves that ϕ is C^p. □

Finally I will show that the restrictive conditions ($E = F$, $x_0 = 0_E = y_0$, etc.) that I imposed do not involve any loss of generality.

For any Banach space B and $\boldsymbol{v} \in B$, let $T^B_{\boldsymbol{v}} : B \to B$ be the translation map $T^B_{\boldsymbol{v}}(\boldsymbol{x}) \doteq \boldsymbol{x} + \boldsymbol{v}$. These maps are obviously C^∞ and invertible and their first derivative, $\mathbf{D}T^B_v(x) = 1_B$, the identity map of B. So if f is as in the Theorem and $U_0 = T^E_{-x_0}(U)$, then $f_0 : U_0 \to E$ defined by $f_0(x) = \mathbf{D}f(x_0)^{-1}\big(f(x + x_0) - y_0\big)$, satisfies the seemingly more restrictive hypotheses under which we have proved the result. This should be thought of as a "normalized" version of f.

The rest of the argument will be more transparent if we refer to the following commutative diagram in which all the horizontal arrows represent C^∞ bijections:

$$\begin{array}{ccccc} U_0 & & \xrightarrow{\quad T^E_{x_0} \quad} & & U \\ \Big\downarrow{\scriptstyle f_0} & & & & \Big\downarrow{\scriptstyle f} \\ E & \xleftarrow[\mathbf{D}f(x_0)^{-1}]{} & F & \xleftarrow[T^F_{-y_0}]{} & F. \end{array}$$

Suppose $V_0 \subset U_0$ is open and (V_0, ψ_0, W_0) is a local inverse for f_0 at 0_E. Let

$$\begin{aligned} W &= T^F_{\boldsymbol{y_0}}(\mathbf{D}f(x_0)(W_0)), \\ V &= T^E_{\boldsymbol{x_0}}(V_0), \quad \text{and} \\ \psi(z) &= \psi_0\Big(\mathbf{D}f(x_0)^{-1}(\boldsymbol{z} - \boldsymbol{y_0})\Big) + \boldsymbol{x_0}. \end{aligned}$$

Then clearly, (V, ψ, W) as defined above is a local inverse of f at x_0. This completes the proof of the Inverse Function Theorem. □

Remark 5.3.2 (Some finer details).

1. The inverse function theorem can only give *local* inverses. For instance, the function

$$f(x, y) = (e^x \cos y, e^x \sin y)$$

has an invertible derivative at every point of $\mathbb{R}^2$ but no global inverse since the trigonometric functions are periodic.

2. The conditions on the derivative *cannot* be relaxed. The following example illustrates this.

Let $f : \mathbb{R}^2 \to \mathbb{R}^2$ be defined by $f(x, y) = (x, \phi(x, y))$ where

$$\phi(x, y) = \begin{cases} y - x^2 & \text{if } x^2 \leq y, \\ \dfrac{(y^2 - x^2 y)}{x^2} & \text{if } 0 \leq y \leq x^2 \\ \dfrac{(y^2 + x^2 y)}{x^2} & \text{if } y \leq 0. \end{cases}$$

Then f is differentiable at every point in $\mathbb{R}^2$ and $\mathbf{D}f(0, 0)$ the identity map of $\mathbb{R}^2$ onto itself, but $\mathbf{D}f$ is not continuous at $(0, 0)$. It can be checked that in any open set containing (0,0) there are distinct points ξ', ξ'' such that $f(\xi') = f(\xi'')$.

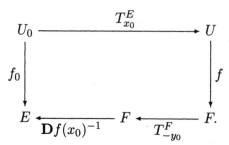

The next item on our agenda is the Implicit Function Theorem, one part of which is equivalent to the Inverse Function Theorem. But the other part (establishing the uniqueness of the implicit function) is a delicate and lengthy business. Before embarking upon this I find it difficult to resist the temptation of showing you one interesting application of the Inverse Function Theorem.

I will present a proof of the fundamental theorem of Algebra which asserts that *every monic polynomial* $A(z) = z^n + a_1 z^{n-1} + \cdots + a_n$ *with complex coefficients has a root.*

The Fundamental Theorem of Algebra

The usual proof of this theorem occupies one line and proceeds via an invocation of Liouville's Theorem, a powerful result in the theory of holomorphic functions. But this proof hides the reasons for the validity of the "fundamental theorem". The proof I present, on the other hand, uses only a very weak form of the Inverse Function Theorem and it is immediately clear where it would break down if we were seeking real roots for polynomials with real coefficients. I would like to thank Anindya Sen for showing me the cute proof which follows.

Theorem 5.3.3 (The fundamental theorem of Algebra).
Let $A(z) = z^n + a_1 z^{n-1} + \cdots + a_k z^{n-k} + \cdots + a_n$ *be a polynomial with complex numbers* a_k *as coefficients. Then* A *has a root* i. e., *there is a* $\zeta \in \mathbb{C}$ *such that* $A(\zeta) = 0$.

Proof. We break up the proof into a few small lemmas. Since these lemmas are not used elsewhere, we simply number them (1) through (5). The strategy is to consider $z \mapsto A(z)$ as a map $A : \mathbb{C} \to \mathbb{C}$ and show that the image of $A : \mathbb{C} \to \mathbb{C}$ is both open and closed. Since $\mathbb{C}$ is connected and obviously nonempty; this will imply that Image $A = \mathbb{C}$.

Lemma (1). *If* $A : \mathbb{C} \to \mathbb{C}$ *is a polynomial of degree* $n \geq 1$ *then given any* $\epsilon > 0$, *there exists a* $\delta > 0$ *such that* $|A(z)| \leq \epsilon$ *implies that* $|z| \leq \delta$.

Proof. This is simply a matter of estimating $|A(z)|$ as follows.

$$\begin{aligned} |A(z)| &= |z^n|.|1 + a_1 z^{-1} + \cdots + a_n z^{-n}| \\ &\geq |z^n|.\left|1 - (|a_1 z^{-1} + \cdots + a_n z^{-n}|)\right| \end{aligned}$$

Given $\varepsilon > 0$, there exists $\delta > 0$ such that, $|z| > \delta$ implies that $|z^n| > 2\varepsilon$ and $|a_1 z^{-1} + \cdots + a_n z^{-n}|$ is strictly less than $1/2$. Hence if $|z| > \delta$

then $|A(z)| > \varepsilon$. This is obviously equivalent to the statement of the lemma. □

Lemma (2). *Let A be as before. Then,* Image $A \subset \mathbb{C}$ *is a closed set.*

Proof. Suppose that (Image A) is not closed. Then there exists a $y \notin$ Image (A) and a sequence $y_n \in$ (Image A) such that $\lim_{n\to\infty} y_n = y$. We can also assume without loss of generality that $|y_n - y| < 1$ for all n. Now, each $y_n = A(x_n)$ for some $x_n \in \mathbb{C}$. Furthermore, since $|y_n| < |y|+1$ Lemma 1 implies that there is a $\delta > 0$ such that each x_n belongs to the closed disc $\overline{B}_\delta(0)$. But $\overline{B}_\delta(0)$ is compact. Hence, by passing to a subsequence, if necessary, and using the continuity of A, we see that $y = \lim A(x_n) = A(x)$. But this contradicts the fact that y is in the complement of (Image A). □

I will now introduce some jargon and notation.

If $E \supset U \xrightarrow{f} F$ is a differentiable map of an open set U of a Banach space E into a Banach space F, then a point $x \in U$ is said to be a *regular point* of f if $\mathbf{D}f(x) : E \to F$ is surjective. If $\mathbf{D}f(x) : E \to F$ is not surjective, x is said to be a *critical point* of f. A point $y \in F$ is said to be a *regular value* of f if every $x \in f^{-1}(y)$ is a regular point of f. If y is not a regular value then it is said to be a *critical value* of f.

In particular, note that *if $y \notin$ (Image f) $\subset F$, then y is a* regular value *of f.*

We now resume our study of the map $A : \mathbb{C} \to \mathbb{C}$. The following notations will be used in the remainder of the proof of the Fundamental Theorem of Algebra.

(a) $\mathcal{C}(A)$ will denote the set of critical values of A;

(b) $\mathbf{R}(A)$ will denote the set of regular values of A and

(c) $\mathcal{R}(A)$ will denote the intersection of (Image A) and $\mathbf{R}(A)$.

Lemma (3). *$\mathcal{C}(A)$ is a finite set and $\mathcal{R}(A)$ is non-empty.*

Proof. The critical points of A are the points, z, for which $A'(z) = 0$. Since A' is a nonzero polynomial of degree $n-1$, there are at most $n-1$ critical points and since, each critical value is the image of a critical point, $\mathcal{C}(A)$ has at most $n-1$ points.

Now each critical value has at most n inverse images. Hence, there are at most $n(n-1)$ points z in $\mathbb{C}$ such that $A(z) \in \mathcal{C}(A)$. Hence, there exists infinitely many $z \in \mathbb{C}$ such that $A(z)$ is a regular value. So, $\mathcal{R}(A)$ is non-empty. □

Lemma (4). *$\mathcal{R}(A)$ is an open set.*

Proof. Let $y \in \mathcal{R}(A)$ and let $x \in f^{-1}(y)$. Since x is a regular point, the Inverse Function Theorem tells us that there are open sets U (resp. V) containing x, (resp. $y = A(x)$) such that $f|U : U \to V$ is bijective. Hence, every point in V has at least one inverse image. So, $V \subset$ (Image A). Also, since $\mathcal{C}(A)$ is finite, we can shrink V so as to exclude any critical values. Then, $V \subset \mathcal{R}(A)$. So, $\mathcal{R}(A)$ is open. □

Lemma (5). *(Image A) is open.*

Proof. Let y be any point in (Image A). If y is a regular value, then $y \in \mathcal{R}(A)$. But, by Lemma 4, $\mathcal{R}(A)$ is open and $\mathcal{R}(A) \subset$ (Image A). So y is in the interior of (Image A).

Now suppose, to the contrary, that y is a critical value. Since, $\mathcal{C}(A)$ is finite, we can choose a $\delta > 0$ such that y is the only critical value inside the open disc $\mathbb{B}_\delta(y)$.

Now consider the punctured disc

$$\mathbb{B}_\delta^\times(y) \doteq \mathbb{B}_\delta(y) - \{y\}.$$

$\mathbb{B}_\delta^\times(y)$ is open and connected. Also by our choice of δ , $\mathbb{B}_\delta^\times(y)$ contains no critical points. So, we can write

$$\mathbb{B}_\delta^\times(y) = (\mathbb{B}_\delta^\times(y) \cap \mathcal{R}(A)) \sqcup \left(\mathbb{B}_\delta^\times(y) \cap (\text{Image } A)^c\right),$$

where "$\sqcup$" stands for "disjoint union" of sets. But, from the previous lemmas, (Image $A)^c$ and $\mathcal{R}(A)$ are both open. So, each of the sets appearing on the right-hand side above are open. But, since $\mathbb{B}_\delta^\times(y)$ is connected, one of them must be empty.

Suppose, $\mathbb{B}_\delta^\times(y) \cap \mathcal{R}(A)$ is empty. Then, $\mathbb{B}_\delta^\times(y) \subset$ (Image $A)^c$. This is impossible; since if $t \in [0,1]$ and $f(x) = y$, then $A(tx) \in$ (Image A) and if $|1-t| \neq 0$ is small enough, then $A(tx) \in \mathbb{B}_\delta^\times(y)$.

Hence, $\mathbb{B}_\delta^\times(y) \cap$ (Image $A)^c$ must be empty. But, then $\mathbb{B}_\delta(y) \subset \mathcal{R}(A)$. This means that $\mathbb{B}_\delta(y) \subset$ (Image A). Thus any point in (Image A) is contained in an open subset of $\mathbb{C}$ which is contained in (Image A). Hence, (Image A) is open. □

Thus (Image A) $\subset \mathbb{C}$ *is a nonempty, open and closed subset of* $\mathbb{C}$*: hence* (Image A) $= \mathbb{C}$. This completes the proof of the Fundamental Theorem of Algebra. □

Remark. It is worth noting that the first four Lemmas hold for polynomial functions $\mathbb{R} \to \mathbb{R}$ as well but, of course, not every real polynomial has a real root.

The crucial point in the proof of Lemma 5 is the fact that while in $\mathbb{C}$ a ball of radius $r > 0$ with its centre removed (the "punctured disc") is connected, a ball in $\mathbb{R}$, which is just an interval $(x - \eta, x + \eta)$, with its centre x removed is not a connected set.

5.4 The Implicit Function Theorem

Although I have, in the preamble to this chapter spoken at some length about the Inverse Function Theorem, in practice, an equivalent result, namely, the Implicit Function Theorem is more often used. It is almost trivial to see that the latter result implies the Inverse Function Theorem. I will use the Inverse Function Theorem to prove the Implicit Function Theorem; this choice is dictated by the extremely elegant and transparent proof (due to Lang, see [Lan65]) of the Inverse Function Theorem that I presented in the previous section. A direct proof of the Implicit Function Theorem is a lot more involved; see for example [Die57].

Theorem 5.4.1 (The Implicit Function Theorem).
Let E, F, G be Banach spaces and $U \subset E \times F$ an open set containing the point (x_0, y_0). Suppose that $f : U \to G$ is a C^p-map satisfying the conditions:

1. $f(x_0, y_0) = z_0$.
2. $\mathbf{D}_2 f(x_0, y_0) : F \to G$ *is an isomorphism of Banach spaces.*

Then there is an open set $U' \subset E$ containing x_0 and a C^p-map (the implicit function) $\phi : U' \to F$ *such that $\phi(x_0) = y_0$ and for all $x \in U'$, the point $\big(x, \phi(x)\big)$ belongs to U and $f\big(x, \phi(x)\big) = z_0$. The derivative of ϕ is given by the formula*

$$\mathbf{D}\phi(x) = -\mathbf{D}_2 f\big(x, \phi(x)\big)^{-1} \circ \mathbf{D}_1 f\big(x, \phi(x)\big).$$

Moreover, if O is a sufficiently small connected open set containing x_0 and $\xi : U \to F$ is any map, $\xi : O \to F$, such that $\xi(x_0) = y_0$ and $f\big(x, \xi(x)\big) = z_0$ then $\xi = \phi|O$.

The complete proof of this result is somewhat lengthy and it might be worthwhile keeping in mind the next few observations.

REMARKS 5.4.1 (ON THE IMPLICIT FUNCTION THEOREM)

Remark 5.4.1a. The function ϕ is said to be "implicitly defined" in a neighbourhood of x_0 by the conditions $f(x, y) = z_0$ and $\phi(x_0) = z_0$.

Remark 5.4.1b. Recall that $\mathbf{GL}(F)$ is open in $\mathbb{L}(F)$ and since f is C^p its partial derivatives are continuous. Hence the entity '$\left(\mathbf{D}_2 f(x, \phi(x))\right)^{-1}$', that features in the formula for the derivative of the implicit function exists if U' is sufficiently small.

Remark 5.4.1c (Implicit implies Inverse).
Let $U \subset E$ be an open subset of a Banach space, x_0 a point in U and suppose, $g : U \to F$ is a C^1-map into a Banach space F such that $g(x_0) = y_0$ and $\mathbf{D}g(x_0) : E \to F$ is an isomorphism of Banach spaces.

Consider the open subset $(F \times U) \subset F \times F$ and the map $f : (F \times U) \to F$ defined by $f(y, x) = y - g(x)$. From the hypotheses on g it is clear that f satisfies the hypotheses of the Implicit Function Theorem at the point (y_0, x_0) with $z_0 = 0_F$.

The function, ϕ, implicitly defined by $f\big(x, \phi(x)\big) = 0_F$ and the condition, $\phi(y_0) = x_0$ is a local inverse of g at y_0.

We now begin the proof of the Implicit Function Theorem.

Existence of an implicit function.

Consider the function $\Phi : U \times F \to E \times G$ defined by $\Phi(u, v) = \big(u, f(u, v)\big)$. Then the derivative of Φ may be written as a "partitioned matrix" of continuous linear transformations in the form

$$\mathbf{D}\widetilde{\Phi}(x, y) = \begin{bmatrix} 1_E & 0 \\ \mathbf{D}_1 f(x, y) & \mathbf{D}_2 f(x, y) \end{bmatrix}.$$

Now it is easy to check that if $T : E \oplus F \to E \oplus G$ is a linear transformation whose partitioned form is

$$\begin{bmatrix} 1_E & 0 \\ T_{21} & T_{22} \end{bmatrix}$$

then T is invertible iff $T_{22} : F \to G$ is invertible and that, in this situation, T^{-1} has the partitioned form:

$$T^{-1} = \begin{bmatrix} 1_E & 0 \\ -T_{22}^{-1} T_{21} & T_{22}^{-1} \end{bmatrix}.$$

It follows that if $(x, y) \in U$ is a point such that $\mathbf{D}_2 f(x,y) : F \to G$ is an isomorphism then Φ has an invertible derivative at (x, y) and

$$(\star) \qquad \big(\mathbf{D}\Phi(x,y)\big)^{-1} = \begin{bmatrix} 1_E & 0 \\ -\mathbf{D}_2 f(x,y)^{-1}\mathbf{D}_1 f(x,y) & \mathbf{D}_2 f(x,y)^{-1} \end{bmatrix}.$$

So the map $\Phi : U \times F \to E \times G$ satisfies the hypotheses of the Inverse Function Theorem at the point (x_0, y_0).

Let (V', Ψ, W'') be a local inverse of Φ at z_0. (Here $V' \subset U \times F$ is an open subset and $W' = \Phi(V')$)

Let $U' = p_1(V')$, where $p_1 : U \times F \to U$ is the projection onto the first factor. Then writing $\Psi(x,y) = \big(\psi_1(x,y), \psi_2(x,y)\big) \in E \times F$, we get for $(x, y) \in W'$:

$$\begin{aligned}(x,y) &= \Phi(\psi_1(x,y), \psi_2(x,y)) \\ &= \Big(\psi_1(x,y), f\big(\psi_1(x,y), \psi_2(x,y)\big)\Big)\end{aligned}$$

which implies that $\psi_1(x,y) = x$ and $f\big(x, \psi_2(x,y)\big) = y$. The implicit function we are after is now on hand; define $\phi : U' \to F$ by setting $\phi(x) = \psi_2(x, z_0)$.

The computation of the derivative is just a matter of using the formula $(\star)$ for the derivative of Ψ; I leave this as an exercise. □

To prove the uniqueness part of the Implicit Function Theorem I will need a lemma which is quite close in spirit to the CMP. It yields not just a fixed point but a "*fixed function*".

Lemma 5.4.2 (The fixed-function lemma).
Let E and F be Banach spaces. Let A (resp. B) be $\mathbb{B}_a(0_E)$ (resp. $\mathbb{B}_b(0_F)$). Suppose $u : A \times B \to F$ is a map which satisfies the following conditions.

1. $u(0_E, 0_F) = 0_F$.

2. *u is uniformly Lipschitz in the second variable, with Lipschitz constant k; that is, for all $x \in A$ and $y_1, y_2 \in B$:*
$$||u(x,y_1) - u(x,y_2)||_F \le k \cdot ||y_1 - y_2||_F.$$
Suppose further that:

3. $||u(x, 0_F)||_F \le (1-k)b$ *and*

4. *we have* $0 \le k < 1$ *and* $b < (1-k)a$.

Then there is a unique map $f : A \to B$ such that for all $x \in A$, $u(x, f(x)) = f(x)$.

During the proof we will no longer be as pedantic as we have been in the statement of the result; the zero vectors in both E and F will be denoted 0.

Proof. For $n \geq -1$, inductively define functions $v_n : A \to F$ by setting:

$$\begin{aligned} v_{-1}(x) &= 0 \\ v_{n+1}(x) &= u(x, v_n(x)). \end{aligned}$$

Of course, for this inductive definition to make sense, we must show that each $v_p : A \to F$ actually takes values in B ; this is our first objective.

Observe that $v_p(x) = v_p(x) - v_{-1}(x)$ and hence,

$$v_p(x) = \sum_{j=0}^{p} \{v_j(x) - v_{j-1}(x)\}.$$

So if we assume that $v_k(x) \in B$ for $-1 \leq k \leq p-1$, then,

$$\begin{aligned} ||v_p(x)||_F &\leq \sum_{j=0}^{p} ||v_j(x) - v_{j-1}(x)||_F \\ &\leq \sum_{j=0}^{p} k^j ||u(x,0)||_F \\ &< (1-k)^{-1} ||u(x,0)||_F < b. \end{aligned}$$

where we are using condition 3 in the last step.

This shows that all the v_n's take values in B and hence the v_n's are indeed properly defined.

Next we will show, inductively, that the v_n's are continuous. Certainly v_{-1} is so, since it is constant. Suppose that $v_p : A \to F$ is continuous for $-1 \leq p \leq n$ and let $\epsilon > 0$ be given.

Then,
$$\begin{aligned} ||v_{n+1}(x) - v_{n+1}(x')||_F &= ||u(x, v_n(x)) - u(x', v_n(x'))||_F \\ &\leq ||u(x, v_n(x)) - u(x, v_n(x'))||_F \\ &\quad + ||u(x, v_n(x')) - u(x', v_n(x'))||_F \\ &\leq k||v_n(x) - v_n(x')||_F \\ &\quad + ||u(x, v_n(x')) - u(x', v_n(x'))) \end{aligned}$$

Since u and v_n (by the induction hypothesis) are continuous, there is a $\delta > 0$, such that if $||\boldsymbol{x} - \boldsymbol{x}'||_E < \delta$, then $||u(x, v_n(x')) - u(x', v_n(x'))||_F < \epsilon/2$. Hence, if $||\boldsymbol{x} - \boldsymbol{x}'||_E < \epsilon = \min(\delta, \epsilon/2)$ then, by the Lipschitz condition, the first term on the right hand side of the last equation is less than $\epsilon/2$ also and we see that if $||\boldsymbol{x} - \boldsymbol{x}'||_E < \epsilon$, then

$$||v_{n+1}(x) - v_{n+1}(x')||_F < \varepsilon, \quad \text{proving that } v_{n+1} \text{ is continuous.}$$

This completes the inductive definition of the maps $v_n : A \to B$.

Now we proceed as in the proof of the CMP.

Using the Lipschitz condition we calculate as follows.

$$||v_n(x) - v_{n-1}(x)||_F = ||u(x, v_{n-1(x)}) - (x, v_{n-2}(x))||_F$$

After n applications of the Lipschitz condition, we get:

$$\begin{aligned} ||v_n(x) - v_{n-1}(x)||_F &\leq (1 + k + \cdots + k^n) \cdot ||u(x, 0)||_F \\ &\leq \tfrac{1}{k-1}||u(x, 0)||_F. \end{aligned}$$

Now these inequalities also show that that the series associated with the sequence $f_n(x) \doteq v_n(x) - v_{n-1}(x)$ is absolutely convergent. In fact, since

$$\sum_m^n f_i(x) = v_n(x) - v_m(x)$$

and the norm of the left side is $\leq k^m b$, $\sum_n f_n(x)$ converges uniformly to a continuous function $x \mapsto f(x)$ of A to B. Now $\sum_{j=-1}^n f_n(x) = v_n(x)$ and $v_{n+1}(x) = u(x, v_n(x))$ for all $x \in A$; this implies that $f(x) = u(x, f(x))$. The uniqueness follows trivially from the Lipschitz condition. □

The uniqueness of the implicitly-defined function.
My strategy (adapted from [Die57]) will be to define a function $u : A \times B \to F$ where $A = \mathbb{B}_a(0) \subset E$ and $B = \mathbb{B}_b(0) \subset F$ satisfying the conditions of the Fixed-Function lemma in such a way that any "fixed function", of the lemma yields an implicit function for the Theorem and vice versa.

To simplify notation, the invertible linear transformation, $\mathbf{D}_2 f(x_0, y_0)$ will be denoted L_0.

Consider the functions

$$g(x, y) = y - L_0^{-1}(f(x, y) - z_0), \tag{5.4.1}$$

$$\text{and } u(x, y) = g(x_0 + x, y_0 + y) - y_0. \tag{5.4.2}$$

Since $(x_0, y_0) \in U$ and U is open in $E \times F$, provided a and b are chosen suitably small, ϕ will be defined at $(x_0 + x, y_0 + y)$ when $(x, y) \in A \times B$, so the definitions of g and u make sense. In the course of the proof, I will, if necessary, readjust a, b to smaller values if thereby some further desirable outcome can be achieved thereby. In such a situation, I will simply say "readjusting $a, b\ldots$" and leave it at that. For instance, for any fixed $0 < k < 1$ the condition 3 of the Fixed-Function lemma can be obtained by readjusting a and b.

We first check that $u : A \times B \to F$ does indeed fit the three requirements of the strategy outlined above.

1. We check that $u(0,0) = 0$:

$$u(0,0) = y_0 - \{y_0 - L_0^{-1}(f(x_0,y_0) - z_0)\} = 0, \text{ since } f(x_0,y_0) = z_0.$$

This shows that u satisfies the first of the hypotheses of the Fixed-Function lemma.

2. Now suppose $\xi : A \to B$ is the "fixed function" arising from u. Then,

$$\begin{aligned}\xi(x) &= u(x, \xi(x)) \\ &= g(x_0 + x, y_0 + \xi(x)) - y_0 \quad \text{from (5.4.1) and (5.4.2)} \\ &= y_0 + \xi(x) \\ &- L_0^{-1}\left(\phi(x_0 + x, y_0 + \xi(x)) - z_0\right) - y_0\end{aligned}$$

from which we see that, $\phi(x_0 + x, y_0 + \xi(x)) = z_0$. This implies that if $\xi : A \to B$ is a "fixed function" for u then $\psi_\xi : \mathbb{B}_a(x) \to F$ defined by

$$\psi_\xi(x) \doteq u(x - x_0) + y_0$$

is the function implicitly defined in a neighbourhood of x_0 by the conditions $f(x,y) = z_0$ and $\psi_\xi(x_0) = y_0$.

3. Conversely, if ψ is a function implicitly defined in a neighbourhood N of x_0 by the conditions $f(x,y) = z_0$ and $\psi(x_0) = y_0$ then provided $a > 0$ is so small that $\mathbb{B}_a(x_0) \subset N$, the function $\xi_\psi : A \to F$ defined by: $\xi_\psi(x) \doteq \psi(x - x_0) - y_0$ is a fixed function for u.

It now remains to check that if $a, b > 0$ are suitably small, then $u : A \times B \to F$ satisfies the conditions of the Fixed Function Lemma (Lemma 5.4.2).

So suppose that $x \in A$ and $y_1 \neq y_2 \in B$. Then, writing $\widetilde{x}$ (resp.$\widetilde{y}$) for $x_0 + x$ (resp. $y_0 + y$) we have

$$\begin{aligned}u(x,y_1) - u(x,y_2) &= g(\widetilde{x},\widetilde{y}_1) - g(\widetilde{x},\widetilde{y}_2) \\ &= (y_1 - y_2) - L_0^{-1}\left(f(\widetilde{x},\widetilde{y}_1) - f(\widetilde{x},\widetilde{y}_2)\right) \\ &= L_0^{-1}\{\mathbf{D}_2\phi(x_0,y_0)(y_1 - y_2) - f(\widetilde{x},\widetilde{y}_1) + f(\widetilde{x},\widetilde{y}_2)\}\end{aligned}$$

Now choose $\varepsilon > 0$ so small that $\varepsilon||L_0^{-1}|| < 1/2$. Since f is C^p, its partial derivatives will be continuous. Hence, by the second Mean-Value theorem (page 169) adjusting a, b suitably we can ensure that

$$||\mathbf{D}_2 f(x_0,y_0)(y_1 - y_2) - f(\widetilde{x},\widetilde{y}_1) + \phi(\widetilde{x},\widetilde{y}_2)|| \leq \varepsilon||y_1 - y_2||$$

which establishes that if $x \in A$ and $y_1, y_2 \in B$ then

$$||u(x, y_1) - u(x, y_2)|| \leq \varepsilon ||L_0^{-1}|| \cdot ||y_1 - y_2|| \leq \frac{1}{2}||y_1 - y_2||.$$

So u satisfies the crucial Lipschitz condition of Lemma 5.4.2.

This establishes the uniqueness part of the Implicit Function Theorem and thus *the proof of the Implicit Function Theorem is now complete.* □

I now state (with brief indications of proofs) particular cases of the Implicit Function Theorem.

First, the formulation which appears in most Calculus textbooks of the traditional variety.

Theorem 5.4.3 (Classical form of the Implicit Function Theorem). *Let $U \subset \mathbb{R}^n \times \mathbb{R}^m$ be an open set and for $k = 1, \dots, n$ let $f_k : U \to \mathbb{R}$ be C^1-functions. Suppose at the point $p = (p_1, \dots, p_m, p_{m+1}, \dots, p_{m+n})$ the following submatrix of the Jacobian matrix, $J(f;p) = \dfrac{\partial(f_1, \dots, f_n)}{\partial(x_1, \dots, x_{m+n})}$, has nonzero determinant.*

$$\begin{bmatrix} \dfrac{\partial f_1}{\partial x_{m+1}} & \cdots & \dfrac{\partial f_1}{\partial x_{m+n}} \\ \vdots & \cdots & \vdots \\ \dfrac{\partial f_n}{\partial x_{m+1}} & \cdots & \dfrac{\partial f_n}{\partial x_{m+n}} \end{bmatrix}$$

Let $f_{m+k}(p) = c_k$ for $k = 1, \dots, n$.

Identifying the linear subspace of $\mathbb{R}^{m+n}$ where the last n coordinates vanish with $\mathbb{R}^m$, there is an open set $U' \subset \mathbb{R}^m$ containing $(p_1, \dots, p_m)$ and n C^1-functions $g_\imath : U' \to \mathbb{R}$ such that for all $(x_1, \dots, x_m) \in U'$ and $i = 1 \dots, n$:

$$\begin{aligned} f_{m+\imath}(x_1, \dots, x_m, g_1(x_1, \dots, x_m), \dots, g_n(x_1, \dots, x_m)) &= c_\imath \\ g_\imath(p_1, \dots, p_m) &= p_{m+\imath}. \end{aligned}$$

Moreover, if the set U' is connected and small enough then the $g_\imath$'s are uniquely determined.

If the functions f_k are C^p then the g_k's will also be C^p.

Proof. This is simply a restatement, using coordinates, of the Implicit Function Theorem we have proved with $E = \mathbb{R}^m$, $F, G = \mathbb{R}^n$. □

For Hilbert spaces there is a more interesting and useful variant of the Implicit Function Theorem which I will soon describe. For its proof, I will need to use the following celebrated result of Functional Analysis.

Theorem (The Open Mapping Theorem).
Let E_1, E_2 be Banach spaces and $T : E_1 \to E_2$ a continuous linear transformation. If T is surjective then T is an open map *i. e., if $U \subset E_1$ is open, then $T(U)$ is open in E_2.*

An immediate corollary is the result we will use in our "Hilbertian Implicit Function theorem".

Proposition 5.4.4. *If $T : E_1 \to E_2$ is a continuous linear transformation between Banach spaces that is bijective, then T is an isomorphism of Banach spaces; that is, $T^{-1} : E_2 \to E_1$ is also continuous.* □

Theorem 5.4.5 (Implicit Functions in Hilbert spaces).
Let $\mathcal{H}, \mathcal{H}'$ be Hilbert spaces, $U \subset \mathcal{H}$ an open set and $f : U \to \mathcal{H}'$ a C^1-map. Let $p \in U$ be a point such that

1. *$f(p) = \xi$ and*
2. *$\mathbf{D}f(p) : \mathcal{H} \to \mathcal{H}'$ is surjective.*

Let K be the kernel of $\mathbf{D}f(p)$ and $L \doteq K^{\perp}$ the orthogonal complement of K.

Let
$$\pi_K : \mathcal{H} \to K$$
$$\pi_L : \mathcal{H} \to L$$
be the orthogonal projections from $\mathcal{H}$. If p_K, p_L are the projections of p onto K and L respectively, then there are open sets, $U_1 \subset K$ containing p_K and $U_2 \subset L$ containing p_L such that, $(U_1 \times U_2) \subset U$ and a C^1-map, $\phi : U_1 \to U_2$ such that $f\big(x, \phi(x)\big) = \xi$ and $\phi(p_K) = p_L$.

Proof. Observe that $\mathbf{D}_2 f(p_K, p_L) : L \to \mathcal{H}'$ is a bijective and continuous linear transformation. Hence by the Open Mapping Theorem is an isomorphism. So the hypotheses of the Implicit Function Theorem are satisfied. The result follows easily. □

Remark 5.4.6. There are very good reasons why a theorem such as Theorem 5.4.5 cannot hold for arbitrary Banach spaces.

1. There is no obvious choice for the supplement of $\ker \mathbf{D}f(p)$ as in the Hilbert space situation.

2. There exist Banach spaces which have closed subspaces which do not have any *closed* supplement. If a C^1 map $f : U \to E$ defined on an open

set, U, of such a Banach space had as kernel, a subspace which did not have a closed supplement then we cannot write a product decomposition as I have done above, using *Banach spaces.* and outside this realm all our tools of Calculus fail.

6

Applications: Stationary values

Large portions of mathematical physics, almost all of Classical Mechanics, the theory of Differential Equations and so on *ad infinitum* may be regarded as applications of what has been done in the last two chapters.

In order to bring some kind of focus to the discussion and (to enable myself to finish in a finite amount of time!) I have decided to only discuss applications of Differential Calculus which in some way involve **`stationary`** points. (Loosely speaking these are points at which a differentiable function attains a local maximum or minimum.) The values attained by the function at such points are called the *extrema* of the function.

For obvious reasons any text will confine attention to finding extrema of "nice" functions, that is, functions whose extrema can be obtained easily by the methods expounded. Because we have discussed the Differential Calculus of functions defined on infinite dimensional Banach spaces, there is a very large range of examples that can be discussed.

Section 2 of this chapter is entirely devoted to discussing the situation of infinite dimensions. However, the "bare bones" account of integration I have provided in Chapter 4 does not provide sufficient machinery for giving a completely satisfactory treatment of some results. The reason we have chosen to present these results in spite of the shortcomings of our treatment is that the lacunae are not in our treatment of "Differential Calculus" but in our sketchy treatment of integration. We will point out the difficulties in greater detail later; for now it suffices to mention that we will be assuming the existence of certain Hilbert spaces of curves without being able to justify all our statements concerning these spaces.

The hope is to stimulate the reader's interest and make him aware that Differential Calculus done in a basis-free context embraces a very wide and exciting range of mathematics.

In § 6.1, I give an account of what is usually called Lagrange's *Method* of Undetermined Multipliers. My treatment is somewhat unorthodox, but the basic ideas presented are well known.

In § 6.2, the only section of the book which is exclusively devoted to infinite-dimensional spaces, I discuss the Calculus of Variations. After deriving the Euler–Lagrange equations I discuss two sample problems of this vast area. The geodesic problem is discussed as an example of a problem of obtaining unconstrained extrema and the classical `isoperimetric` problem as an instance of constrained extrema. Two other well-known examples appear as exercises.

6.1 Lagrange's Theorem

Suppose that $f : U \to \mathbb{R}$ is a differentiable function, defined on an open set U of a Banach space and $p \in U$ is a point where f attains a local maximum, (resp. minimum). This means that $f(x) \leq f(p)$ (resp. $f(x) \geq f(p)$) for all x belonging to the punctured ball, $\mathbb{B}_\delta(p) - p$, where $\delta > 0$ is some real number. Then we have already seen that $\mathbf{D}f(p) \equiv \mathbf{d}f(p) : E \to \mathbb{R}$ is zero. Of course this is only a necessary, but by no means, sufficient condition for an extremum. Points at which the derivative of a real-valued function is zero are called stationary points and if x is a stationary point of $E \supset U \xrightarrow{f} \mathbb{R}$ then $f(x)$ will be called a *stationary value* of f. Clearly the first derivative of a function at a point cannot be expected to give any information about what happens to the value of the function outside a small neighbourhood of this point. So it really is not correct to call stationary values 'extrema'. I belabour this point because the theorem we are about to discuss is generally regarded as a method of finding "maxima/minima". Under favourable circumstances one may, by means other than a study of the first derivative, be able to establish that the value attained by a function at a given stationary point is a local maximum or minimum. But our concern will mainly be to look for stationary points and stationary values.

In mathematics, as in ordinary life, one's efforts to maximize the value of a 'function' is often subject to some conditions: for instance, one would like to maximize one's wealth but usually our efforts are constrained by conditions such as "Thou shalt not steal". Lagrange's "Method of Undetermined Multipliers" examines the following situation that often arises in Physics.

U is an open set in a Hilbert space E, $f : U \to \mathbb{R}$ a C^1-map and $\phi : U \to F$ a C^1-map into another Hilbert space. Let $c \in F$ be a point and $M_c \doteq \phi^{-1}(c)$.

A minimum of f *subject to the constraint* $\phi(x) = c$ (or a minimum of $f|M_c$) is a point $x_0 \in M_c$ such that there is some open set O containing x_0 and for all $x \in M \cap O$, $f(x_0) \leq f(x)$. Maxima of $f|M_c$ are defined analogously.

More generally we define stationary values of f restricted to the closed subset $M_c = \phi^{-1}(c)$ in the following manner which is a *very slight* extension of ideas which are by now quite familiar to you.

Definition 6.1.1 (Stationary points under constraints).
Let f, U, E, F, ϕ, c, M_c be as in the discussion above. Let $\mathcal{C}(p, \phi)$ be the collection of C^1-curves, $\alpha : I \to U$ satisfying the following conditions.

1. I is an open interval containing 0.
2. $\alpha(0) = p$ and $\alpha(t) \in M_c$ for all $t \in I$.

The point p is said to be a *stationary point of f subject to the constraint*, $\phi(p) = c$, if the following condition obtains:

$$\text{For every curve } \alpha \in \mathcal{C}(p, \phi), \ \left.\left(\frac{d(f \circ \alpha)}{dt}\right)\right|_{t=0} = 0.$$

We will, for brevity, say that p is a stationary point of $f|M_c$.

We wish to find stationary points of f when the argument of f is restricted to the closed subset M_c of U. Lagrange's method enables us, under suitable hypotheses, to determine, the *stationary points* of $f|M_c$

Theorem 6.1.2 (Lagrange's Theorem).
Let E, F be Hilbert spaces, $V \subset E$ an open set, and suppose that $f : V \to \mathbb{R}$ and $\phi : V \to F$ are C^1 maps. Let $c \in F$ be a regular value *of ϕ, that is, if $x \in M_c \doteq \{x \in V : \phi(x) = c\}$ then $\mathbf{D}\phi(x) : E \to F$ is surjective.*

Under these conditions, $p \in V$ is a stationary point of f, subject to the constraint $\phi(p) = c$, iff the differential, $\mathbf{d}f(p) \in E^*$, belongs to the image of the dual map, induced by the derivative of ϕ at the point p.

This necessary and sufficient condition, for a stationary point subject to the constraint, $\phi(p) = c$, henceforth called the **Lagrange condition** is:

$$\mathbf{d}f(p) \in \text{Image}\left\{ \left(\mathbf{D}\phi(p)\right)^* : F^* \to E^*.\right\}$$

Notice that I am using * to indicate continuous duals. Hopefully there is no need any longer to use a special notation to remind you that in the context of Banach spaces, only continuous duals are used.

Sufficiency of the Lagrange Condition.

I will begin by studying the implications of $c \in F$ being a *regular value* of $\phi : V \to F$. This means that $\mathbf{D}\phi(p) : E \to F$ is surjective whenever $p \in M_c$.. Since E and F are Hilbert spaces we can use the special form of the Implicit Function Theorem, (Theorem 5.4.5) for each $p \in M_c$.

Suppose $p \in M_c$ and $K = \ker \mathbf{D}\phi(p)$. Let $L = K^{\perp}$, be the orthogonal complement of K. I will write E as the direct sum $E \cong K \oplus L$ and denote by π_K (resp. π_L) the orthogonal projections of E onto K (resp. L). Using this decomposition, points of E will be written in "(K, L)-coordinates"; that is, we will write $q = (x_q, y_q)$ where $\boldsymbol{x}_q = \pi_K(\boldsymbol{q})$ and $\boldsymbol{y}_q = \pi_L(\boldsymbol{q})$. Then clearly, $\boldsymbol{q} = \boldsymbol{x}_q + \boldsymbol{y}_q$. In particular, the point p of Lagrange's theorem will be denoted (x_p, y_p).

Now Theorem 5.4.5 assures us that there is an open set $\mathcal{U} \subset E$ containing p, a connected, open set $U \subset K$ containing x_p and a unique C^1-map $g : U \to L$ such that, the following hold:

$$\begin{aligned}
&(1_g) \cdots\cdots & (M_c \cap \mathcal{U}) &= \{(x, g(x)) : x \in U\}, \\
&(2_g) \cdots\cdots & g(x_p) &= y_p, \text{ and} \\
&(3_g) \cdots\cdots & \mathbf{D}g(x) &= -\mathbf{D}_2\phi(x, g(x))^{-1}\mathbf{D}_1\phi(x, g(x)).
\end{aligned}$$

Suppose that ω is a differentiable curve in $U \subset K$, defined on an open interval J containing 0, and passing through x_p at time $t = 0$. Then the curve $\Omega(t) = \big(\omega(t), g\big(\omega(t)\big)\big)$ is a curve belonging to the collection $\mathcal{C}(p, \phi)$.

In fact, every curve in $\mathcal{C}(p, \phi)$ arises in this way. To see this, simply note that if Ω is a differentiable curve in E lying entirely in $(M_c \cap \mathcal{U})$, then $\Omega(t) = \big(\omega(t), g\big(\omega(t)\big)\big)$, where $\omega(t) = \pi_K\big(\Omega(t)\big)$.

Suppose that

$$(\dagger) \cdots\cdots\cdots\cdots \quad \mathbf{d}f(p) \in \text{Image}\,\{\mathbf{D}\phi(p)^* : F^* \to E^*\}.$$

To establish that p is a stationary point of $f|M_c$, we have to show that

$$\frac{d}{dt}f(\Omega(t))\Big|_{t=0} = 0$$

for every C^1-curve that lies entirely in $M_c \cap \mathcal{U}$ and which passes through $p \equiv (x_p, y_p)$ at time $t = 0$. So suppose $\Omega : (-\varepsilon, \varepsilon) \to M_c \cap \mathcal{U}$ is such a curve. Then we can write $\Omega(t) = \big(\omega(t), g(\omega(t))\big)$.

Writing $\boldsymbol{v}$ for $\omega'(0)$ we compute as follows.

$$\begin{aligned}\frac{d}{dt}f(\Omega(t))\Big|_{t=0} &= \langle \mathbf{d}f(p), \Omega'(0)\rangle \\ &= \langle \mathbf{d}f(p), \boldsymbol{v} + \mathbf{D}g(x_p)(\boldsymbol{v})\rangle \\ &= \langle \mathbf{d}f(p), \boldsymbol{v} - \big(\mathbf{D}_2\phi(p)\big)^{-1}\mathbf{D}_1\phi(p)(\boldsymbol{v})\rangle.\end{aligned}$$

By hypothesis, $\mathbf{d}f(p) = \Xi \circ \mathbf{D}\phi(p)$ for some $\Xi \in F^*$.

Since $\mathbf{D}\phi(p)|K = \mathbf{D}\phi(p$ and $\mathbf{D}\phi(p)|L = \mathbf{D}_2\phi(p)$, we get:

$$\frac{d}{dt}f(\Omega(t))\Big|_{t=0} \langle \Xi, \mathbf{D}\phi(p)\big(\boldsymbol{v} - \big(\mathbf{D}_2\phi(p)\big)^{-1}\mathbf{D}_1\phi(p)(\boldsymbol{v})\big)\rangle = 0. \qquad \square$$

To prove the necessity of the Lagrange condition, we need a simple result from Linear Algebra.

Lemma. *Let E, F be Hilbert spaces and $T : E \to F$ a continuous linear transformation that is surjective. Let $K = \ker T$ and $A \doteq \operatorname{Ann}(K) \subset E^*$, the annihilator of K. Then $A = \operatorname{Image}\{T^* : F^* \to E^*\}$.*

Proof. Suppose $\alpha \in A$. Then define $\beta \in F^*$ by setting $\beta(\boldsymbol{y}) = \alpha(T(\boldsymbol{x}))$ for any $\boldsymbol{x} \in T^{-1}(y) \subset E$. β is well defined: if $T(\boldsymbol{x}') = \boldsymbol{y}$ also, then $\boldsymbol{x} - \boldsymbol{x}' \in \ker T$ and hence $\alpha\big(T(\boldsymbol{x}')\big) = \alpha\big(T(\boldsymbol{x}' - \boldsymbol{x}) + T(\boldsymbol{x})\big) = \alpha\big(T(\boldsymbol{x})\big)$ since $T(\boldsymbol{x} - \boldsymbol{x}') = 0$. Now for all $\boldsymbol{e} \in E$, we have:

$$\begin{aligned}T^*(\beta)(\boldsymbol{e}) &= \beta\big(T(\boldsymbol{e})\big) \qquad \text{(by the definition of } T^*,\text{)} \\ &= \alpha(\boldsymbol{e}) \qquad \text{(by the definition of } \beta,\text{)}\end{aligned}$$

showing that $\alpha = T^*(\beta) \in \operatorname{Image} T^*$. Thus $\operatorname{Ann}(K) \subset \operatorname{Image} T^*$.

Conversely, suppose $\beta \in \operatorname{Image} T^*$, say $\beta = \alpha \circ T$. Then for any $\boldsymbol{x} \in K$, $\beta(\boldsymbol{x}) = \alpha(T(\boldsymbol{x})) = 0$, since $K = \ker T$, showing that $\operatorname{Image} T^* \subset \operatorname{Ann}(K)$. So, $\operatorname{Ann}(K) = \operatorname{Image} T^*$. $\square$

The necessity of the Lagrange condition.

We will be using the direct sum decomposition $E \approxeq K \oplus L$ described earlier. In particular, the relations $1_g, 2_g, 3_g$ will prove useful.

Suppose p is a stationary point of $f|M_c$ and $\Omega : (-\varepsilon, +\varepsilon) \to M_c \subset \mathcal{U}$ a curve in the collection $\mathcal{C}(p, \phi)$.

Let $\widetilde{g} : U \to E$ be the map $x \mapsto (x, g(x))$ and $\omega : (-\varepsilon, +\varepsilon) \to U$ be a curve such that $\omega(0) = x_p$ and $\Omega(t) = \widetilde{g}(\omega(t))$.

Define $G : U \to \mathbb{R}$ to be the composition: $U \xrightarrow{\tilde{g}} M_c \xrightarrow{f} \mathbb{R}$. Then it is clear that if p is a stationary point of $f|M_c$, then x_p is a stationary point of $G : U \to \mathbb{R}$. This means that $\langle \mathbf{d}G(x_p), \omega'(0)\rangle = 0$ for every C^1-curve in U passing through x_p at time $t = 0$. Since every vector in K can be represented as $\omega'(0)$ for some such curve it follows that, $\mathbf{d}G(x_p) \in \mathrm{Ann}(K)$. Now $K = \ker \mathbf{D}\phi(p)$ and by the above lemma, $\mathrm{Ann}(\ker \mathbf{D}\phi(p)) = \mathrm{Image}\,\{(\mathbf{D}\phi(p))^* : F^* \to E^*\}$. The proof of Lagrange's Theorem is complete. □

Remarks 6.1.2 (On Lagrange's Theorem)

Remark 6.1.2a. Obviously there is a certin amount of inaccuracy involved in attributing this particular result to Lagrange, since Lagrange lived more than a century before Hilbert spaces were formally defined. Moreover, I have never seen this result stated or proved as a necessary and sufficient condition for the existence of stationary points subject to constraints. An exercise in volume 3 of Dieudonné's *Treatise on Analysis* [Die80] is the nearest thing to Theorem 6.1.2, that I have seen in the literature.

Usually one half or the other of the theorem is presented as a *useful recipe* for finding the "constrained extrema of a function" and this is referred to as "Lagrange's method of Undetermined Multipliers". But, in fact, it has limited usefulness as a tool in actually determining the constrained extrema, in problems of Optimization Theory.

On the other hand it is a beautiful and satisfying result which should be part of any course on Advanced Calculus. Clearly it ought to be attributed to its originator.

In the succeeding remarks I will elucidate the connection between the theorem as stated above and the classical 'Method' as expounded in textbooks of Advanced Calculus.

Remark 6.1.2b (Comparing Lagrange's Theorem and Method).
I wish to compare "Lagrange's Theorem" with "Lagrange's Method" which I will describe in a moment. (While describing Lagrange's Method I have used the phrase "stationary point" though the standard texts in Calculus employ the words "extremum", "miximum" or "minimum". As I have already pointed out the first derivative cannot possibly find points at which a function attains a maximum or minimum.) Here is a "complete statement" of Lagrange's 'Method'.

The Method of Undetermined Multipliers

Let $U \subset \mathbb{R}^n$ be an open set and suppose that for $j = 1, \ldots, m < n$,
$\phi_j : U \to \mathbb{R}$ *and* $f : U \to \mathbb{R}$ *are* C^1*-maps.*
Suppose further that for some $c = (c_1, \ldots, c_m) \in \mathbb{R}^m$, whenever $x \in U$ and $\phi_j(x) = c_j$, the matrix:

$$\frac{\partial(\phi_1, \ldots, \phi_m)}{\partial(x_1, \ldots, x_n)}(x) \doteq \begin{bmatrix} \frac{\partial \phi_1}{\partial x_1} & \frac{\partial \phi_1}{\partial x_2} & \cdots & \frac{\partial \phi_1}{\partial x_n} \\ \frac{\partial \phi_2}{\partial x_1} & \frac{\partial \phi_2}{\partial x_2} & \cdots & \frac{\partial \phi_2}{\partial x_n} \\ \vdots & \vdots & \cdots & \vdots \\ \frac{\partial \phi_m}{\partial x_1} & \frac{\partial \phi_m}{\partial x_2} & \cdots & \frac{\partial \phi_m}{\partial x_n} \end{bmatrix}_x$$

has rank m.

Then a necessary and sufficient condition for $p \in U$ to be a stationary point of f (subject to the conditions $\phi_j(p) = c_j$, for $1 \leq j \leq m$), is that there are constants $\lambda_1, \ldots, \lambda_m \in \mathbb{R}$ ("the undetermined multipliers") *such that:*

(Lagrange A)
$$\begin{aligned} \lambda_1 \frac{\partial \phi_1}{\partial x_1} + \cdots + \lambda_m \frac{\partial \phi_m}{\partial x_1} &= \frac{\partial f}{\partial x_1} \\ \vdots \quad + \cdots + \quad \vdots \quad &= \quad \vdots \\ \lambda_1 \frac{\partial \phi_1}{\partial x_n} + \cdots + \lambda_m \frac{\partial \phi_m}{\partial x_n} &= \frac{\partial f}{\partial x_n} \end{aligned}$$

where all the partial derivatives are evaluated at the point $p \in U$.

Thus to find a stationary point $p \in U$ of f subject to the constraints:

$$\begin{aligned} \phi_1(x_1, \ldots, x_n) &= c_1 \\ \vdots \quad & \quad \vdots \\ \phi_m(x_1, \ldots, x_n) &= c_m \end{aligned}$$

one needs to find $(m+n)$ real numbers, $p_1, \ldots, p_n$ (which are the coordinates of $p \in U$) and the "multipliers" $\lambda_1, \ldots, \lambda_m$, from the n equations (Lagrange A) supplemented by the m "constraining equations:

(Lagrange B)
$$\begin{aligned} \phi_1(p_1, \ldots, p_n) &= c_1 \\ \vdots \quad &= \vdots \\ \phi_m(p_1, \ldots, p_n) &= c_m \end{aligned}$$

Here is how one derives the method of undetermined multipliers from Lagrange's Theorem. To conform to the notation of the theorem I will denote $\mathbb{R}^n, \mathbb{R}^m$ by E, F respectively.

We are trying to find stationary points of the C^1-function $E \supset U \xrightarrow{f} \mathbb{R}$ when the arguments satisfy $\phi(x_1, \ldots, x_m) = \mathbf{c} = (c_1, \ldots, c_m) \in \mathbb{R}^m$ where

$$\begin{aligned}&\phi : U \to F \text{ is the } C^1\text{-map,}\\ &(x_1, \ldots, x_m) \mapsto (\phi(x_1, \ldots, x_n), \ldots, \phi_m(x_1, \ldots, x_n))\\ &\text{and } \mathbf{c} = (c_1, \ldots, c_m).\end{aligned}$$

The rank condition on the matrix $\dfrac{\partial(\phi_1, \ldots, \phi_m)}{\partial(x_1, \ldots, x_n)}(p)$ simply means that $\mathbf{c}$ is a regular value of ϕ. Thus the hypotheses of the "theorem" and the "method" are the same.

With respect to the standard onb's of E and F, the dual spaces are just row vectors rather than column vectors. The condition for $p \in U$ to be a stationary point subject to the constraint $\phi(p) = \mathbf{c}$ according to Lagrange's Theorem is that there is a $\Lambda \in F^*$ (that is a row vector $[\lambda_1, \ldots, \lambda_m]$) such that

$$\mathbf{d}f(p) = [\lambda_1, \ldots, \lambda_m] \cdot \begin{bmatrix} \frac{\partial \phi_1}{\partial x_1} & \frac{\partial \phi_1}{\partial x_2} & \cdots & \frac{\partial \phi_1}{\partial x_n} \\ \frac{\partial \phi_2}{\partial x_1} & \frac{\partial \phi_2}{\partial x_2} & \cdots & \frac{\partial \phi_2}{\partial x_n} \\ \vdots & \vdots & \cdots & \vdots \\ \frac{\partial \phi_m}{\partial x_1} & \frac{\partial \phi_m}{\partial x_2} & \cdots & \frac{\partial \phi_m}{\partial x_n} \end{bmatrix}_p$$

Unfortunately, the right-hand side of this equation is a row vector each of whose entries is a sum of n terms. For ease of type-setting (and comprehension) I will write out the transpose of both sides. This yields the system of n equations

$$\text{(Lagrange A)} \qquad \begin{aligned} \lambda_1 \frac{\partial \phi_1}{\partial x_1} + \cdots + \lambda_m \frac{\partial \phi_m}{\partial x_1} &= \frac{\partial f}{\partial x_1} \\ \vdots \quad + \cdots + \quad \vdots \quad &= \quad \vdots \\ \lambda_1 \frac{\partial \phi_1}{\partial x_n} + \cdots + \lambda_m \frac{\partial \phi_m}{\partial x_n} &= \frac{\partial f}{\partial x_n} \end{aligned}$$

where all the partial derivatives are being evaluated at the point $p = (p_1, \ldots, p_n)$. *This completes the derivation of the "method" from Lagrange's Theorem.* □

Remark 6.1.2c. Clearly, in general, there can be little hope of solving the equations marked Lagrange A and B. However, when the functions ϕ and f have nice symmetry properties, (that is they are left invariant when the subscripts of the variables x_k are permuted by some large subgroup of $\mathfrak{S}_n$), it may be possible to apply this method successfully.

In fact, many famous results such as the AM–GM inequality, Hölder's and Minkowski's inequalities can be established quite easily using this technique.

This, perhaps, is the reason why this method has acquired a certain reputation for being a useful tool for finding maxima/minima of functions.

Remark 6.1.2d (An analytic formulation).
Let me translate Lagrange's theorem into the language of "tangent spaces" introduced in § 4.4. The condition that c be a regular value of ϕ ensures that at each point $x \in M_c$, the closed set M_c has a well-defined tangent space and the Lagrange theorem asserts that $p \in M_c$ is a stationary point of $f|M_c$ iff the differential, $\mathbf{d}f(p)$, annihilate $T_p(M_c)$.

Remark 6.1.2e (The geometric formulation).
The theorem may also be reformulated in a more 'geometric' manner:

> $p \in M_c$ is a constrained stationary point iff the gradient of f at p is orthogonal to the kernel of $\mathbf{D}\phi(p)$, i. e. $\nabla f(p) \in L$.

The equivalence of the two formulations is an immediate consequence of the lemma that was proved in establishing the necessity of the condition. Indeed, the proof amounted to showing that $\nabla f(p)$ was orthogonal to the kernel of $\mathbf{D}\phi(p)$.

Remark 6.1.2f (The geometric version of the 'Method').
I have already deduced the classical form of the multiplier condition from the statement of Theorem 6.1.1. This classical formulation looks really attractive in the geometric setting of gradients:

> *Let* $U \subset \mathbb{R}^n$ *be an open set and let* $f, \phi_1, \ldots, \phi_m : U \to \mathbb{R}$ *be* C^1*-functions satisfying the conditions laid down in the statement of the "Method of Undetermined Multipliers".*
>
> *Then* $p \in U$ *is a stationary point of* f *subject to the constraint* $p \in M_c$ *iff there are constants* $\lambda_1, \ldots, \lambda_m$, *so that*

$$\nabla f(p) = \sum_{j=1}^{m} \lambda_j \nabla \phi_j(p);$$

In other words, the gradient of f at p must be a linear combination of the gradients (at p) of the constraining functions. □

6.1.1 *Applications of Lagrange's Theorem*

1. AM–GM inequality

This is a classic application of undetermined multipliers.

Proposition 6.1.3 (The AM–GM inequality).
The arithmetic mean is never less than the geometric mean. *More precisely given nonnegative numbers, $x_1, \ldots, x_n$, we have the inequality:*

$$\frac{1}{n}(x_1 + \cdots + x_n) \geq \Big(\prod_{j=1}^{n} x_j\Big)^{1/n}$$

Proof. I will show that if $\sum_{j=1}^{n} x_j = M$, then the right-hand side of the inequality of the proposition cannot exceed $\left(\frac{M}{n}\right)$. For economy of notation, I will write $P(\boldsymbol{x}) = (\prod_{j=1}^{n} x_j)$ and $S(\boldsymbol{x}) = x_1 + \cdots + x_n$, where $\boldsymbol{x} = (x_1, \ldots, x_n) \in \mathbb{R}^n$.

The left side of the AM–GM inequality is obviously nonnegative and the right-hand side vanishes if any of the x_j's are zero. So it suffices to look for the maxima of $P(\boldsymbol{x})$ restricted to the open set

$$U \doteq \{\boldsymbol{x} \in \mathbb{R}^n : x_j > 0 \text{ for all } j\}.$$

subject to the constraint, $S(\boldsymbol{x}) = M$.

From Lagrange's theorem it follows that if $\boldsymbol{\xi} = (\xi_1, \ldots, \xi_n) \in U$ is a stationary point of $f|(S^{-1}\{M\})$ then there is a constant λ such that for each $j = 1, \ldots, n$ we have the relation:

$$(1_j) \qquad \left.\frac{\partial P}{\partial x_j}\right|_{\boldsymbol{\xi}} = \lambda \left.\frac{\partial S}{x_j}\right|_{\boldsymbol{\xi}} \quad \text{or,}$$

$$(2_j) \qquad \frac{P(\boldsymbol{\xi})}{\xi_j} = \lambda.$$

Multiplying the equation (2_j) by ξ_j and adding all the n relations resulting thereby, we get: $n \cdot P(\boldsymbol{\xi}) = \lambda S(\boldsymbol{\xi})$ or, $P(\boldsymbol{\xi}) = M/n$.

Using this relation (2_j) now takes the form:

$$(*) \qquad \lambda = \frac{nP(\boldsymbol{\xi})}{\xi_j}.$$

On the left-hand side we have a constant and the numerator of the right-hand side is independent of j; hence, all the ξ_j's are equal.

It follows that the AM and the GM coincide at $p_0 = (\frac{M}{n}, \ldots, \frac{M}{n}) \in U$ and p_0 is the unique stationary point of $P : U \to \mathbb{R}$ restricted to $S^{-1}\{M\}$. It remains then to prove that $P|(S^{-1}\{M\})$ attains a maximum at p_0.

To do this by directly using the Hessian condition would be rather tedious, so we employ a trick!

For $x_1, \ldots, x_n > 0$ consider the function ϕ defined by $\phi(x_1, \ldots, x_n) \doteq \log\bigl(P(x_1, \ldots, x_n)\bigr)$. Clearly, since the logarithm is strictly increasing, the point p_0 is a local maximum (resp. minimum) for ϕ iff it is a local maximum (resp. minimum) for P. The differential of ϕ is given by:

$$\mathbf{d}\phi(x_1, \ldots, x_n) = \textstyle\sum_{i=1}^{n} \left(\dfrac{\mathbf{d}x_i}{x_i}\right).$$

and hence the Hessian of ϕ is the $(n \times n)$ *diagonal* matrix with k^{th} diagonal entry equal to $(-x_k^{-2})$.

Obviously $\mathbf{D}^2\phi(p_0)(\boldsymbol{v}, \boldsymbol{v}) < 0$ for every $\boldsymbol{v} \neq 0 \in \mathbb{R}^n$. Hence, p_0 is a local (and global) maxima for $\phi|(S^{-1}\{M\})$. The proof is complete. □

2. Diagonalization of self-adjoint matrices

I have used the phrase "self-adjoint matrices" to cover both real symmetric and complex Hermitian matrices.

What I now discuss is the most elementary version of the Spectral Theorem: one of the most beautiful and useful results of functional analysis.

Theorem 6.1.4 (Diagonalization of self-adjoint matrices).
Let A *be a real (resp. complex)* $(n \times n)$ *matrix which is symmetric (resp. Hermitian). Then there is an orthogonal (resp. unitary) matrix* X *such that* XAX* *is a diagonal matrix.*

Here the superscript * denotes transpose in the real case and conjugate transpose in the complex case.

Proof. I will deal with the more general case of complex Hermitian matrices. Let A be an $(n \times n)$ Hermitian matrix. I will denote the linear transformation, $T(\mathrm{A}) : \mathbb{C}^n \mathbb{T} \mathbb{C}^n$, more simply by $A : \mathbb{C}^n \to \mathbb{C}^n$.

Since A is Hermitian $A : \mathbb{C}^n \to \mathbb{C}^n$ is self-adjoint, that is:

$$\langle A(\boldsymbol{x}), \boldsymbol{y}\rangle = \langle \boldsymbol{x}, A(\boldsymbol{y})\rangle \text{ for all } \boldsymbol{x}, \boldsymbol{y} \in \mathbb{C}^n.$$

Consider the function $q : \mathbb{C}^n \times \mathbb{C}^n \to \mathbb{C}$ defined by $q(\boldsymbol{x}, \boldsymbol{y}) = \langle A(\boldsymbol{x}), \boldsymbol{y}\rangle$. This is an $\mathbb{R}$-bilinear function and hence a C^∞-map on $\mathbb{R}^{2n} \times \mathbb{R}^{2n} \equiv$

$\mathbb{C}^n \times \mathbb{C}^n$. So, $Q(\boldsymbol{x}) = \langle A(\boldsymbol{x}), \boldsymbol{x}\rangle$ is a C^∞-map on $\mathbb{R}^{2n}$.

$$\text{Now, } \overline{\langle A(\boldsymbol{x}), \boldsymbol{x}\rangle} = \langle \boldsymbol{x}, A(\boldsymbol{x})\rangle = \langle A^*(\boldsymbol{x}), \boldsymbol{x}\rangle = \langle A(\boldsymbol{x}), \boldsymbol{x}\rangle \quad \text{because } A \text{ is self-adjoint.}$$

This shows that Q is a real-valued C^∞-map on $\mathbb{R}^{2n}$.

The unit sphere, $\mathbb{S}_1(0)$, of $\mathbb{R}^{2n}$ is closed and bounded, hence compact. So Q attains a supremum and infimum on this set, which we will denote $\mathbf{S}$. So stationary points of $Q|\mathbf{S}$ exist; let us determine these stationary points.

First observe that $\mathbf{S}$ is precisely the set on which the function $N(x) \doteq ||\boldsymbol{x}||^2$ is equal to 1. The derivative of N has, of course, been computed already: $\mathbf{D}N(x)(\boldsymbol{h}) = 2\langle \boldsymbol{x}, \boldsymbol{h}\rangle$.

Now we compute the derivative of Q.

$$\begin{aligned} Q(x+h) - Q(x) &= \langle A(\boldsymbol{x}+\boldsymbol{h}), \boldsymbol{x}+\boldsymbol{h}\rangle - \langle A(\boldsymbol{x}, \boldsymbol{x}\rangle, \\ &= \langle A(\boldsymbol{x}), \boldsymbol{h}\rangle + \langle A(\boldsymbol{h}), \boldsymbol{x}\rangle + o(\boldsymbol{h}), \\ &= 2\langle A(\boldsymbol{x}, \boldsymbol{h})\rangle + o(\boldsymbol{h}), \text{since } A \text{ is self-adjoint.} \\ \text{Hence } \mathbf{D}Q(x)(\boldsymbol{h}) &= 2\langle A(\boldsymbol{x}), \boldsymbol{h}\rangle. \end{aligned}$$

Since stationary points of $Q|\mathbf{S}$ are precisely the stationary points of $Q : \mathbb{R}^{2n} \to \mathbb{R}$ subject to the constraint, $N(\boldsymbol{x}) = 1$, Lagrange's theorem tells us that $\boldsymbol{x} \in \mathbb{R}^{2n}$ is a stationary point of $Q|\mathbf{S}$ iff there is a $\lambda \in \mathbb{R}$ such that $\lambda\langle \boldsymbol{x}, \boldsymbol{h}\rangle = \langle A(\boldsymbol{x}), \boldsymbol{h}\rangle$ for every $\boldsymbol{h} \in \mathbb{R}^{2n}$. Then the nonsingularity of the inner product implies that $A(\boldsymbol{x}) = \lambda\boldsymbol{x}$. Since $\boldsymbol{x}$ is a nonzero vector (it has length 1!) it is an eigenvector of A corresponding to the eigenvalue λ.

So $\boldsymbol{x}$ is a stationary point of $Q|\mathbf{S}$ iff $\boldsymbol{x}$ is an eigenvector of A and the associated stationary value is the eigenvalue corresponding to $\boldsymbol{x}$.

If $E \subset \mathbb{R}^{2n}$ is an A-stable subspace, that is $A(E) \subset E$, then for all $\boldsymbol{e} \in E$ and all $\boldsymbol{f} \in E^\perp$, the orthogonal complement E, we have:

$$0 = \langle A(\boldsymbol{e}), \boldsymbol{f}\rangle = \langle \boldsymbol{e}, A(\boldsymbol{f})\rangle,$$

since E is A-stable and A is self-adjoint.

This shows that $E^\perp$ is also A-stable.

Clearly $A|E^\perp : E^\perp \to E^\perp$ is a self-adjoint linear transformation.

To obtain the unitary matrix X reaquired by the theorem we proceed as follows.

Let $\boldsymbol{x}_1$ be a point where $Q|\mathbf{S}$ attains its supremum. Then $\boldsymbol{x}_1$ is an eigenvector of A of unit length having the highest eigenvalue. Let E_1 be the $\mathbb{C}$-linear subspace spanned by $\boldsymbol{x}_1$. This is obviously an A-stable subspace in $\mathbb{C}^n$. Let $Q_1 = Q|E_1^\perp$. Proceeding as before, we now find a eigen vector of $A|E_1^\perp$ of unit length corresponding to the highest eigenvalue of $A|E_1^\perp$. This is, of course, a stationary point of Q_1 restricted to $\mathbf{S} \cap E_1^\perp$. This gives an eigenvector $\boldsymbol{x}_2$ of A that has unit length and is orthogonal to $\boldsymbol{x}_1$. Let E_2 be the $\mathbb{C}$-linear subspace spanned by $\boldsymbol{x}_1$ and $\boldsymbol{x}_2$. This is clearly an A-stable subspace of $\mathbb{C}$-dimension 2.

Now determining the point at which Q restricted to the unit sphere of $E_2^\perp$ will yield an eigenvector of unit length of A which is orthogonal to E_2.

Proceeding in this manner, after n steps we will have an *orthonormal basis*, $\mathcal{X} = \{\boldsymbol{x}_1, \ldots, \boldsymbol{x}_n\}$, of $\mathbb{C}^n$, consisting of eigenvectors of A.

The matrix of A with respect to the basis $\mathcal{X}$ is obviously diagonal, having λ_k as the kth diagonal entry where the eigenvalues are ordered by magnitude, that is, $\lambda_1 \geq \ldots \lambda_n$.

If we regard the orthonormal basis, $\mathcal{X}$, as a collection of n column vectors we get an $(n \times n)$ matrix X. The orthonormality condition implies that $\mathrm{XX}^* = \mathrm{X}^*\mathrm{X} = \mathrm{I}_n$, that is, X is a unitary matrix. Then from Theorem 1.3.10 mentioned on page 16 it follows that XAX^* is the diagonal matrix with entries $\lambda_1, \ldots, \lambda_n$, the eigenvalues of A. This completes the proof of the theorem. □

Remark 6.1.5 (On the Spectral Theorem).
What I have presented is a quick and geometrically attractive proof of the the most elementary version of the Spectral Theorem. For a proper treatment you should read Halmos' account; see pages 153–8 of [Hal61].

It is also worth noting that the above proof shows that the eigenvalues of Hermitian matrices are real.

Exercises for § 6.1

6.1.1 A TYPICAL APPLICATION OF LAGRANGE'S THEOREM

Find the maximum of $\log x + \log y + 3\log z$ on that portion of the sphere $\mathbb{S}_r(0) \subset \mathbb{R}^3$ for which $x > 0, y > 0, z > 0$. Use this result to prove that if a, b, c, are positive real numbers, then

$$abc^3 \leq 27\left(\frac{a+b+c}{5}\right)^5.$$

Use Lagrange's method to establish the triangle inequality for the L^p-norm on $\mathbb{R}^n$, for $p \geq 1$.

> HINT: Fix a point $x = (x_1, \ldots, x_n) \in \mathbb{R}^n$ and maximize the function $F : \mathbb{R}^n_+ \to \mathbb{R}$ defined by
>
> $$F(y_1, \ldots, y_n) = (x_1 + y_1)^p + \cdots + (x_n + y_n)^p.$$
>
> In particular, show that the maximum is attained at points of the form $c(x_1, \ldots, x_n)$.

6.2 The Calculus of Variations

Phrases in common parlance, such as "path of least resistance" or "water finds its own level" testify to our unstated, but firm, conviction (based usually upon experience, sometimes on optimism) that many natural processes evolve in such a way that some real-valued function of "physical significance", such as `potential energy` or `entropy` is minimized or maximized. This line of thought has proved extremely fruitful in physics. For example, even before Differential Calculus had been formalized, Fermat's `Principle of least action` had been formulated and this was used to derive Snell's Law which governs the refraction of light.

Such ideas have, over the centuries, been refined into a flourishing branch of mathematics: the Calculus of Variations. This is a vast field and there is no question of even attempting a survey (however cursory) of this topic. I will discuss two problems: that of `geodesics` and the `isoperimetric problem`; corresponding respectively to free and constrained stationary points respectively. Other examples of each genre will be presented as exercises. These will give glimpses of other famous classical results.

Since this chapter is essentially concerned with maxima/minima of functions defined on infinite-dimensional Hilbert spaces I would like to reiterate a few facts that have appeared earlier.

- Although I deal with functions defined on infinite-dimensional spaces, the basic idea is something you learnt in your *first* Calculus course:

  if a C^1 *function* f *that is real-valued has a* local maximum or minimum *at a point* p *in its domain, then* $\mathbf{D}f(p) = 0$.

- The spaces involved will be infinite-dimensional Hilbert spaces. Now we have already seen in Chapter 3 (Example 3.1.28) that closed and bounded subsets of such spaces are not necessarily compact. This makes it difficult in such situations to ensure that *local stationary points* exist. We will not investigate the conditions for the existence of stationary points. We will simply try to develop the *consequences* of the existence of stationary points. These are of interest in themselves.

- Recall that the condition on the second derivative (see Proposition 4.3.12) which helps decide the nature (maximum or minimum) of the stationary point was rather subtle in the infinite-dimensional case. So we will not generally investigate the nature of the stationary point.

I will now describe the mathematical setup for a kind of situation that arises frequently in Physics. I begin by specifying what 'a curve' will mean in this context.

Definition (Curves).
By a C^1*-curve* taking values in a Banach space E (resp. an open set $U \subset E$) I will mean a map $\alpha : [0,1] \to E$ (resp. $\alpha : [0,1] \to U$) whose restriction to $(0,1)$ is C^1 and has a left (resp. right) derivative at 0 (resp. 1). The set of C^1-curves taking values in U is denoted $C^1([0,1];U)$. If $p,q \in E$, then $C^1(p,q;U)$ will denote the set:

$$\{\alpha \in [0,1] : \alpha(0) = p \text{ and } \alpha(1) = q\}.$$

The set up:

The best way to understand the next set of conditions is to imagine that we are trying to move an object from one point to another (through a substance which offers some resistance to movement) while expending a minimum of effort.

1. E is a Euclidean space, $U \subset E$ an open subset and $p,q \in U$ two points.

2. $f : U \times E \times \mathbb{R} \to \mathbb{R}$ is a C^2-function which is of some mathematical or physical interest.

3. For any $\alpha \in C^1(p,q;U)$ a 'functional', depending on α and f, is defined by the formula

$$L(\alpha; f) = \int_0^1 f\big(\alpha(t), \alpha'(t), t\big)\, dt.$$

(The word 'functional' is being used to conform with traditional nomenclature.)

The problem of Calculus of Variations is to determine the *stationary curves* for the functional L, that is to find curves $\alpha_* \in C^1(p,q;U)$ such that for all "neighbouring" curves, $\beta \in C^1(p,q;U)$, $L(\alpha_*; f) \geq L(\beta; f)$; or, $L(\alpha_*; f) \leq L(\beta; f)$.

We will consider L to be a function defined on $C^1(p,q;U)$ in the hope of using the methods of Differential Calculus to find necessary conditions for a curve to be a "stationary curve" for the functional l.

The first (apparent) difficulty is that the set $C^1(p,q;U)$ does not form a linear space under the "natural" operations of pointwise addition and scalar multiplication of curves. However $C^1(p,q;E)$ is an *affine space.*

Consider the set, $C^1(0,0;E)$. This certainly is closed under pointwise addition of curves and (pointwise) multiplication of curves by real numbers. Obviously, it is a linear space. Given any $\gamma \in C^1(p,q;E)$, there is an injective function $A_\gamma : C^1(p,q;E) \to C^1(0,0;E)$ defined by: $\alpha \mapsto \alpha - \gamma$.

(This is a well-known trick: the set of solutions of a non-homogeneous linear equation do not form a linear space but given *any* solution of the non-homogeneous equation, sometimes called a "particular solution", *all* solutions of the non-homogeneous equation can be obtained by adding solutions of the associated homogeneous equation to the "particular solution").

The restriction of functions such as A_γ embed $C^1(p,q;U)$ as a subset of $C^1(0,0;E)$. So if we define a Banach space structure on $C^1(0,0;E)$ (that is define a complete norm) then we can think of $C^1(p,q;U)$ as a subset of a Banach space and the notion of "neighbouring curves" is then obvious.

Since I intend to use Lagrange's theorem to solve , the "isoperimetric problem" I will have to define a Hilbert space structure on $C^1(0,0;E)$.

And herein lies the difficulty.

The obvious starting point would be to use the Euclidean inner product on E and for any two $\alpha, \beta \in C^1(0,0;E)$ define their inner product, (α, β), by setting:

$$(\alpha, \beta) = \int_0^1 \langle \alpha(t), \beta(t) \rangle dt + \langle \alpha'(t), \beta'(t) \rangle dt.$$

This certainly defines an inner product on $C^1(0,0;E)$ but unfortunately, the resulting metric is **not complete** and therefore useless for our present purpose!

In the exercises to § 3.1, I had shown how a metric space, X, which is not complete, can be isometrically embedded in a 'smallest' complete metric space: the completion of X and that this completion is essentially unique. An analogous procedure, using convergent Cauchy sequences can be used to define a Hilbert space starting from $C^1(0,0;E)$ and the inner product defined above. But this abstract existence theorem is of little use since we do not have any information about how to deal with the "limits of Cauchy sequences of curves" that are the vectors of the Hilbert space thus constructed!

If we had the theory of the Lebesgue integration and the resulting theory of distributional derivatives at our disposal, then we could have identified this completion and given a completely satisfactory account of the Calculus of Variations from the point of view adopted for this book.

Given the above prerequisites, we can identify the completion of the space $C^1(0,0;E)$ (with the above inner product) as the Hilbert space, generally called the *Sobolev space*, $W^{1,2}(0,0;E)$, whose elements are curves defined on $[0,1]$ which start and end at the origin of E and whose coordinate functions in terms of an orthonormal basis of E are L^2 functions whose distributional derivatives are also in $L^2[0,1]$.[1]

Similarly we can define the set $W^{1,2}(p,q;U)$ of L^2 functions on $[0,1]$ which take the value p at 0 and q at 1 and whose distributional derivatives are also L^2 functions.

This will be embedded as an open subset in $W^{1,2}(0,0;E)$ by the functions A_γ for any $\gamma \in W^{1,2}(p,q;U)$.

In what follows I will refer to elements of $W^{1,2}(p,q;U)$ as curves from p to q lying in U and elements of $W^{1,2}(0,0;E)$ as "loops".

I will also have to assume that the functionals under consideration only have C^1-curves as "stationary points". This hypothesis is unnecessary. We will derive a necessary condition for stationary behaviour known as the Euler–Lagrange equations and by a celebrated theorem

[1] L^2 functions or elements of $L^2[0,1[$ are real-valued functions defined on [0,1] which are square integrable with respect to the Lebesgue integral.

The distributional derivative is a strengthening (made possible by the theory of the Lebesgue integral) of the notion of weak derivative which I have discussed in Chapter 4, (see Remark 4.1.25b).

(The Elliptic Regularity Theorem) all solutions of the Euler–Lagrange equations are smooth.

* * * * *

Let us suppose that $\alpha_* \in C^1(p, q; E)$ is a stationary point for the functional L described in the "set up". Then given any loop, $\omega \in W^{1,2}(0, 0; E)$ if $s \in \mathbb{R}$ has sufficiently small absolute value, the curve, ω_{α_*} defined by

$$t \mapsto \alpha_*(t) + s\cdot\omega(t)$$

is again a curve in $W^{1,2}(p, q; U)$. Thus for $\varepsilon > 0$ sufficiently small, we may regard $\omega_{\alpha_*} : (-\varepsilon, +\varepsilon) \to W^{1,2}(p, q; U)$ as a differentiable curve which represents $\omega \in W^{1,2}(0, 0; E)$ as a tangent vector at α_*.

Since α_* is a stationary point for L, we have $\mathbf{d}L(\alpha_*) = 0$ which means that for any C^1-loop $\omega \in W^{1,2}(0, 0; E)$ of sufficiently small norm, we must have:

$$L(\alpha_* + \omega) - L(\alpha_*) = o(\omega).$$

We now compute the left-hand side of the relation above.

$$\begin{aligned} &L(\alpha_* + \omega) - L(\alpha_*) \\ &= \int_0^1 \{f(\alpha_* + \omega, \alpha'_* + \omega', t) - f(\alpha_*, \alpha'_*, t)\}dt \\ &= \int_0^1 \{\mathbf{D}_1 f(\alpha_*, \alpha'_*, t)(\omega) + \mathbf{D}_2 f(\alpha_*, \alpha'_*, t)(\omega')\}dt + o(\omega) \end{aligned}$$

After applying "integration by parts" to the second term inside the last integral we get:

$$L(\alpha_* + \omega) - L(\alpha_*) \int_0^1 \{\ \mathbf{D}_1 f(\alpha_*, \alpha'_*, t) - \tfrac{d}{dt}\big(\mathbf{D}_2 f(\alpha_*, \alpha'_*, t)\big)\ \}(\omega)dt + o(\omega)$$

Since this holds for *every* $W^{1,2}$-loop the entity enclosed by the braces in the above equation must be identically zero. This gives the necessary condition we are seeking:

$\dagger$E-L $\cdots\cdots\cdots\cdots$ $\mathbf{D}_1 f(\alpha_*, \alpha'_*, t) = \tfrac{d}{dt}\big(\mathbf{D}_2 f(\alpha, \alpha'_*, t)\big).$

This is a basis-free version of the celebrated Euler–Lagrange equations

Remark 6.2.1 (On the Euler–Lagrange equation).
I will now use traditional notation so that you will recognize the Euler–Lagrange equation(s) in their more usual form in standard texts.

Usually the the Euclidean space is taken to be $\mathbb{R}^n$, the coordinates of a point are $(x_1, \ldots, x_n)$, the path is written $\alpha(t) = \big(x_1(t), \ldots, x_n(t)\big)$ and the velocity of α at time t is denoted by $\dot{\alpha}(t) = \big(\dot{x}_1(t), \ldots, \dot{x}_n(t)\big)$.

Now it is simply a matter of writing the differentials (row vectors) as gradients (column vectors) to see that the single Euler–Lagrange equation, $\dagger_{\text{E-L}}$, now reappears as a set of n equations:

$$\frac{d}{dt}\left(\frac{\partial f}{\partial \dot{x}_i}\right) - \frac{\partial f}{\partial x_i} = 0 \qquad (i = 1, \ldots, n.)$$

6.2.1 *Applications of the Euler–Lagrange equations*

1. The Geodesic Problems

In discussing applications of the Calculus of Variations, I face a difficulty since I wish to avoid assuming any knowledge of physics or mechanics. So I have chosen to discuss the problem of finding "length minimizing curves" or *geodesics.* This topic really belongs to Differential Geometry, but by making a few assumptions one can reduce this question, at least in some important instances, to a problem in Calculus.

In order to avoid introducing a host of new ideas in a sketchy manner, I will confine myself mainly to open subsets of Euclidean spaces, in fact, $\mathbb{R}^2$, though the definition of "arc length" which follows, obviously carries over to other spaces, (like spheres in Euclidean spaces) where one has a well-defined notion of the tangent space at each point. Perhaps it should be mentioned that the definition I have used for `geodesics` is unconventional, but the results that I derive are consistent with results obtained using conventional tools of differential geometry.

I have already remarked that if $\alpha : [a, b] \to E$ is a C^1 curve (that is α is C^1 in (a, b) and has a left (resp. right) derivative at a (resp. b)) then its derivative at $t_0 \in (a, b)$, denoted either by $\alpha'(t_0)$ or $\dot{\alpha}(t_0)$ is a tangent vector belonging to $T_{\alpha(t_0)}E$ and thus if we think of t as time, then it is natural to think of $||\alpha'(t_0)||_E$, as the *speed* (or the magnitude of the velocity) of α at t_0. Then, intuitively, the distance traversed by $\alpha(t)$ as t ranges over the interval $[a, b]$ ought to be taken as the "length of the curve" between the points $\alpha(a)$ and $\alpha(b)$.

. From elementary examples one learns in the first course of Calculus, it is clear that this length should be the integral from a to b of the speed of the curve. This motivates the next definition.

Definition (Arc Length).
Given two points $p, q \in U$, and $\alpha : [a, b] \to U$ a C^1 curve such that $\alpha(a) = p$ and $\alpha(b) = q$, the *arc length*, $L(\alpha)$, of α is defined by the integral:

$$L(\alpha) = \int_a^b ||\alpha'(t)||dt.$$

One can define arc-length in Euclidean spaces more directly, but I have chosen this method since this is how one proceeds in Differential Geometry.

Let $C^1(p, q)$ be the set of C^1 maps from [0,1] to U which have a left (resp. right) derivative at (resp. 1). As before this can be given the C^1-metric. As pointed out this does not yield a subset of a Banach space, but can be embedded in $W^{1,2}(E)$ which is a Hilbert space. We are interested in curves for which the function $L : W^{1,2}(p, q) \to \mathbb{R}$ attains a stationary value. Again we assume that this will be attained at a C^1 curve. Moreover our interest is in geometry not analysis; what I mean is that we are interested in the "locus traced out" in E by $\alpha(t)$ as t varies over the domain of definition of α and not the function α itself. Our hope is that such curves will be of some geometric significance.

The trouble is that the arc-length function *cannot* have isolated minima! Consider for example curves in $\mathbb{R}^2$ joining $(0, 0)$ and $(1, 1)$. The curve

$$\alpha(t) = (t, t)$$

is obviously the simplest straight line curve joining them. But the curves

$$\alpha_s(t) = (t^{1+s}, t^{1+s}), \text{ for } s > 0$$

also trace the same locus and these curves have the same length!

In Differential Geometry, geodesics are defined to be curves with zero acceleration and these turn out to be constant speed curves at which the arc-length function takes a stationary value. But since I want to avoid explicit use of new notions such as acceleration, I follow a somewhat different method of attacking the problem. Since the integral which defines arc-length is not changed when t ranges over an interval on which the speed is zero, we assume without loss of generality that the curves we discuss are regular (i. e. $\alpha'(t) \neq 0$) except at a finite number of points in its domain of definition. To simplify the notation, I deal with curves in the plane and use x, y as the standard coordinates. Thus a curve

$\alpha : [a,b] \to \mathbb{R}^2$ will be written as

$$\alpha(t) = \big(x_\alpha(t), y_\alpha(t)\big),$$

or, when there is no danger of confusion, more simply, by $\alpha(t) = \big(x(t), y(t)\big)$.

Let $\alpha : [a,b] \to \mathbb{R}^2$ be a C^1 curve as above and suppose that $\alpha'(t) = 0$ only at a finite number of points $t_1, \ldots, t_k \in [a,b]$. If t_0 is not one of these points and $\alpha(t_0) = P_0 = (x_0, y_0)$ then one of $\dot{x}_\alpha(t_0)$ or $\dot{x}_\alpha(t_0)$ is nonzero. Suppose $\dot{x}_\alpha(t_0) \neq 0$. Then in some open set containing P_0 the curve can be represented as the graph of a C^1 function $y : (t_0 - \delta, t_0 + \delta)$. Now it is easy to see that for $\xi \in (t_0 - \delta, t_0 + \delta)$,

$$y'(\xi) = \frac{\dot{x}(\xi)}{\dot{y}(\xi)}$$

and hence over this interval the integral (*) defining L takes a more convenient form, as the following calculation shows.

$$\text{Since } \big(\dot{\alpha}(t)\big)^2 = (\dot{x}(t))^2 + (\dot{y})^2,$$

$$(\dagger) \qquad \int_{t_0-\delta}^{t_0+\delta} \sqrt{1 + \frac{\dot{y}(\xi)^2}{\dot{x}(\xi)^2}}\,\dot{x}(\xi)dt = \int_{t_0-\delta}^{t_0+\delta} \sqrt{1 + (y')^2}dx.$$

(If, on the other hand, we have $\dot{y}(t_0) \neq 0$ we can get an analogous representation with the roles of x and y interchanged.)

The advantage here is that the integral on the right-hand side of (†) does not depend on the time parameter t! So the arc-length integral $L(\alpha)$ can be decomposed into a sum of integrals of the form

$$\int_{a_i}^{b_i} \sqrt{1 + (y'(x))^2}dx \quad \text{or,} \quad \int_{a_i}^{b_i} \sqrt{1 + (x'(y))^2}dy.$$

Thus we are now faced with "Euler–Lagrange" problems for functionals of the form:

$$L(y; f) = \int_a^b f(y, y', x)dx \quad \text{where, } f = \sqrt{1 + (y')^2}.$$

The Euler–Lagrange equation yields:

$$\frac{d}{dx}\left(\frac{\partial}{\partial y'}\sqrt{1 + (y')^2}\right) = 0 \quad \text{since } f \text{ is independent of } y,$$

$$\text{or,} \quad \frac{d}{dx}\left(\frac{y'}{(1 + y'^2)^{\frac{1}{2}}}\right) = 0.$$

This leads to,

$$(\ddagger) \qquad \frac{y''}{(1+y'^2)^{\frac{3}{2}}} = 0.$$

$$\text{Hence, } y'' = 0$$

and finally, we get $y(x) = k_1 + k_2$ where k_1, k_2 are constants.

Thus in each of the subintervals, $[a_i, b_i]$ the curve has the form of a straight line segment. Thus a geodesic in an open subset of $\mathbb{R}^2$ consists of a piece-wise straight arc.

The quantity on the left hand side of ($\ddagger$), is a celebrated geometric invariant, called the curvature, at x, of the plane curve $y = y(x)$ and is usually denoted by, $\kappa_y(x)$.

All this might seem a rather unnecessary use of the very powerful version of the Differential Calculus we have built up: after all we have merely verified Euclid's definition *the shortest distance between two points is a straight line.* The power of the idea of geodesics will be better illustrated by the next example which concerns the hyperbolic plane. This also indicates the usefulness of the rather abstract treatment of tangent spaces in §4.4.

Geodesics of the hyperbolic plane, $\mathfrak{H}^2$

In the last section of Chapter 4, I had indicated that if $U \subset \mathbb{R}^n$ is an open set, then for any $p, q \in U$, $T_pU \approxeq T_qU$ but that this is only an isomorphism of linear spaces. T_pU is the linear space of derivations on the germs of C^∞ functions at p; so if $p \neq q$ then T_pU and T_qU are totally different mathematical entities. In particular, they *do not* inherit the inner product of $\mathbb{R}^n$. I am now going to exploit this to describe a very interesting situation, where our notion of geodesics leads to a **non-Euclidean geometry**.

Following Poincaré (see [O'N97]) we define a different notion of lengths of tangent vectors in upper half-plane:

$$\mathfrak{H}^2 = \{(x, y) \in \mathbb{R}^2 : y > 0\}.$$

Note that since $\mathfrak{H}^2$ is an open subset of $\mathbb{R}^2$ the tangent space at any point is isomorphic to $\mathbb{R}^2$. We define the inner product of $\boldsymbol{v}_1, \boldsymbol{v}_2 \in$

$T_{(x,y)}\mathfrak{H}^2$ by the formula:

$$\langle \boldsymbol{v}_1, \boldsymbol{v}_2 \rangle_{(x,y)} = \frac{[\boldsymbol{v}_1, \boldsymbol{v}_2]}{y^2}$$

where [,] is the usual inner product. Thus lengths of tangent vectors depend not just on their components but on their *location.*

Now the problem of finding geodesics is, of course, completely different. If $\alpha : [0,1] \to \mathfrak{H}^2$ is a C^1-curve $\alpha(t) = (\alpha_1(t), \alpha_2(t))$, then we are dealing here with the problem of finding stationary paths for the functional

$$L(\alpha) = \int_0^1 \frac{\sqrt{\dot{\alpha}_1^2 + \dot{\alpha}_2^2}}{\alpha_2} dt.$$

It is easier to work with a curve written in the form $y = y(x)$; but this would automatically exclude a very simple and important class of geodesics of $\mathfrak{H}^2$. So let me start by describing these.

Consider curves $\gamma(t) = (c, g(t))$ with $g'(t) > 0$. These are straight lines parallel to the y-axis.

For such curves the "f" which appears in the Euler–Lagrange equations is $f(\gamma, \dot{\gamma}, t) = \dot{g}/g$. In this case, since the first coordinate does not play a role, there is only one Euler–Lagrange equation

$$\frac{d}{dt}\left(\frac{1}{g(t)}\right) + \frac{\dot{g}}{g^2} = 0.$$

and this is an identity satisfied by all C^1-functions $g : [a, b] \to \mathbb{R}$. So all such paths are geodesics.

I now turn to classifying those curves of the form $y = y(x)$ that are geodesics. In this case

$$L(y) = \int_0^1 \frac{\sqrt{1 + (y')^2}}{y} dx$$

and the Euler–Lagrange equation is:

$$\frac{d}{dx}\left(\frac{y'}{y\left(1 + y'^2\right)^{1/2}}\right) + \frac{\left(1 + y'^2\right)^{1/2}}{y^2} = 0$$

which (after some elementary manipulations) yields the differential equation:

$$y\frac{d^2y}{dx^2} + \left(\frac{dy}{dx}\right)^2 + 1 = 0.$$

To solve this, first observe that

$$\frac{d}{dx}\left(y\frac{dy}{dx}\right) = y\frac{d^2y}{dx^2} + \left(\frac{dy}{dx}\right)^2.$$

So we get

$$y\frac{dy}{dx} = c_1 - x.$$

Integrating a second time, we get:

$$y^2 + (x - c_1)^2 = c_2 \quad \text{which is very familiar!}$$

Thus (sub-arcs of) semicircles with centres on the x-axis are geodesics of the hyperbolic plane. These and straight lines parallel to the y-axis are, in fact, the only geodesics in $\mathfrak{H}^2$.

Remark 6.2.2 (Non-Euclidean Geometry in $\mathfrak{H}$).
Using elementary high school geometry one can check that geodesics in the hyperbolic upper half-plane satisfy the following properties:

- Given any two points $p, q \in \mathfrak{H}^2$ there is a unique geodesic $G(p,q)$ connecting them.

- If G is a geodesic and p is a point not lying on the obvious "completion" of G then there are infinitely many complete geodesics through p which do not intersect the completion of L. (If L is a straight line segment parallel to the y-axis and passing through, say (x_0, y) then the completion is the "half line" $\{(x_0, y) \in \mathbb{R}^2 \ : \ y > 0\}$ and similarly if L is part of a semicircle orthogonal to the x axis, then the completion is the entire semicircle, without the endpoints, of course!)

Thus if we define a *geometry* in $\mathfrak{H}$ in which a geodesic between two points is the equivalent of a straight line then the resulting system satisfies all the axioms of Euclid's plane geometry *except the parallel postulate.*

I started giving examples of geodesics after saying that they are the curves which are "arc-length minimizing"; but in all the examples (i. e. $\mathbb{R}^2$ and $\mathfrak{H}^2$) I only showed that the geodesics were curves for which the arc-length functional L attained a stationary value. I will partially redeem myself by showing that in some special cases geodesics do, indeed, represent local minima of the arc-length functional.

I begin with the case of $\mathbb{R}^2$. As explained earlier, in this case geodesics can always be represented (locally) as curves of the form $y = y(x)$ or $x = x(y)$. So suppose, that the curve $y = y(x)$ (that is the graph of the function $y : [0,1] \to \mathbb{R}$) is a geodesic in $\mathbb{R}^2$ joining the points $p = (0, y(0))$ and $q = (1, y(1))$. Then as we have seen, the equation for the curve is $y(x) = y(0) + x(y(1) - y(0))$ and $L(\alpha) = \{1 + (y(1) - y(0))^2\}^{1/2}$. Now suppose that, $\xi(x) = y(x) + \omega(x)$ is another C^1 curve joining p and q. Then, clearly $\omega(0) = \omega(1) = 0$. The length of ξ is easily estimated:

$$L(\xi) = \int_0^1 \{1 + \xi'(s)^2\}^{1/2}\, ds \tag{6.2.1}$$

$$= \int_0^1 \{1 + (\alpha' + \omega')^2\}^{1/2}\, ds, \tag{6.2.2}$$

Now using Taylor's formula (for functions from [0,1] to $\mathbb{R}$) the integrand on the second line can be written:

$$\{1 + \alpha'^2\}^{1/2} + \frac{\alpha'\omega'}{\{1 + \alpha'^2\}^{1/2}} + \frac{\omega'^2}{2\{1 + (\alpha' + \theta\omega')^2\}^{3/2}}.$$

where $0 < \theta < 1$. From this it follows that:

$$L(\xi) - L(\alpha) = \int_0^1 \frac{\alpha'\omega'}{\{1 + \alpha'^2\}^{1/2}}\, ds + \int_0^1 \frac{\omega'^2}{2\{1 + (\alpha' + \theta\omega')^2\}^{3/2}}\, ds.$$

Since α' is a constant and ω vanishes at $s = 0$ and 1, the first integral is zero. The second integral is ≥ 0 since the integrand is never negative and vanishes iff $\omega(s) = 0$ identically. *Thus, straight lines are minimum-length curves between points* in $\mathbb{R}^2$. The proof for $\mathbb{R}^n$ is essentially the same except that to deal with the second order term we have to show that the Hessian matrix of $F(x_1, \ldots, x_n) = \{1 + x_1^2 + \cdots + x_n^2\}^{1/2}$ is positive-definite. This is something for specialists and enthusiasts of matrix theory and should not detain us from moving ahead.

Remark 6.2.3. Essentially the same argument would show that the geodesics of $\mathfrak{H}^2$ are arc-length minimizing curves between points. (This is particularly easy for the straight segments parallel to the y-axis.)

I hope I have conveyed to you the richness of the seemingly innocent query "What is the shortest route between two points?"

Let us move on and consider problems of the Calculus of Variations which are subject to constraints.

2. Isoperimetric Problems

This is a class of problems concerning stationary values of a function defined on an infinite-dimensional Hilbert space, where the function is required to satisfy some constraints. Typically, the object is to find a function $y : [a, b] \to \mathbb{R}$ such that the 'functional':

$$F(y) = \int_a^b f(y(x), y'(x), x)\, dx$$

achieves a stationary value subject to a constraint of the form:

$$\int_a^b g(y(x), y'(x), x)\, dx = K, \text{ a constant.}$$

Clearly if we can provide a Hilbert space structure on the set of functions from $[a, b] \to \mathbb{R}$, we may obtain a situation where Lagrange's theorem may be applied.

The 'original' isoperimetric problem is the only one I will discuss. It may be stated as follows:

Among all simple, closed C^1 curves $\alpha : [0, 1] \to \mathbb{R}^2$ in the plane, having a fixed length, say L, which one encloses the greatest possible area?

The first difficulty is that such a curve can *never* be the graph of a function $f : [0, 1] \to \mathbb{R}$. While you know from your first course in Calculus how to find the area "under the graph" of $f : [0, 1] \to \mathbb{R}$ and between the lines $x = 0$ and $x = 1$, that is of little help in the present instance. The following theorem (which is proved in the Appendix) is what we need.

Theorem 6.2.4 (Green's Theorem).
Let $\alpha(t) = (x(t), y(t))$ for $a \le t \le b$, be a simple, closed C^1 curve in the plane; that is α is a C^1 curve defined on $[a, b]$ such that $\alpha(a) = \alpha(b)$ and that if $t \ne t'$ then $\alpha(t) = \alpha(t')$ iff $\{t, t'\} = \{a, b\}$. Then the area $A(\alpha)$ enclosed by α is given by the formula:

$$A(\alpha) = \frac{1}{2} \int_0^1 (\dot{x}y - \dot{y}x)\, dt.$$

Given this result the problem becomes one of finding the stationary point of

$$A(\alpha) = \frac{1}{2}\int_0^1 (\dot{x}y - \dot{y}x)\, dt.$$

subject to the constraint

$$\int_0^1 \left(\dot{x}^2 + \dot{y}^2\right)^{1/2} dt = L.$$

We will consider, (without loss of generality,) only curves $\alpha : [0,1] \to \mathbb{R}^2$ such that $\alpha(0) = (0,0) = \alpha(1)$. As before we can start with the C^1-metric on such curves and go to the Hilbert space completion, the Sobolev space $W^{1,2}([0,1];\mathbb{R}^2)$ and as before, assume that the stationary points are C^1 curves.

Since the constraint is $\mathbb{R}$-valued, the "Lagrange multiplier" ($\in \mathbb{R}^*$) is simply multiplication by a constant, λ, and we are reduced to finding a stationary path for the functional

$$\int_0^1 \left\{ \frac{1}{2}(\dot{x}y - \dot{y}x) - \lambda \left(\dot{x}^2 + \dot{y}^2 \right)^{\frac{1}{2}} \right\} dt.$$

To find a stationary path we write down the Euler–Lagrange equations which are:

$$\begin{aligned} \frac{d}{dt}\left(\frac{1}{2}y - \lambda\frac{\dot{x}}{(\dot{x}^2+\dot{y}^2)^{1/2}}\right) &= -\frac{1}{2}\dot{y} \\ \frac{d}{dt}\left(-\frac{1}{2}x - \lambda\frac{\dot{y}}{(\dot{x}^2+\dot{y}^2)^{1/2}}\right) &= +\frac{1}{2}\dot{x} \end{aligned}$$

So after integration we get:

$$\lambda\frac{\dot{x}}{(\dot{x}^2+\dot{y}^2)^{\frac{1}{2}}} = y - c_2, \qquad \lambda\frac{\dot{y}}{(\dot{x}^2+\dot{y}^2)^{1/2}} = -x + c_1.$$

From these we directly get:

$$(y - c_2)^2 = \lambda^2\frac{\dot{x}^2}{\dot{x}^2+\dot{y}^2} \qquad \text{and} \quad (x - c_1)^2 = \lambda^2\frac{\dot{y}^2}{\dot{x}^2+\dot{y}^2};$$

hence a desired curve must satisfy the equation:

$$(x - c_1)^2 + (y - c_2)^2 = \lambda^2.$$

So the curve we are seeking must be a circle. Notice that we have not yet used the initial condition $\alpha(0) = \alpha(1) = (0,0)$; nor have we determined the multiplier! This is because we have not yet used the condition that the length of the curve is L. The curve is a circle of radius λ; hence for the length (circumference) of the curve to be L, $\lambda = L/2\pi$. Now the numbers c_1, c_2 can be chosen arbitrarily as long as they satisfy the condition $c_1^2 + c_2^2 = L^2/4\pi^2$.

Exercises for § 6.2

6.2.1 The Catenary Problem

The problem is to determine the shape assumed by a uniformly dense string of given length with fixed end-points under the influence of gravity. (Of course, we assume that the length of the string is greater than the distance between the fixed end-points.)

Suppose gravity acts in the negative y-direction. The equilibrium position is that for which the centre of gravity be as low as possible. Show that this leads to the following variational problem: Find the function $y = y(x)$ for which

$$P(y) = \int_{x_0}^{x_1} y\sqrt{1 + y'^2}dx,$$

is as small as possible while

$$L = \int_{x_0}^{x_1} \sqrt{1 + y'^2}dx \text{ is constant.}$$

and the boundary values $y(x_0) = y_0$ and $y(x_1) = y_1$ are given.

Show that the solution is:

(The catenary) $$y + \lambda = c\cosh\left(\frac{x}{c} + c_1\right).$$

6.2.2 The Brachistochrone Problem

Two points $A = (x_0, y_0)$ and $B = (x_1, 0)$, with $y_0 > 0$ are to be connected by a curve along which a frictionless particle moves, in the shortest possible time, from B to A under gravity acting in the negative y-direction.

The key to this problem is the law of *conservation of energy.* The initial velocity of the particle is zero. After falling a distance y it has the velocity $\sqrt{2gy}$ where g is the acceleration due to gravity. It follows that the time of transit is the integral

$$T = \int_{x_0}^{x_1} \frac{\sqrt{1+y'^2}}{2gy} dx$$

The function y must in addition satisfy the two condition: $y(x_0) = y_0$, $y(x_1) = 0$.

Show that the solution is:

(The cycloid) $$y = \frac{1}{c}\cosh(cx + d),$$

where c, d are constants. It is said that Newton solved this problem in response to a challenge from Bernoulli: remember he did not have the Euler–Lagrange equations at hand!

Appendix: Green's Theorem

In this short section I hope to convince you of the validity of the formula for the area enclosed by a simple, closed C^1 curve lying in $\mathbb{R}^2$. (This was the key to our solution of the isoperimetric problem.) A meticulous proof of Green's Theorem would force us to stray into deep waters. These finer points are raised, though not resolved, in a series of remarks following the 'proof'.

A piece-wise C^1 curve in $\mathbb{R}^2$ consists of a a partition $\mathcal{P} = \{a = t_0 < t_1 < \cdots < t_n = b\}$ of a closed interval, $[a, b]$, and a map $\alpha : [a, b] \to \mathbb{R}^2$ such that for each $i = 1, \ldots, n$, $\alpha|[t_{i-1}, t_i]$ is a C^1-curve.

Theorem 6.2.5 (Area enclosed by a simple closed curve).
Let $\alpha : [0, 1] \to \mathbb{R}^2$ be a piece-wise C^1 curve $\alpha(t) = (x(t), y(t))$ which is simple and closed i. e. $\alpha(t) = \alpha(t')$ iff $\{t, t'\} = \{0, 1\}$. Let us suppose that as t goes from 0 to 1, α moves in the plane in a counter-clockwise direction. Then the area $A(\alpha)$ enclosed by α is given by

$$(\dagger) \qquad A(\alpha) = \frac{1}{2}\int_0^1 (x\dot{y} - y\dot{x})\, dt$$

Proof. Consider Figure 6.2.1 which shows the image of a piece-wise C^1 curve of constant speed describing a rectangle ABCD whose sides are parallel to the coordinate axes and whose left hand bottom corner (resp. right hand top corner) is situated at (x_0, y_0) (resp. (x_1, y_1)). This curve is defined by the equations:

$$\alpha(t) = \begin{cases} (x_0, y_0) + 4t(x_1 - x_0, 0) & \text{for } 0 \le t \le \frac{1}{4}, \\ (x_1, y_0) + 4(t - \frac{1}{4})(0, (y_1 - y_0)) & \text{for } \frac{1}{4} \le t \le \frac{1}{2}, \\ (x_1, y_1) - 4(t - \frac{1}{2})(x_1 - x_0, 0) & \text{for } \frac{1}{2} \le t \le \frac{3}{4}, \\ (x_0, y_1) - 4(t - \frac{3}{4})(0, (y_1 - y_0) & \text{for } \frac{3}{4} \le t \le 1. \end{cases}$$

The area of this rectangle is $(x_1 - x_0)\cdot(y_1 - y_0)$. Let us compute the integral on the right-hand side of (†). The integral around the curve breaks up into four obvious steps.

Integrating from 0 to $\frac{1}{4}$ one gets Area($PABQ$) with $-$ve sign,
integrating from $\frac{1}{4}$ to $\frac{1}{2}$ one gets Area($SBCR$) with +ve sign,
integrating from $\frac{1}{2}$ to $\frac{3}{4}$ one gets Area($CDPQ$) with +ve sign, and
integrating from $\frac{3}{4}$ to 1 one gets Area($DASR$) with $-$ve sign.

Hence the final answer comes out to be twice the area of $ABCD$.

This verifies the formula for a particularly simple curve.

Remark. It can be checked that if $\beta(t) = (x'(t), y'(t))$ is another simple curve whose image is the same as that of α, then the integral on the right side does not change its value.

We now look at *slightly* more complicated situations.

For example, let us look at the following grid of rectangles shown in (b) of Figure 6.2.1. We can write equations for each of the squares and use the formula (†) to compute the area of each of the individual squares; adding up the integrals along the various paths, (the arrows in the squares indicate the sense in which the line integrals are performed when computing the area of the square) it is clear that the contributions arising from the sections of the grid in the 'interior' of ADPM will cancel out. (For instance, when computing the area of FGKJ, one integrates along the segment GK, going from G to K, whereas when computing the area of GHLK one integrates from K to G: clearly these cancel out each other.) As a result, we are left with the integral of $(x\dot{y} - y\dot{x})$ along ADPM and this gives the sum of the areas of the nine rectangles.

Now given an arbitrary curve satisfying the conditions of the theorem, we cover it by a rectangular grid (see Figure 6.2.1(c)). If we compute the areas of each of the rectangles and add, the area we get is the same as the line integral along the outer edge of the grid. By choosing a fine enough grid, we will not only get arbitrarily near the area enclosed by

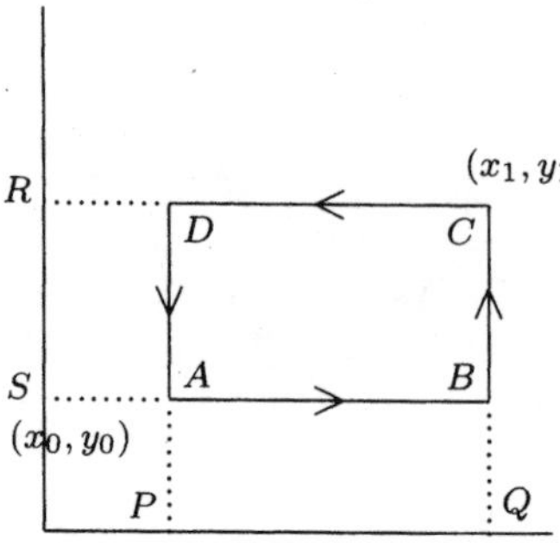

(a) Integrating around a rectangle

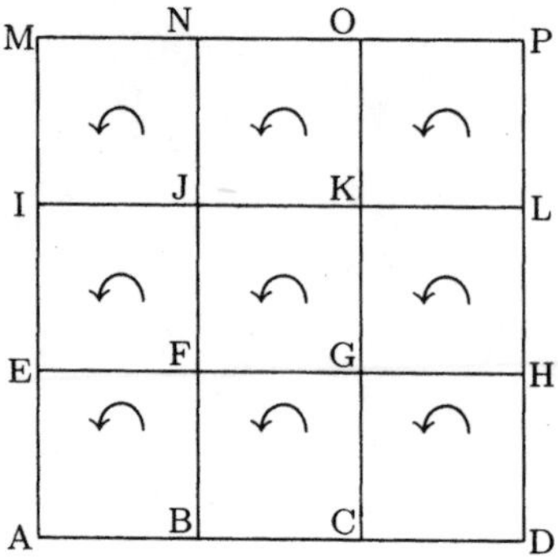

(b) Finding the area of a grid of rectangles

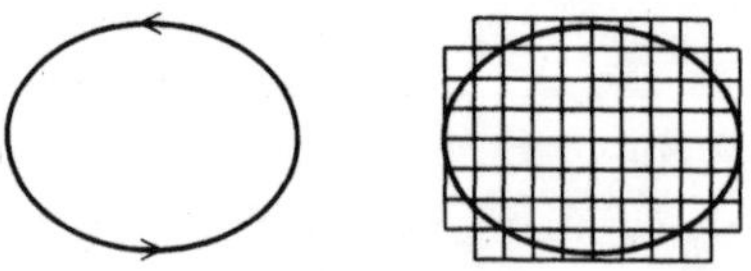

(c) Approximating the area enclosed by a curve with a grid

FIGURE 6.2.1. Proof of Green's Formula

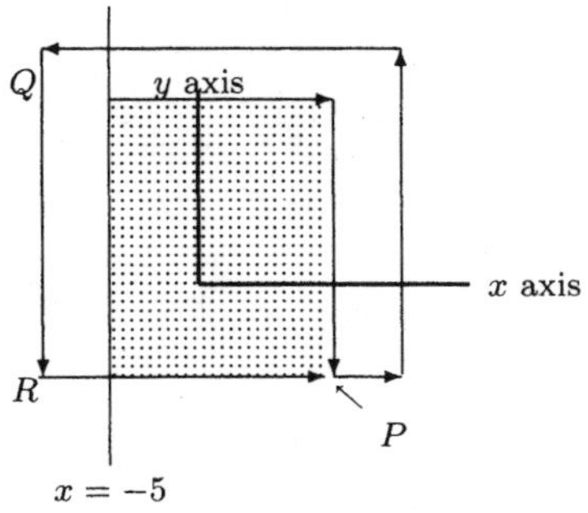

FIGURE 6.2.2. An indecisive curve

the curve the line integral around the outer edge of the grid will converge to the integral (†) of the theorem. □

Remark 6.2.5g (The hidden difficulties).
The adjective 'simple' qualifying the curve C means that $\alpha(t_1) = \alpha(t_2)$ iff t_1, t_2 is a permutation of $0, 1$. But really I have, in the proof, assumed the curve to be 'simple' in the vulgar sense of the word. The difficulty lies in clearly defining what is meant by a parametrization $\alpha : [a, b] \to \mathbb{R}^2$ of a curve describing the curve in a *counter-clockwise fashion.*

This is clearly not describable in terms of differential calculus alone because it is a property dependent on the behaviour of the function α on its entire domain.

Consider, for example, an "indecisive curve" shown in Figure 6.2.2 whose nature is explained below.

α starts off from $(20, -10) \in \mathbb{R}^2$, moves parallel to the x-axis to the point $(25, -10)$, then goes up to $(25, 20)$ along a line parallel to the y-axis, takes a left turn to $Q = (-10, 20)$, then moves down to $R = (-10, -10)$ and moves directly towards its initial point. When at a very small distance ε (say the diameter of a hydrogen atom!) from $(20, -10)$ it "changes its mind" moves up by ε and proceeds left (parallel to the x axis). Again when at a distance ε from the line $x = -5$, it moves up by ε and proceeds to go right till within ε of the line $x = 20$ and continues in this fashion till after a lot of to and fro movements, when it is within ε of the point $(-5, 20)$ it actually goes to the line $y = 20$ and moves right along this line to the point (20,20) and then moves down along the line $x = 20$ and returns to $(20, -10)$. In the figure the "decisive sections" of this piece-wise C^1 curve are shown using arrows, the haphazard motion indicated by an array of dots.

I hope you will agree that it is not so clear whether this curve completes its circuit in a clockwise or counter-clockwise fashion nor is it clear how the "grid method" outlined above will work for such a curve.

There is an elementary (yet formal) definition available in terms of an integral over the domain of α, but this involves the choice of a point in the *the interior* of the simple closed curve.

Remark 6.2.5h (What is the 'interior' of a curve?).
The notion of the interior of a curve presupposes the validity of the celebrated *Jordan Curve Theorem*:

> Given any simple closed curve in the plane, the complement of the image of the curve has exactly two connected components; one of these components is bounded, the other unbounded. The bounded component is called the *interior* of the curve.

The Jordan curve theorem requires a fairly involved proof. The easy proofs are not elementary; requiring some use of algebraic topology. An elementary, though not easy proof can be found in [Die57].

Remark 6.2.5i (Oriented areas, volumes etc.).
Notice that the formula in Green's theorem would give a negative answer if the curve went around in a clockwise fashion. Thus we now seem to be dealing with areas *with sign* or oriented areas. This is actually implicit in the process of integration itself: if $c > 0$ is a constant, then $\int_a^b c\,dt > 0$ iff $b > a$. If we look at the "reverse curve" $\overline{\alpha}$ defined by:

$$\overline{\alpha} : t \mapsto (x(1-t), y(1-t)),$$

then it is clear that $A(\alpha) = -A(\overline{\alpha})$.

Finally notice that while in Statistics, Physics etc. you have been told that to compute areas you must perform a 'multiple' or 'double' integration, Green's theorem computes area in terms of an integral over an interval. This is the 2-dimensional analogue of the Fundamental Theorem of Calculus. (The fundamental theorem computes an integral (of a derivative) over an interval by evaluating a function at two points (which is a kind of 0-dimensional integral.) There is a very general result which relates integration over a special kind of n-dimensional regions with integration over the $(n-1)$ dimensional 'boundary' of the region. This result, known as Stokes' Theorem (see [GP74], [ST76]) combines notions from Algebra, Analysis and Topology to yield a wonderfully flexible and powerful tool for the study of spaces of arbitrary dimension.

References

[Car67] H. Cartan. *Differential Calculus.* Hermann, Paris, first edition, 1967.

[Die57] J. Dieudonné. *Foundations of Modern Analysis.* Academic Press, New York, first edition, 1957.

[Die80] J. Dieudonné. *Treatise on Analysis*, volume 3. Academic Press, New York, 1980.

[GP74] V. Guillemin and A. Pollack. *Differential Topology.* Prentice-Hall Inc., Engelwood Cliffs, N. J., 1974.

[Hal61] P. R. Halmos. *Finite Dimensional Vector Spaces.* Van Nostrand, New York, first edition, 1961.

[Hel94] H. Helson. *Linear Algebra.* Hindustan Book Agency, New Delhi, 1994.

[HS74] M. Hirsch and S. Smale. *Differential Equations, Dynamical Systems and Linear Algebra.* Academic Press, New York, first edition, 1974.

[Inc56] E. L. Ince. *Ordinary Differential Equations.* Dover Press, 1956.

[Lan65] S. Lang. *Introduction to Differentiable Manifolds.* Interscience Publishers, first edition, 1965.

[O'N78] Barrett O'Neill. *Semi-Riemannian Geometry.* Academic Press, New York, 1978.

[O'N97] Barrett O'Neill. *Elementary Differential Geometry.* Academic Press, New York, 1997.

[Par80] K. R. Parthasarathy. Notes on analysis. Mimeographed notes, 1980.

[Rud61] Walter Rudin. *Principles of Mathematical Analysis.* McGraw Hill, first edition, 1961.

[Sim63] G. F. Simmons. *Introduction to Topology and Modern Analysis.* McGraw Hill, 1963.

[ST76] I. M. Singer and J. A. Thorpe. *Lecture Notes on Elementary Topology and Geometry.* UTM. Springer-Verlag, New York-Heidelberg- Berlin, 1976.

Index

Texts and Readings in Mathematics

1. R. B. Bapat: Linear Algebra and Linear Models (Second Edition)
2. Rajendra Bhatia: Fourier Series (Second Edition)
3. C. Musili: Representations of Finite Groups
4. H. Helson: Linear Algebra (Second Edition)
5. D. Sarason: Notes on Complex Function Theory
6. M. G. Nadkarni: Basic Ergodic Theory (Second Edition)
7. H. Helson: Harmonic Analysis (Second Edition)
8. K. Chandrasekharan: A Course on Integration Theory
9. K. Chandrasekharan: A Course on Topological Groups
10. R. Bhatia (ed.): Analysis, Geometry and Probability
11. K. R. Davidson: C* – Algebras by Example
12. M. Bhattacharjee et al.: Notes on Infinite Permutation Groups
13. V. S. Sunder: Functional Analysis — Spectral Theory
14. V. S. Varadarajan: Algebra in Ancient and Modern Times
15. M. G. Nadkarni: Spectral Theory of Dynamical Systems
16. A. Borel: Semisimple Groups and Riemannian Symmetric Spaces
17. M. Marcolli: Seiberg – Witten Gauge Theory
18. A. Bottcher and S. M. Grudsky: Toeplitz Matrices, Asymptotic Linear Algebra and Functional Analysis
19. A. R. Rao and P. Bhimasankaram: Linear Algebra (Second Edition)
20. C. Musili: Algebraic Geometry for Beginners
21. A. R. Rajwade: Convex Polyhedra with Regularity Conditions and Hilbert's Third Problem
22. S. Kumaresan: A Course in Differential Geometry and Lie Groups
23. Stef Tijs: Introduction to Game Theory
24. B. Sury: The Congruence Subgroup Problem
25. R. Bhatia (ed.): Connected at Infinity
26. K. Mukherjea: Differential Calculus in Normed Linear Spaces
27. Satya Deo: Algebraic Topology: A Primer
28. S. Kesavan: Nonlinear Functional Analysis: A First Course
29. S. Szabó: Topics in Factorization of Abelian Groups
30. S. Kumaresan and G. Santhanam: An Expedition to Geometry
31. D. Mumford: Lectures on Curves on an Algebraic Surface (Reprint)
32. J. W. Milnor and J. D. Stasheff: Characteristic Classes (Reprint)
33. K. R. Parthasarathy: Introduction to Probability and Measure (Corrected Reprint)
34. A. Mukherjee: Topics in Differential Topology
35. K. R. Parthasarathy: Mathematical Foundations of Quantum Mechanics
36. K. B. Athreya and S. N. Lahiri: Measure Theory
37. Terence Tao: Analysis I
38. Terence Tao: Analysis II
39. W. Decker and C. Lossen: Computing in Algebraic Geometry

40. A. Goswami and B. V. Rao: A Course in Applied Stochastic Processes
41. K. B. Athreya and S. N. Lahiri: Probability Theory
42. A. R. Rajwade and A. K Bhandari: Surprises and Counterexamples in Real Function Theory
43. G. H. Golub and C. F. Van Loan: Matrix Computations (Reprint of the Third Edition)
44. Rajendra Bhatia: Positive Definite Matrices
45. K. R. Parthasarathy: Coding Theorems of Classical and Quantum Information Theory